Oakland Community College
Orchard Ridge Campus Library
27055 Orchard Lake Road
Farmington Hills, MI 48334

DATE DUE			
APR 0 5 2004			
OCT 2 0 2007			

HIGHSMITH #45114

OAKLAND COMMUNITY COLLEGE
ORCHARD RIDGE LIBRARY
27055 ORCHARD LAKE ROAD
FARMINGTON HILLS, MI 48334

DEMCO

The Geometric Viewpoint

A Survey of Geometries

Thomas Q. Sibley

St. John's University

ADDISON-WESLEY

An imprint of Addison Wesley Longman, Inc.

Reading, Massachusetts · Menlo Park, California · New York · Harlow, England
Don Mills, Ontario · Sydney · Mexico City · Madrid · Amsterdam

Sponsoring Editor: Jennifer Albanese
Associate Production Supervisor: Kim Ellwood
Senior Marketing Manager: Andrew Fisher
Cover Design: Susan Carsten
Manufacturing Supervisor: Ralph J. Mattivello
Compositor: Windfall Software
Art Coordinator: Sandra Selfridge

Credits for illustrations and text materials can be found after the index.

This text is in the Addison Wesley Longman Higher Mathematics Series.
For more information about Addison Wesley Longman Mathematics books, visit our World Wide Web site at http://www.aw.com/he/.

Library of Congress Cataloging-in-Publication Data

Sibley, Thomas Q.
    The geometric viewpoint : a survey of geometries / Thomas Q. Sibley.
       p.   cm.
    Includes bibliographical references (p. –) and index.
    ISBN 0-201-87450-4
     1. Geometry.   I. Title
QA445.S54   1998
516—dc21   97-14024
                                                                                                     CIP

Copyright © 1998 by Addison Wesley Longman.

All rights reserved. No part of this publication may be reproduced, stored in a retrieval system, or transmitted, in any form or by any means, electronic, mechanical, photocopying, recording, or otherwise, without the prior written permission of the publisher. Printed in the United States of America.

ISBN   0-201-87450-4

1 2 3 4 5 6 7 8 9 10 ML 99 98 97

Preface

> I begin to understand that while logic is a most excellent guide in governing our reason, it does not, as regards stimulation to discovery, compare with the power of sharp distinction which belongs to geometry. —*Galileo Galilei*

Geometry combines visual delights and powerful abstractions, concrete intuitions and general theories, historical perspective and contemporary applications, and surprising insights and satisfying certainty. In this textbook, I try to weave together these facets of geometry. I also want to convey the multiple connections that different topics in geometry have with each other and that geometry has with other areas of mathematics. These connections link different chapters together without sacrificing the survey nature of the whole text.

The enduring appeal and importance of geometry stem from its synthesis of intuition and reasoning. Mathematical intuition is as hard for mathematics majors to develop as is the art of proving theorems. Geometry is an ideal subject for developing both. However, geometry courses and texts for mathematics majors usually emphasize reasoning. This book strives to build students' intuition and reasoning through the text and the variety of problems and projects. The dynamic geometry software now available provides one valuable way for students to build their intuition and prepare them for proofs. Thus problems dependent on technology join hands-on explorations, proofs, and other problems. I see this book as building on the momentum of the NCTM *Standards* for primary and secondary education and on the calculus reform movement in its goal of helping students internalize mathematical concepts and thinking.

- Proofs end with the symbol ∎. Examples end with the symbol ●. We abbreviate "if and only if" with "iff." In a definition, the word being defined is italicized.

Prerequisites.

The major prerequisite for using this text is the ability to read proofs and other mathematical content at the level of an undergraduate mathematics major. The book assumes a familiarity with high school algebra and geometry, although a detailed memory of geometry theorems isn't expected. (Appendix A supplies many of them in Euclid's language and in sometimes more modern phrasing.) Calculus is needed for Section 2.4 and, as noted, occasionally in examples, problems, and projects. An understanding of vectors is needed for Sections 2.3 and 2.5. Linear algebra is needed for Sections 4.3, 4.4, 4.5, 6.3, 6.4, 6.5, 6.6, and 7.4. We discuss groups in Chapters 4, 5, and 6, and fields in Section 7.4, but students aren't expected to be familiar with them.

TO THE STUDENT

Everyday speech equates intuitive with easy and obvious. However, recent psychology research confirms what mathematicians have always understood: People build their own intuitions through reflection on new experiences. My students often describe the process as learning to think in a new geometry. I hope the text's explanations are clear and provide new insights. I also hope you find many nonroutine problems you can solve after an extended effort; such challenges enrich your intuition.

Problems.

Learning mathematics centers on doing mathematics, so problems are the heart of any mathematics textbook. I hope that you (and your teachers) will enjoy spending time pondering and discussing the problems. Make lots of diagrams and, when relevant, physical or computer models. The problems are varied and include some that encourage hands-on experimentation and conjecturing, as well as more traditional proofs and computations. Answers to selected problems appear in the back of the book.

Projects.

Essays, paper topics, and more extended and open-ended problems appear at the end of each chapter as projects. Too often textbooks shift to a new topic just when students are ready to make their own connections. These projects encourage extensions of the ideas discussed in the text. Many of these projects benefit from group efforts.

Visualization.

This text has an unusually high number of figures because I believe that students benefit greatly from visual thinking. In addition, you will be asked to draw your own designs, to use a computer, or to build three-dimensional models. Visualization not only builds your intuition about the topic at hand, it can also lead you to new insights.

History.

Geometry reveals the rich influences over the centuries between mathematics and other fields. Students in geometry, even more than other areas of mathematics, benefit from historical background, which I have included in the first section of each chapter and the biographies. The biographies give a personal flavor to some of the work discussed in the text. One common thread I noted in reading about these geometers was the importance of intuition and visual thinking. As a student, I sometimes questioned my mathematical ability because I needed to visualize and to construct my own intuitive understanding instead of grasping abstract ideas directly. Now I realize that great mathematicians also built on intuition and visualization for their abstract proofs and theories. Perhaps this understanding will help you as well.

Readings and Media.
Geometry is a field blessed with many wonderful and accessible expository writings and a long tradition of visual materials. I hope that you find the list of readings and media at the end of each chapter valuable while you are studying from this book and later as a resource.

TO THE TEACHER

Suggested Course Outlines.
This survey text supports various courses. If time permits, one option is to do the entire book in sequence. Even an entire year devoted to this course would require that you pick carefully among the problems and projects in order to cover all the chapters. One-quarter courses could cover two chapters—for example, Chapters 1 and 3, Chapters 4 and 5, or shortened versions of the following semester course options.

- Course 1, teacher preparation: Chapters 1, 3, and 4, and, as needed, Sections 2.1 and 2.2. Sections 3.5 and 4.6 are optional. If time permits, topics from Chapters 2, 5, and 7 may be included.
- Course 2, Euclidean geometry: Chapters 1, 2, 4 (except Section 4.6), and as much of 5 as time permits.
- Course 3, transformational geometry: Chapters 4, 5, and 6, and, as needed, Sections 1.1, 1.2, 1.5, 1.6, 2.1, 2.2, and 2.5.
- Course 4, axiomatic systems and models: Chapters 1 and 3, Sections 2.1, 6.1, 6.2, 6.3, 7.1, 7.2, 7.4, and parts of Sections 2.3 and 6.5.

More than one class period should be devoted to most of the sections. In the first period I usually discuss the main points of the section, model one or more of the problems or relevant projects, and/or have the students do one of the group projects. I often use part of the next class (or classes) for group presentations of problems or projects, which can spark discussion about different approaches. I hold class in our computer lab several times during the semester. I encourage students to use their graphing calculators for matrix computations in Chapters 4 and 6.

The projects serve several pedagogical purposes. Some of them are intended as group explorations to introduce topics in class. Others are suitable for essay topics or classroom discussion. Others are open-ended or extended problems. The most succinct ones, of the form "Investigate . . . ," are leads for papers. I have assigned papers in my geometry class for more than 10 years. Students uniformly regard these papers as the high point of the class.

The chapters are largely independent, although they are often interconnected. Chapter 3 depends on the first four sections of Chapter 1. Section 4.6 relies on Section 3.1.

Chapter 5 requires the first two sections of Chapter 4. Sections 6.3–6.6 depend on Chapter 4, except Section 4.6. Sections 6.5 and 6.6 also rely on Section 3.1. Section 7.4 requires Chapters 4 and 6.

Instructor's Resource Manual.
Because of the unusual variety of problems and projects, answers to the problems and selected projects and suggestions for their use can be found in the *Instructor's Resource Manual,* available from Addison Wesley Longman. Instructor support for integrating Geometer's Sketchpad into the geometry course will be included.

ACKNOWLEDGMENTS

I appreciate the support and helpful suggestions that many people made to improve this book, starting with the students who studied from the rough versions. My editors, Marianne Lepp and Jennifer Albanese, and my production supervisor, Kim Ellwood, have provided much needed direction, encouragement, and suggestions for improvement. I am grateful to the reviewers, whose insightful and careful critiques helped me greatly. They are: Bradford Findell, University of New Hampshire; Yvonne Greeleaf, Rivier College; Daisy McCoy, Lyndon State University; Jeanette Palmiter, Portland State University; and Diana Venters, University of North Carolina, Charlotte. These people have pointed out many mistakes and unclear passages; any remaining faults are, naturally, my full responsibility.

I particularly want to thank three people. Paul Krueger not only painstakingly made many of the fine illustrations that are an integral part of this book, but he also pondered with me over the years the connections of geometry with other fields. Connie Gerads Fournelle helped greatly with her valuable perspective on the text, both as a student using an early version and later as my assistant writing answers to selected problems. My wife, Jennifer Galovich, provided the unswerving support and love I needed to keep chipping away at this project, even when no end to it seemed discernible.

Contents

1 Euclidean Geometry 1

1.1 Overview and History 2

1.2 Constructions, Congruence, and Parallels: Euclid's Approach to Geometry 9

Archimedes 15

1.3 A Critique of Euclid–Modern Axiomatics 22

David Hilbert 24

1.4 Axiomatic Systems, Models, and Metamathematics 29

1.5 Similar Figures 35

1.6 Three-Dimensional Geometry 39

Projects for Chapter 1 52

Suggested Readings 57

Suggested Media 58

2 Analytic Geometry 59

2.1 Overview and History 60

René Descartes 61

2.2 Conics and Locus Problems 65

Pierre Fermat 67

2.3 Further Topics in Analytic Geometry 72

2.4 Curves in Computer-Aided Design 78

2.5 Higher Dimensional Analytic Geometry 85

Gaspard Monge 87

Projects for Chapter 2   93
Suggested Readings   95
Suggested Media   96

3 Non-Euclidean Geometries   97

3.1   Overview and History   98
  Nikolai Lobachevsky and János Bolyai   100
  Carl Friedrich Gauss   101
  Georg Friedrich Bernhard Riemann   102
3.2   Properties of Lines and Omega Triangles   106
3.3   Saccheri Quadrilaterals and Triangles   112
3.4   Area and Hyperbolic Designs   116
3.5   Spherical and Single Elliptic Geometries   124
  Projects for Chapter 3   128
  Suggested Readings   129
  Suggested Media   129

4 Transformational Geometry   131

4.1   Overview and History   132
4.2   Isometries   137
  Felix Klein   143
4.3   Algebraic Representation of Transformations   145
4.4   Similarities and Affine Transformations   153
  Sophus Lie   155
4.5   Transformations in Higher Dimensions; Computer-Aided Design   161
4.6   Inversions and the Complex Plane   167
  Augustus Möbius   172
  Projects for Chapter 4   175
  Suggested Readings   177
  Suggested Media   178

5 Symmetry   179

5.1   Overview and History   180
5.2   Finite Plane Symmetry Groups   185
5.3   Symmetry in the Plane   189
5.4   Symmetries in Higher Dimensions   203
5.5   Symmetry in Science   206

　　　　　Marjorie Senechal　209
　5.6　Fractals　214
　　　　　Benoit Mandelbrot　216
　　　　　Projects for Chapter 5　222
　　　　　Suggested Readings　223
　　　　　Suggested Media　224

6　Projective Geometry　225

　6.1　Overview and History　226
　6.2　Axiomatic Projective Geometry　231
　　　　　Jean Victor Poncelet　233
　6.3　Analytic Projective Geometry　238
　6.4　Projective Transformations　243
　6.5　Subgeometries　247
　　　　　Arthur Cayley　248
　6.6　Projective Space　253
　　　　　Projects for Chapter 6　260
　　　　　Suggested Readings　261
　　　　　Suggested Media　261

7　Finite Geometries　263

　7.1　Overview and History　264
　　　　　Leonhard Euler　265
　7.2　Affine and Projective Planes　267
　7.3　Design Theory　272
　　　　　Sir Ronald A. Fisher　273
　7.4　Finite Analytic Geometry　278
　　　　　Projects for Chapter 7　283
　　　　　Suggested Readings　285

Appendix A　Definitions, Postulates, Common Notions, and Propositions from Book I of Euclid's *Elements*　287

Appendix B　Hilbert's Axioms for Euclidean Plane Geometry　293

　　Selected Answers　297

　　Index　305

1
Euclidean Geometry

Geodesic domes, such as the U.S. Pavilion at Expo '67 pictured here, satisfy pragmatic engineering demands with an elegant geometric shape. Buckminster Fuller used elementary two- and three-dimensional geometry, familiar to mathematicians for more than 2000 years, to fashion the first geodesic domes in the 1950s.

ΑΓΕΩΜΕΤΡΗΤΟΣ ΜΗΔΕΙΣ ΕΙΣΙΤΩ
"Let no one unversed in geometry enter here."
(the inscription Plato is said to have
placed over the entrance to his Academy)

1.1 Overview and History

Geometry has a rich heritage as well as contemporary importance. We begin this chapter with a historical examination, focusing on the fundamental contributions of the ancient Greeks. We discuss modern critiques of this foundation and newer geometry stemming from that early work. We also investigate axiomatic systems and models.

Geometric understanding developed in all ancient cultures, consisting largely of geometric patterns and empirical methods for finding areas and volumes of various shapes. The best preserved and most developed pre-Greek mathematics came from Egypt and Babylonia. An Egyptian papyrus dated 1850 B.C. gave an exact procedure for finding the volume of a truncated square pyramid. However, the Egyptians were probably not aware that it was exact or which of their other methods, such as finding the area of a circle, were not exact. A century earlier (by 1950 B.C.) the Babylonians possessed a sophisticated number system and methods to solve problems that we would describe as first- and second-degree equations in one and two variables. The Babylonians, among others, used what we call the Pythagorean theorem. Although Egyptian and Babylonian mathematics dealt with specific numbers rather than general formulas, the variety of examples that survived convinces scholars that these peoples understood the generality of their methods.

1.1.1 Euclid, the Pythagoreans, and Zeno

The heritage of deductive mathematics started in ancient Greece and was built on the work of the Babylonians and Egyptians. The Greeks discovered and proved many mathematical properties, including the familiar ones of high school geometry. They also organized this knowledge into an axiomatic system, now known as Euclidean geometry, which honors the Greek mathematician Euclid. We know little about Euclid (circa 300 B.C.) except his mathematics, including the most influential mathematics book of all time: the *Elements* [13]. In it he organized virtually all the elementary mathematics known at the time into a coherent whole. The *Elements* contains definitions, axioms, and 465 theorems and their proofs—but no explanations or applications. For centuries this format represented the ideal for mathematicians and influenced many other areas of knowledge. The Greeks called an axiomatic approach *synthetic* because it synthesizes (proves) new results from statements already known. The Greeks often used a process they called *analysis* to discover new results that they then proved. They analyzed a problem by assuming the desired solution and worked backward to something known. We mimic this procedure in analytic geometry and algebra by assuming that there is an answer, the unknown x, and solving for it. In modern times synthetic geometry has come to mean geometry without coordinates because coordinates are central to analytic geometry.

The Pythagoreans, followers of Pythagoras (580 to 500 B.C.), were among the first groups to focus on theoretical mathematics. Although the Pythagorean theorem had been known at least in numerical form in many cultures, the Pythagoreans are credited with proving this key link between geometry and numbers. The Pythagoreans built their mathematics and their mystical musings on positive whole numbers and their ratios, proportions, and properties. The Pythagoreans developed the theory of positive whole numbers, investigating prime numbers, square numbers, and triangular numbers, among others. They also developed geometric proofs. There is evidence that the Pythagoreans found the proof of Theorem 1.1.1, a theorem as fundamental as the Pythagorean theorem. (See Heath [13, vol. I, 317–320].)

Notation. We use the following notation. The *line segment* between two points A and B is denoted $\overline{AB}$, the *length* of $\overline{AB}$ is denoted AB, and the *line* through A and B is denoted $\overleftrightarrow{AB}$. The *triangle* with vertices A, B, and C is denoted $\triangle ABC$, and the *angle* of that triangle with vertex at B is denoted $\angle ABC$. We abbreviate "sum of the measures of the angles" as *angle sum*.

Theorem 1.1.1 In Euclidean geometry the angle sum of a triangle is $180°$.

Proof. Let $\triangle ABC$ be any triangle and construct the line $\overleftrightarrow{DE}$ parallel to $\overline{BC}$ through A (Fig. 1.1). Then (by Euclid's proposition I-29) $\angle DAB \cong \angle CBA$ and $\angle CAE \cong \angle ACB$. Thus the three angles of the triangle are congruent to the three angles $\angle DAB$, $\angle BAC$, and $\angle CAE$ that comprise a straight angle. Hence the angle sum of the triangle equals the measure of a straight angle, $180°$. ∎

The Pythagorean attempt to ground mathematics on numbers ran into an irreconcilable conflict with their discovery of *incommensurables* (irrational numbers, in modern terms). Commensurable lengths have a common measure—a unit length such that the given lengths are integer multiples of the unit. For example, $\frac{2}{3}$ and $\frac{4}{5}$ have $\frac{1}{15}$ as a common measure. The Pythagoreans proved that the diagonal and side of a square were incommensurable. We say that the diagonal is $\sqrt{2}$ times as long as the side and that $\sqrt{2}$ is not a rational number. Recall that a rational number can be written as a fraction.

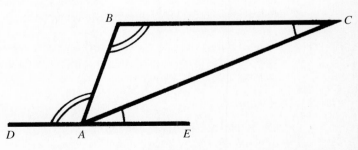

Figure 1.1

Theorem 1.1.2 No rational number equals $\sqrt{2}$. (The diagonal of a square is incommensurable with the side.)

Proof. Suppose, for a contradiction, that there were two integers p and q such that $p/q = \sqrt{2}$. Without loss of generality (WLOG), assume that p and q are not both even: otherwise we could factor out any common factors of 2. Then $(p/q)^2 = 2$, or $p^2 = 2q^2$. Thus p^2 must be an even number, which in turn, forces p to be even. (To illustrate, suppose for a moment that p is odd, say, $p = 2k + 1$. Then $p^2 = 4k^2 + 4k + 1$, an odd number.) If we rewrite the even number p as $2r$, for some integer r, then $(2r)^2 = 2q^2$, or more simply $2r^2 = q^2$. As before, we deduce that q must be even. However, p and q are not both even, giving us a contradiction. Thus our initial supposition is invalid, and $\sqrt{2}$ is not a rational number. ∎

Exercise 1 Modify the proof of Theorem 1.1.2 to prove that $\sqrt{3}$ is not rational. Explain where the corresponding argument fails when you try to show that $\sqrt{4}$ is not rational.

Theorem 1.1.2 ruined the Pythagorean's philosophical commitment to explain everything in terms of whole numbers and their ratios. This and other philosophical problems led later Greek mathematicians to base their mathematics on geometry. For example, they no longer thought of lengths, areas, and volumes as numbers because these values could be irrational. The lack of rational numbers for measurement ruled out geometric formulas. Nevertheless, the Greeks made impressive advances in geometry and developed careful, well-founded proofs. The theoretical, abstract nature of Greek mathematics separated it from practical and computational mathematics. However, modern scientists and mathematicians have found important applications of Greek discoveries and the theoretical approach.

Zeno's paradoxes—and the irrationality of $\sqrt{2}$—spurred a careful study of the foundations of geometry. Zeno (circa 450 B.C.) proposed four paradoxes purporting to disprove obvious facts about motion. Zeno's most famous paradox, Achilles and the Tortoise, continues to puzzle people. (See Salmon [26].)

Achilles and the Tortoise.

Achilles, the swiftest human, gives a tortoise a head start in a race. Zeno argued that Achilles can never pass the tortoise. For Achilles to catch the tortoise, he must first run to where the tortoise started, but by then the tortoise will have crawled a bit farther. Achilles must now run to this new place, but the tortoise will then be a tiny bit farther along. No matter how often this process is repeated and no matter how small the tortoise's lead, Achilles always remains behind.

Exercise 2 Zeno's argument is paradoxical because each step seems reasonable, yet we know that faster things regularly pass slower ones. Discuss this paradox, trying especially to find an error in Zeno's argument.

1.1.2 Plato and Aristotle

The school of philosophy founded by Plato (429–348 B.C.) next took the lead in the study of geometry. One of Plato's pupils, Eudoxus, developed a theory of proportion and a way to give careful proofs of sophisticated results that applied equally to commensurables and incommensurables. Furthermore, Eudoxus's work, like limits in calculus, avoided Zeno's paradoxes altogether. Another pupil, Theaetetus, developed a classification of incommensurable lengths and gave the first proof that there are five regular polyhedra.

Plato viewed geometry as vital training for philosophy. He thought that only those who understood the truths of geometry could grasp philosophical truths. In his view, mathematics was certain because it was about ideal, eternal truths, and mathematics was applicable because the physical world was an imperfect reflection of the ideal truth.

Now mathematics is often viewed as part of science with its emphasis on physical reality rather than Plato's ideal view. But mathematics, with its astounding certainty that surpasses any other subject's reliability, seems to have a different content than any science. After all, no one can physically measure π to one hundred place accuracy, let alone the more than one billion places that have been found with the aid of computers. Obviously, though, mathematics is not isolated from the real world, as its sophisticated and varied applications reveal.

Aristotle (384–322 B.C.), Plato's most famous student, established his own school of philosophy. Aristotle considered mathematics to be an abstraction of concrete experience. Thus for him the applicability of mathematics derived from its origin in the world. Aristotle thought that mathematics owed its certainty to its careful proofs. He emphasized the necessity of starting with simple, unquestionable truths (axioms or postulates) and carefully proving all other truths from them. His work on logic set the standards of reasoning for two thousand years just as his contributions to many other areas—science, law, ethics and esthetics—profoundly influenced Western culture. (See Kline [18, Chapters 1–3] for more information on ancient, including Greek, mathematics.)

PROBLEMS FOR SECTION 1.1

In any problems requiring proofs, you may assume any common geometric properties you know, as long as you make your assumptions explicit.

1. The Egyptians used the square of $\frac{8}{9}$ of the diameter for the area of a circle. In measuring a cylindrical granary with a height of 20 ft and a radius of 10 ft, what error in cubic feet and as a percentage would the Egyptians have made? Find the value of k in $(8d/9)^2 = kr^2$. (The Egyptian method effectively approximates π by k, but it is historically misleading to talk about an "Egyptian value of π.")

2. Figure 1.2 shows a truncated pyramid. The following is the Egyptian recipe for the volume of a truncated square pyramid with a height of 6 and lengths of 4 at the base and 2 at the top: "You are to square this 4; result 16. You are to double 4; result 8. You are to square 2; result 4. You are to add the 16, the 8, and the 4; result 28. You are to take one third of 6; result 2. You are to take 28 twice; result 56. See, it is 56. You will find it right." Verify that this recipe corresponds

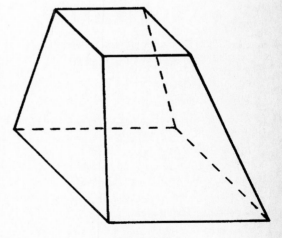

Figure 1.2

to the modern formula $V = \frac{1}{3}h(a^2 + ab + b^2)$. Then derive this formula from $V = \frac{1}{3}HA$ for the volume of a pyramid with height H and base area A. [*Hint:* A truncated pyramid is the difference between an entire pyramid and a smaller pyramid removed from the top. Use proportions to relate the bases and heights of these two pyramids. Note that $a^3 - b^3 = (a^2 + ab + b^2)(a - b)$.]

3. A diagram like that shown in Fig. 1.3 appears on a Babylonian tablet, but written in their base 60 notation. Convert these fractions to decimals and discuss what the numbers tell you about Babylonian mathematics.

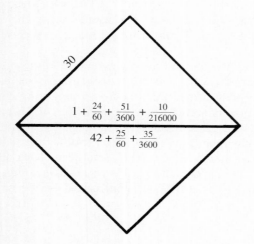

Figure 1.3

4. An (abbreviated) Babylonian problem reads " . . . I added the length and the width and [the result is] 6.5 GAR. . . . [The area is] 7.5 SAR. . . . What are the length and width?" (1 GAR is almost 20 ft, and a SAR is a square GAR.)

   a) Solve this problem with modern methods. Explain your approach.

   b) Verify the quadratic formula $(-b \pm \sqrt{b^2 - 4ac})/2a$ reduces to $-b/2 \pm \sqrt{(b/2)^2 - c}$ for $a = 1$.

   c) Use part (b) to discuss the scribe's recipe for solving this problem:

   Halve the length and width which I added together, and you will get 3.25. Square 3.25 and you will get 10.5625. Subtract 7.5 from 10.5625, and you will get 3.2625. Take its square root, and you will get 1.75. Add it to the one, subtract it from the other, and you will get the length and the width. 5 GAR is the length; 1.5 GAR is the width. . . . Such is the procedure.

5. The Pythagoreans thought that the pentagram, a pentagon with its diagonals, as shown in Fig 1.4, had mystical qualities.

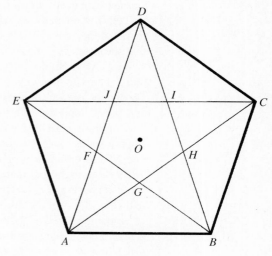

Figure 1.4

   a) Find the measures of the following angles. $\angle AOB$, $\angle OBA$, $\angle ABC$, $\angle BAC$, $\angle AGB$, $\angle CGB$, $\angle ABH$, and $\angle CAD$.

   b) Verify that $\triangle ABC$ and $\triangle AGB$ are isosceles and similar. List other similar triangles. If $AB = 1$ and $AG = x$, explain why $AC = 1 + x$. [*Hint:* Use $\triangle BCG$.]

   c) Explain why in part (b) x satisfies $1 + x = 1/x$. Find the exact value of $1 + x$, the diagonal of the pentagon. Verify that the ratio of AG to GH is also $1 + x$. The number $1 + x \approx 1.618$, the *golden ratio,* appears in many natural settings and applications. (See Huntley [16].)

6. Assume that any diagonal of a convex polygon is inside it.

   a) Find the angle sum of a convex quadrilateral, pentagon, and hexagon.

   b) Find the angle sum of a convex n-gon. Prove it with induction.

   c) What happens if the polygon in part (b) is not convex?

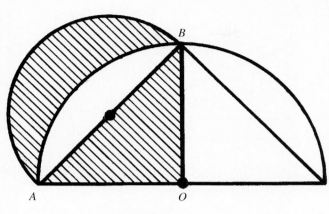

Figure 1.5

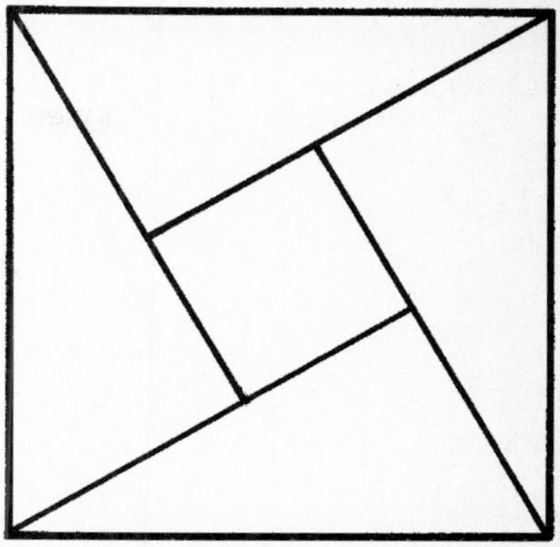

Figure 1.6 BEHOLD!

7. **a)** Find all rectangles with sides of integer lengths whose perimeters equal their areas. (The Pythagoreans considered this problem, but later Greeks did not because they didn't consider areas and lengths as numbers.)

   b) Find a formula for all rectangles whose perimeters equal their areas. Use this formula to explain your answer in part (a).

   c) Repeat part (a) for rectangular boxes whose surface areas equal their volumes.

8. Eratosthenes (circa 284–192 B.C.) made the most famous and accurate of the Greeks' estimates of the circumference of the earth. He found that at noon on the summer solstice, the sun was directly overheard at Syene, Egypt; at the same time 5000 stadia (approximately 500 mi) north, in Alexandria, Egypt, the sun was $\frac{1}{50}$ of a circle off vertical. Make a diagram, compute the circumference of the earth based on his data and explain your procedure. [*Hint:* Assume that the sun's rays at Syene and Alexandria are parallel lines.]

9. In the only surviving Greek mathematical text from before Euclid, Hippocrates (circa 440 B.C.) investigated the areas of *lunes*, which are regions bounded by two circles. Explain his result that the shaded lune shown in Fig. 1.5 has the same area as $\triangle ABO$. [*Hint:* Include the unshaded area between A and B.]

10. Recall the Pythagorean theorem: In a right triangle, the square on the hypotenuse has the same area as the squares on the sides: that is, $a^2 + b^2 = c^2$. Give a proof of the Pythagorean theorem based on Fig 1.6. The Indian mathematician Bhaskara (1114–1185) devised such a proof but provided only this diagram with the word "BEHOLD!" written below it.

11. In Fig 1.7 $\angle ACB$ is a right angle and $\overline{CD}$ is perpendicular to $\overline{AB}$. Why are $\triangle ABC$, $\triangle ACD$, and $\triangle CBD$ all similar? Show that $cy = a^2$ and $cx = b^2$ and develop a proof of the Pythagorean theorem.

12. Study Euclid's proof of the Pythagorean theorem (proposition I-47 in Heath [13, 349]) and compare it with the proofs in Problems 10 and 11. Discuss each proof's assumptions and how convincing and how easy it is to follow.

13. The Greeks proved that a cylinder has three times the volume of a corresponding cone. Use each of the following methods to verify this fact and discuss their advantages, disadvantages, and assumptions.

    a) (Empirical) Find or make a cylinder and a cone with the same height and radius (Fig. 1.8). Use the cone three times to fill the cylinder with sand. How close does the volume of sand in the cylinder come to exactly filling it?

    b) (Calculus) Recall that $\int_a^b \pi(f(x))^2 dx$ gives the volume of revolution generated by rotating

8 Chapter 1 Euclidean Geometry

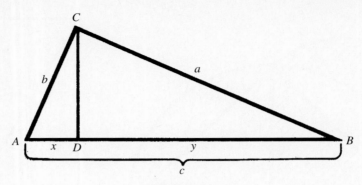

Figure 1.7

around the x-axis the curve $y = f(x)$ from $x = a$ to $x = b$ (See Fig. 1.9). Use calculus to compare the volumes of a cone and a cylinder.

c) (Stacked disks) You can approximate the volume of a cone with the volume of a stack of disks (Fig. 1.10). If h is the height of the cone and n is the number of disks, the height of each disk is h/n. Find a formula for the radius and the volume of the ith disk from the top in terms of the radius.

Find the volume of the stack of disks and simplify it to get the approximation formula

$$\text{Volume} \approx \frac{\pi h r^2}{n^3} \sum_{i=1}^{n} i^2.$$

Assume (or better prove by induction) that $\sum_{i=1}^{n} i^2 = (2n^3 + 3n^2 + n)/6$. What happens to this approximation as n approaches infinity?

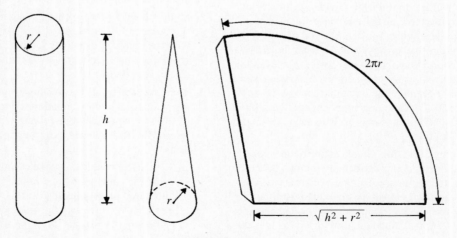

Figure 1.8

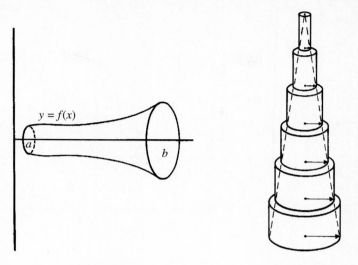

Figure 1.9   **Figure 1.10**

1.2 CONSTRUCTIONS, CONGRUENCE, AND PARALLELS: EUCLID'S APPROACH TO GEOMETRY

Euclid's masterpiece, *Elements*, exemplifies Greek mathematics; it is an axiomatic system about ideal geometric forms. The modern understanding of axiomatic systems (Sections 1.3 and 1.4) and non-Euclidean geometries (Chapter 3) arose from careful reflection on Euclid's work. Euclid's axiomatic system and these developments have greatly influenced modern mathematics and thus provide ample reason to study Euclid. High school geometry courses, based on Euclid's approach, provide another reason to look at the *Elements* in some detail. In this section we consider constructions, congruence, and parallel lines—all familiar topics of high school geometry found in the *Elements*. In Section 1.5 we consider similarity, another high school topic based on Euclid's work.

Euclid united his own work with that of his predecessors. However, he didn't indicate which of the 465 theorems he discovered, and his text was so successful that no prior geometry text was preserved. Scholars credit Euclid with the organization, the choice of axioms (his postulates and common notions), and some of the theorems and proofs. Euclid sought to achieve Aristotle's goal of starting with self-evident truths and proving all other properties from these assumptions.

1.2.1 Constructions

Euclid's first three postulates and many of his propositions reflect the growth of formal geometry from constructions—figures built from line segments and circles. (See Appendix A for the postulates, definitions and propositions of Book I of Euclid's *Elements*; proposition *n* of book I is denoted I-*n*.) Euclid assumed the construction of a line segment given the end points (postulate 1, in Appendix A), the extension of a line segment (postulate 2, in Appendix A), and the construction of a circle given the center

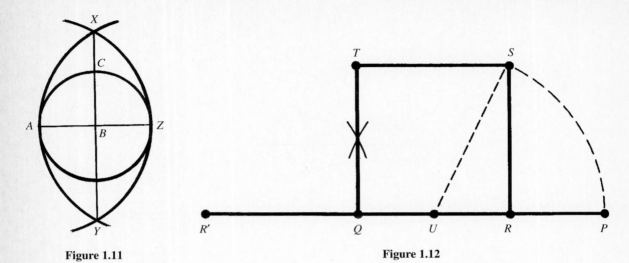

Figure 1.11

Figure 1.12

and the radius (postulate 3, in Appendix A). By long tradition from Greek to modern times only an unmarked straightedge and a compass can be used in constructions.

Example 1 Construct a square if you know one side of it, say, $\overline{AB}$.

Solution. First construct (following Euclid I-11) $\overline{BC}$, the perpendicular to $\overline{AB}$ at B (Fig. 1.11). Construct the circle with center B and radius $\overline{AB}$ (postulate 3, in Appendix A). Extend $\overline{AB}$ (postulate 2, in Appendix A) until it intersects the circle again, say, at Z. Construct circles with centers A and Z and radii $\overline{AZ}$. Let X and Y be their intersections and construct $\overline{XY}$ (postulate 1, in Appendix A). Explain why $\overline{XY}$ is the perpendicular bisector of $\overline{AZ}$. Let C be an intersection of $\overline{XY}$ and the circle centered at B. Then $\overline{BC}$ is a second side of the square. Construct the rest of the square similarly. ●

Example 2 Construct a regular pentagon if you know one side of it.

Solution. Construct a segment $\overline{PQ}$ such that $PQ = (1+\sqrt{5})AB/2$ (Fig. 1.12). By Problem 5 of Section 1.1, PQ is the length of the diagonal of the pentagon. Let $QRST$ be a square whose sides are the same length as $\overline{AB}$. Find the midpoint U of $\overline{QR}$, using the perpendicular bisector of $\overline{QR}$. Construct the circle with center U and radius $\overline{US}$. One of its intersections with $\overleftrightarrow{QR}$ is the desired point P. The Pythagorean theorem (I-47) shows that $\overline{PQ}$ is as long as claimed.

Fig. 1.13 shows the construction of the pentagon. The circle with center A and radius $\overline{AB}$ intersects the circle with center at B and radius $\overline{PQ}$ at E. Similarly, points C and D are the intersections of circles centered at A and B of appropriate radii. Construct $\overline{AB}, \overline{BC}, \overline{CD}, \overline{DE}$, and $\overline{AE}$. Explain why $ABCDE$ is a regular pentagon. ●

Example 3 Construct the tangents to a circle from an outside point.

Solution. Let P be outside the circle with center C (Fig. 1.14). Construct the midpoint M of $\overline{CP}$ and the circle with center M and radius $\overline{MC}$. This circle intersects the original circle in two points, A and B. Then $\overline{PA}$ and $\overline{PB}$ are tangent to the circle through P. Problem 9 provides a way to justify that $\angle PAC$ and $\angle PBC$ are right angles. ●

1.2 Constructions, Congruence, and Parallels: Euclid's Approach to Geometry

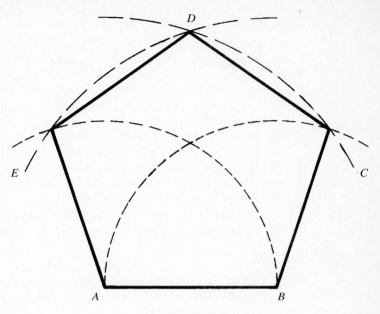

Figure 1.13

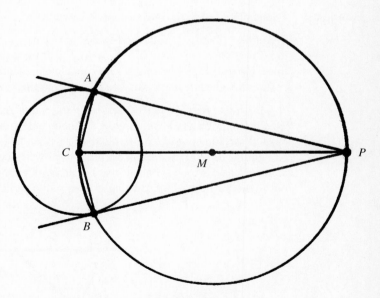

Figure 1.14

1.2.2 Congruence and equality

Euclid's fourth postulate, "All right angles are equal," introduces a second theme, the equality of figures. He used two senses of equal: congruent and equal in measure (length, angle, area, or volume). Intuitively, we know that *congruent* figures coincide if one is placed on the other. Moving figures leads naturally to transformational geometry, which we discuss in Chapter 4. However, Euclid focused on showing that two figures are congruent from the equal measures of various parts. Thus Euclid proves the three well-known triangle congruence theorems *side-angle-side* or *SAS* (I-4), *side-side-side* or *SSS* (I-8), and *angle-angle-side* or *AAS* (I-26).

Euclid didn't think of lengths, areas, and volumes as numbers, so he studied them without formulas. For example, he showed that for parallelograms with the same height, their areas were proportional to the lengths of their bases. Then to compare the areas of any two polygons, he constructed two parallelograms with the same areas (I-45) and compared the parallelograms.

1.2.3 Parallels

Euclid proved the familiar theorems of high school geometry about parallel lines cut by *transversals*, that is, lines intersecting both parallel lines. Euclid's first four postulates allow construction of parallel lines.

Definition 1.2.1 Two lines k and m in the Euclidean plane are *parallel* ($k \| m$) if and only if (iff) they have no points in common or they are equal.

Example 4 Using a line k and a point P not on k, construct a line m parallel to k with P on m.

Solution. Construct l, the perpendicular to k on P, and construct m, the perpendicular to l on P. Then m is parallel to k by I-27. We can generalize this construction with any line l' through P intersecting k, rather than the perpendicular (Fig. 1.15). Proposition I-8 ensures that the two angles at P and Q are congruent, and I-28 guarantees that $m \| k$. ●

Euclid's fifth postulate was essential to proving the most important of his theorems. This assumption dissatisfied many mathematicians because it seemed far from self-evident and too complicated to qualify as a postulate. Many mathematicians from 200 B.C. until A.D. 1800 tried unsuccessfully to prove Euclid's fifth postulate from his other assumptions. Historians believe that Euclid wasn't completely comfortable with his fifth postulate, for he postponed using it until proposition I-29. Playfair's axiom, an easier

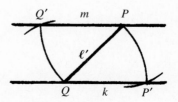

Figure 1.15

1.2 Constructions, Congruence, and Parallels: Euclid's Approach to Geometry

statement about parallel lines, is equivalent to the fifth postulate if we accept Euclid's first 28 propositions. (Two statements A and B are logically *equivalent* provided that we can prove both "If A, then B" and its *converse* "If B, then A".)

Euclid's Fifth Postulate That, if a straight line falling on two straight lines makes the interior angles on the same side less than two right angles, the two straight lines, if produced indefinitely, meet on the side on which are the angles less than the two right angles.

Playfair's Axiom If a point P is not on a line k, then there is on P at most one line m that does not intersect k.

Exercise 1 Illustrate Euclid's fifth postulate and Playfair's axiom. Explain how they relate to each other.

Theorem 1.2.1 Euclid's fifth postulate is equivalent to Playfair's axiom, assuming that Euclid's first 28 propositions hold.

Proof. (Euclid ⇒ Playfair) Suppose that the fifth postulate holds and that we are given a point P not on a line k. Example 4 gives us one parallel, say, $\overleftrightarrow{PR}$, where $\overline{PQ}$ is perpendicular to $\overleftrightarrow{PR}$ and k, as shown in Fig. 1.16. For Playfair's axiom we need to show that $\overleftrightarrow{PR}$ is the only parallel to k through P. We let $\overleftrightarrow{WV}$ be any other line on P with P between V and W. As $\overleftrightarrow{WV}$ is not $\overleftrightarrow{PR}$, either $\angle VPQ$ is acute or $\angle WPQ$ is acute. Then WLOG we assume $\angle WPQ$ to be acute. We now have fulfilled the hypothesis of the fifth postulate: $\angle WPQ$ and $\angle PQU$ together measure less than 180°. Hence $\overleftrightarrow{WV}$ must meet k, showing there is only one parallel to k on P.

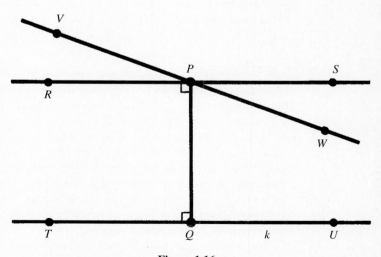

Figure 1.16

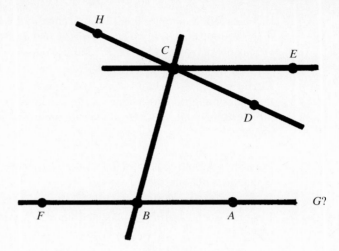

Figure 1.17

(Playfair ⇒ Euclid) Suppose that Playfair's axiom holds and $\overline{BC}$ falls on $\overleftrightarrow{AB}$ and $\overleftrightarrow{CD}$ so that $m\angle ABC + m\angle BCD < 180°$. We must show that $\overleftrightarrow{AB}$ and $\overleftrightarrow{CD}$ meet on the side of A and D. We construct the line $\overleftrightarrow{CE}$, as shown in Fig. 1.17, such that $\angle BCE \cong \angle CBF$. By Example 4, $\overleftrightarrow{CE} \parallel \overleftrightarrow{AB}$. Playfair's axiom states that there is only one parallel, meaning that $\overleftrightarrow{CD}$ must intersect $\overleftrightarrow{AB}$, say, at G. However, Playfair's axiom does not tell us directly on which side of B this point G lies. Note that we have a triangle $\triangle BCG$. Euclid I-17 guarantees that the measures of any two angles of $\triangle BCG$ add to less than $180°$. Our original assumption about angles $\angle ABC$ and $\angle BCD$ fits I-17 perfectly, but that isn't enough. We must show that the angles on the other side, $\angle FBC$ and $\angle HCB$, do not satisfy I-17. The measures of all four of these angles must add to $360°$, and we assumed that the first two add to less than $180°$. Hence the last two must add to more than $180°$, implying that they are not part of the triangle $\triangle BCG$. Hence G is indeed on the same side of B as A is. ∎

1.2.4 The Greek legacy

Three unsuccessful Greek constructions also inspired the development of mathematics. The first, called *doubling the cube*, was to construct (the side of) a cube with twice the volume of a given cube. The second was *trisecting an angle*. The third, *squaring the circle*, was to construct a square with the same area as a given circle. The Greeks did find exact constructions for these problems using methods beyond a straightedge and compass. In fact, the development of conics (ellipses, parabolas, and hyperbolas) was connected with doubling the cube and trisecting an angle. Related to the third problem, Archimedes, the greatest mathematician of the ancient world, proved that the area of a circle equaled the area of a triangle whose height was the radius of the circle and whose base was the circumference of the circle.

Not until the nineteenth century were mathematicians able to prove that these three constructions were impossible with only an unmarked straightedge and compass. Many

ARCHIMEDES

Archimedes (287–212 B.C.) was the greatest mathematician of the ancient world, and its outstanding engineer and physicist. Many legends attest to his engineering feats, as well as to his absent-mindedness. Upon understanding his law of the lever, as illustrated by the figure, Archimedes is supposed to have claimed, "Give me a place to stand and I will move the world."

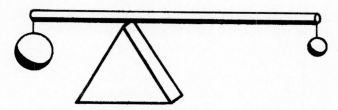

Archimedes' law of the lever: The mass of one weight times its distance from the fulcrum equals the mass of the other weight times its distance from the fulcrum.

The story continues with the King of Syracuse asking for a practical demonstration of mechanical advantage. So, by himself, Archimedes pulled a fully loaded ship up a beach by using a sophisticated arrangement of pulleys. Another time the King asked Archimedes to determine, without harming it, if his new crown was made entirely of gold. In a bath Archimedes grasped the principle of the buoyancy of water and so found a solution to the King's problem. In the excitement of discovery, Archimedes ran naked from the bath shouting "Eureka!", meaning "I have found it!" He apparently forgot the world about him when he worked on a problem—drawing figures in ashes or on the oil rubbed on him after a bath. During the extended Roman siege of Syracuse, Archimedes designed various machines that greatly helped in its defense and intimidated the Roman soldiers. He was killed by a soldier when the Romans finally won.

Archimedes brought great imagination and supreme mathematical expertise to his mathematical investigations. He found and proved many theorems, including the exact area of a circle and a parabolic section. His most famous results give the surface area and the volume of a sphere. He also proved theorems concerning tangents and the centers of gravity for various shapes. A number of problems in this chapter concern other results of Archimedes. He brilliantly used the particular geometric properties of each shape he considered. Archimedes' mathematics went beyond his elegant geometric proofs. He gave upper and lower estimates for π. In the "Sand Reckoner" he devised a number system that could handle huge numbers, including his estimate of the number of grains of sand in the earth. In "The Method," a treatise rediscovered in 1906 after being lost for more than a thousand years, he explained how he used his law of the lever to discover new mathematical results that he later proved rigorously.

Archimedes' geometry epitomized Greek mathematical thought. His flawless proofs of difficult theorems were unsurpassed for more than 1500 years. His work also showed the limitations of Greek mathematics. Each result required its own ingenious approach for its proof, unlike calculus and other modern mathematics. Archimedes' writings in physics and mathematics inspired scholars for centuries, especially during the Renaissance. Even today the beauty of Archimedes' mathematics reminds us of what is best in mathematics.

people misinterpret this impossibility, thinking that mathematicians just haven't been clever enough to find constructions. Thus people still propose solutions for these ancient problems. These proofs require abstract algebra and are beyond the scope of this text. (See Gallian [11].) In brief, mathematicians converted the geometric problem of what lengths were constructible with straightedge and compass into an algebraic problem about what irrational numbers could be written in a particular form by using rational numbers, repeated square roots, and arithmetic operations. Both doubling the cube and trisecting an angle involve cubic equations whose roots cannot in general be written in that particular form. Pierre Wantzel proved the impossibility of these constructions in 1837. To square the circle requires the construction of π, which also cannot be written in the form just described. Ferdinand Lindemann proved this assertion in 1882 by proving that π was *transcendental*; that is, it is not the root of any polynomial whose coefficients are rational numbers.

The decomposition of a figure, another aspect of Euclid's geometry, led to other modern investigations. Euclid proved that certain shapes had the same area by decomposing one shape into smaller pieces that could be reassembled to form the other shape. Decomposition puzzles have been popular for centuries, especially the Chinese Tangram puzzle. W. Bolyai in 1832 and P. Gerwien in 1833 independently showed that two polygons in the plane with the same area could be decomposed into one another by using finitely many smaller polygons. (See Boltyanskii [3].) In Chapter 3, we use decomposition to examine area in hyperbolic geometry.

David Hilbert posed the corresponding problem for three-dimensions in a famous talk in 1900, when he presented a list of 23 important, unsolved problems. The same year Max Dehn proved that a regular tetrahedron (triangular pyramid) could not be cut into finitely many polyhedra and then reassembled to form a cube. This and other results showed that a theory of volumes of polyhedra needed limit arguments for rigorous proofs. (See Boltyanskii [3].)

Euclid's *Elements* included results now considered to be part of number theory, algebra, and irrational numbers. Mathematicians learned from Euclid's text for two thousand years, and many important developments in mathematics stem from it. The *Elements* richly deserves its reputation as the most important mathematics book ever written. (See Kline [18, Chapters 4 and 5] for more historical information.)

PROBLEMS FOR SECTION 1.2

1. a) Identify which of Euclid's first 48 propositions (Appendix A) concern constructions.
   b) Repeat part (a), replacing *constructions* with *congruence*.
   c) Repeat part (a), replacing *constructions* with *equality of measure*.
   d) Repeat part (a), replacing *constructions* with *parallels*.
2. a) For lengths a and b, with $a > b$, construct $a + b$ and $a - b$.
   b) In Fig. 1.18, $\overleftrightarrow{QS} \| \overleftrightarrow{RT}$. Explain how PQ, PR, PS, and PT are related.
   c) Use part (b) to construct the lengths $a \cdot b$ and a/b, given a unit length and the lengths a and b.
3. a) In Fig. 1.19, let $AD = 1$ and $BD = x$. Explain why $CD = \sqrt{x}$.
   b) Let M be the midpoint of $\overline{AB}$ and use algebra to show that $CM = AM = BM$ and so A, B, and C are on a circle centered at M.
   c) Use segments of length 1 and x to construct a segment of length $\sqrt{x}$.

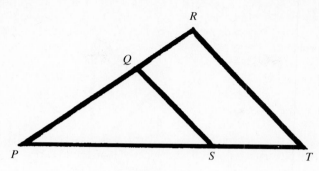

Figure 1.18

d) Construct segment of lengths $\sqrt[4]{x}$ and $\sqrt[8]{x}$. Generalize.

e) Construct a segment of length $\sqrt{1+\sqrt{2}}$.

4. Begin with a circle and construct the following regular inscribed polygons. You may use earlier constructions in later ones. (An *n*-sided *polygon*, or *n-gon*, is a set of distinct vertices P_1, P_2, ..., P_n in a plane and the *edges* (line segments) $\overline{P_1P_2}$, $\overline{P_2P_3}$, ..., $\overline{P_nP_1}$, with the condition that two edges intersect only at their endpoints. A polygon is *inscribed* in a figure only if the vertices of the polygon are on that figure and the rest of the polygon is in the interior of the figure. A polygon is *regular* only if all the edges and angles are congruent.)

a) An equilateral triangle
b) A square
c) A hexagon
d) An octagon
e) A dodecagon (12-gon)
f) Explain how to inscribe a regular 2*n*-gon from one with *n* sides.

Remarks Carl F. Gauss discovered in 1801 which regular polygons could be constructed with straightedge and compass. He showed that, if the number of sides is a product of a power of 2 and distinct primes of the form $(2^{2^k}+1)$, the regular polygon is constructible. The only known primes of this form, called Fermat primes, are 3, 5, 17, 257, and 65,536. Gauss actually constructed a regular 17-gon in 1796. He conjectured and Pierre Wantzel proved in 1837 that no other regular polygons are constructible by straightedge and compass alone.

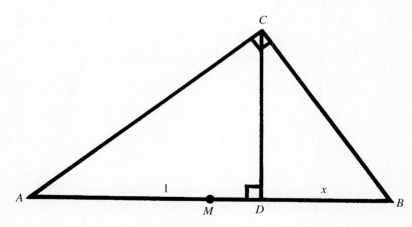

Figure 1.19

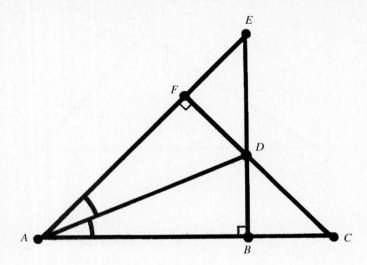

Figure 1.20

5. a) Construct the circumscribed circle for a triangle.
   b) Construct the inscribed circle—the circle tangent to the three sides—for a triangle.
6. Use SAS, SSS, and AAS to find and prove the congruence of the four pairs of triangles shown in Fig. 1.20.
7. In Fig. 1.21 assume that $\overline{AD} \cong \overline{BC}$ and $\angle ADC \cong \angle BCD$. Use only Euclid's first 28 propositions in your proofs.
   a) Prove that $\angle DAB \cong \angle CBA$. [*Hint:* Find two pairs of congruent triangles formed with the given sides and the two diagonals.]
   b) If E is the midpoint of $\overline{CD}$ and F is the midpoint of $\overline{AB}$, show that $\overline{EF}$ is perpendicular to both $\overline{AB}$ and $\overline{CD}$. [*Hint:* Draw $\overline{DF}$ and $\overline{CF}$.]
8. Recall that a *parallelogram* is a quadrilateral whose opposite sides are parallel. Use only Euclid's first 29 propositions in your proofs.
   a) Prove that opposite sides of a parallelogram are congruent.

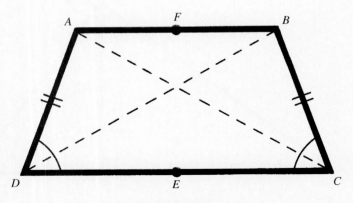

Figure 1.21

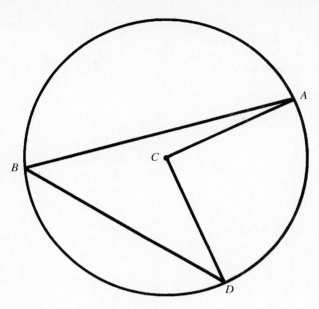

Figure 1.22

b) If $\overline{AB} \parallel \overline{CD}$ and $\overline{AB} \cong \overline{CD}$, prove that $ABCD$ is a parallelogram. If $\overline{AB} \parallel \overline{CD}$ and $\overline{AC} \cong \overline{BD}$, must $ABCD$ be a parallelogram? Explain.

c) Prove that a quadrilateral is a parallelogram iff the diagonals bisect each other.

9. Prove that in a circle, the central angle is twice the inscribed angle. That is, in Fig. 1.22, $m\angle ACD$ is twice $m\angle ABD$, where C is the center of the circle. [*Hint:* Draw the diameter through B and use the isosceles triangles $\triangle BCA$ and $\triangle BCD$.]

10. a) Rewrite Euclid II-13 and explain why it is equivalent to the law of cosines for an acute angle. [*Hint:* Examine Fig. 1.23.]

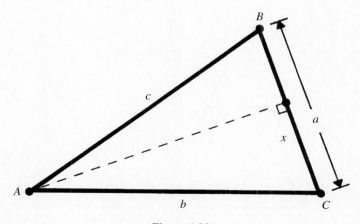

Figure 1.23

Proposition II-13: In a triangle, the square on the side subtending an acute angle is less than the squares on the sides containing the acute angle by twice the rectangle contained by one of the sides about the acute angle, namely, that on which the perpendicular falls, and the straight line cut off within by the perpendicular toward the acute angle.

The law of cosines: In any triangle with sides of length a, b, and c, $c^2 = a^2 + b^2 - 2ab\cos(C)$, where C is the angle opposite side c.

    b) Prove your reformulation of Euclid II-13 by using the Pythagorean theorem.

    c) Does your proof in part (b) hold if the angle is obtuse? If so, explain why; if not, modify it to hold. (Euclid II-12 handles the obtuse case.)

11. Two quadrilaterals $ABCD$ and $EFGH$ are clearly congruent if all corresponding sides, angles, and diagonals are congruent. Look for smaller sets of these correspondences, which are sufficient for convex quadrilaterals.

    a) Give an example to show that the congruence of the four pairs of sides (SSSS) is not sufficient. This insufficiency illustrates a basic engineering property: triangles are rigid (SSS); other polygons need triangular bracing to be rigid.

    b) A diagonal brace (SSSSD) ensures congruence for convex quadrilaterals. State SSSSD clearly and completely; prove it.

    c) State AAAS clearly and completely. Either prove that AAAS is a congruence theorem for convex quadrilaterals or find a counterexample.

    d) Repeat part (c) for SASAA, SAASA, and SASSS.

    e) In Fig. 1.24, $ABCD$ and $EFGH$ satisfy $\overline{AB} \cong \overline{EF}$, $\overline{BC} \cong \overline{FG}$, $\overline{CD} \cong \overline{GH}$, $\overline{DA} \cong \overline{HE}$, and $\overline{BD} \cong \overline{FH}$, yet they are not congruent. Explain why this situation does not contradict part (b). If they exist, find similar examples for SASAA and SASSS.

12. Begin with a unit length and investigate what other lengths are constructible with a straightedge and compass.

    a) Use Problem 2 to describe how to construct integer and rational lengths.

    b) Use Problems 2 and 3 to describe how to construct lengths corresponding to numbers built from rational numbers by using square roots and the four arithmetic operations ($+, -, \times, \div$). Call such numbers *constructible*.

    c) Recall that a straight line has a first-degree equation $ax + by + c = 0$ and that a circle has a second-degree equation $x^2 + y^2 + dx + ey + f = 0$. Suppose that the coefficients of two lines, two circles, or a line and a circle are all constructible numbers. Explain why the coordinates of the intersections of these lines and circles must also be constructible numbers. [*Hint:* Remember the quadratic formula.]

    d) Let (s, t) and (u, v) be the coordinates of two points in the plane. Find the equation of the line through these two points. Explain why the coefficients of this equation can be written

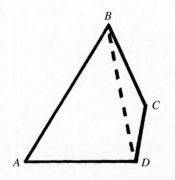
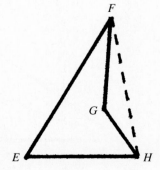

Figure 1.24

in terms of s, t, u, v, and the four arithmetic operations.

e) Repeat part (d) for the circle with center (s, t) and passing through (u, v).

f) Explain why parts (c), (d), and (e) ensure that, if the coordinates of the given points are constructible numbers, the coordinates of any points that you can construct with lines and circles through these points are constructible numbers.

13. Trisecting an angle generally involves a cubic equation. An angle is *constructible* only if given a length, you can construct two other lengths such that the three form a triangle with the desired angle.

    a) Show that an angle is constructible iff the cosine of that angle is constructible. Draw a diagram.

    b) Use trigonometry to show that $\cos(3z) = 4\cos^3(z) - 3\cos(z)$.

    c) If the cosine of a constructible angle is b, show that you can trisect that angle only if you can construct a segment whose length x satisfies the equation $4x^3 - 3x = b$. [Incidentally, $x = \left(\sqrt[3]{b + \sqrt{b^2 - 1}} + \sqrt[3]{b - \sqrt{b^2 - 1}}\right)/2$.]

14. A well-known method for trisecting an angle requires the marking of a particular length on the straightedge, a slightly stronger condition than an unmarked straightedge. Explain why the following method trisects any angle. [*Hint:* Draw $\overline{DM}$, use Problem 3(b), and look for isosceles triangles.]

    In Fig. 1.25 $\angle ABC$ is the angle to trisect, D is any point on $\overline{AB}$, $\overline{DE}$ is perpendicular to $\overline{BC}$, and $\overline{DF}$ is parallel to $\overline{BC}$. On the straightedge mark a length YZ twice BD. Place the straightedge so that it goes through B and so Y is on the segment $\overline{DE}$. Slide the straightedge around, keeping the previous two conditions satisfied, until point Z is on the ray $\overrightarrow{DF}$. In Fig. 1.25, points Y' and Z' indicate the correct positions of Y and Z. M is the midpoint of $\overline{Y'Z'}$. Claim: $m\angle CBZ'$ is one-third of $m\angle ABC$.

15. Prove that the following are equivalent to Playfair's axiom, using Euclid's first 28 propositions and Theorem 1.2.1. Draw diagrams.

    a) If a straight line cuts one of two parallel lines, it cuts the other.

    b) Given two parallel lines and a transversal, the alternate interior angles are congruent (Euclid I-29).

    c) If $k \| l$ and $l \| m$, then $k \| m$ (Euclid I-30).

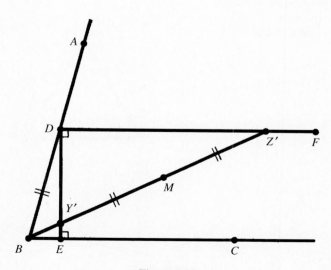

Figure 1.25

1.3 A CRITIQUE OF EUCLID–MODERN AXIOMATICS

At the end of the nineteenth century, Euclid's work received scrutiny far surpassing the preceding 2000 years' efforts. This examination revealed how much more precise and explicit mathematics has become, without calling into question any of Euclid's results. For example, consider Euclid's proof of his first proposition, given in shortened form (from Heath [13, volume I, 241]. Can you find the logical gap in his proof, given his postulates and common notions? (Figure 1.26.)

Proposition I-1 On a given finite straight line to construct an equilateral triangle.

Proof. Let $\overline{AB}$ be the given finite straight line. . . . With center A and radius $\overline{AB}$ let the circle BCD be described; again, with center B and radius $\overline{BA}$ let the circle ACE be described (postulate 3, in Appendix A). From point C, in which the circles cut each other, to points A and B, join the straight lines $\overline{CA}$ and $\overline{CB}$ (postulate 1, in Appendix A). Now, point A is the center of the circle CDB, so AC is equal to AB (definition 15, in Appendix A). Again, . . . BA is equal to BC. . . . And things equal to the same thing are also equal to one another . . . (common notion 1, in Appendix A). Therefore . . . AC, AB, BC are equal to one another. Therefore $\triangle ABC$ is equilateral. ∎

Euclid's construction is straightforward, and he followed it with a proof that the three sides are congruent. However, Euclid never showed, nor could he show from his assumptions, that the circles must intersect. This logical gap indicates one of the many implicit assumptions Euclid made that were very hard to detect because they were "obvious." (We made the same assumption in Example 3 of Section 1.2.) In modern terms Euclid assumed that lines, circles, and other figures are continuous. In the fifth postulate and numerous other places Euclid assumes an order to points on lines with no justification or discussion. (In Example 3 of Section 1.2 we assumed such ordering to

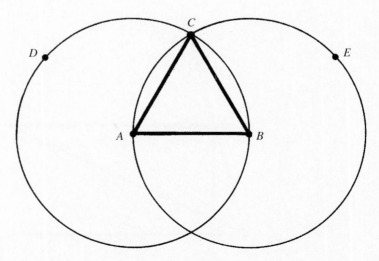

Figure 1.26

talk about a point "outside" a circle.) David Hilbert and others recognized that rigorous proofs require explicit axioms about continuity, order, and other assumptions.

1.3.1 Hilbert's axioms

Hilbert's axiomatic system, first presented in 1898 [15], corrects the logical deficiencies found in Euclid's work. A careful comparison of Euclid's definitions and assumptions (Appendix A) with Hilbert's system (Appendix B) shows some of the differences between a modern axiomatic system and the system Euclid developed. In particular, Hilbert's axioms provide enough explicit properties to prove all the theorems of Euclidean geometry without any logical gaps. Also, Hilbert uses undefined terms, whereas Euclid starts his text with a list of definitions, some having little mathematical value.

The first three groups of Hilbert's axioms codify elementary properties, many previously overlooked. Axioms I-1 and I-2 modernize Euclid's first two postulates. Axiom I-3 guarantees that there are some points in the geometry. The first three axioms of order give the most elementary properties of betweenness on a line. Pasch's axiom (II-4), which says that a line that *enters* a triangle must *exit* as well, deserves more explanation. To help explain Pasch's axiom we derive it from the separation axiom, an alternative (II-4′) that Hilbert gives.

Theorem 1.3.1 The separation axiom (II-4′) implies Pasch's axiom (II-4).

Exercise 1 Illustrate the proof of Theorem 1.3.1.

Proof. Suppose that the separation axiom holds: A line m separates the points that are not on m into two sets such that if X and Y are in the same set, $\overline{XY}$ does not intersect m, and if X and Y are in different sets, $\overline{XY}$ does intersect m. To show Pasch's axiom, suppose that a line l enters $\triangle ABC$ at D on side $\overline{AB}$ but that A, B, and C are not on l. By the separation axiom, A and B are on opposite sides of l; that is, they are in different sets of points. Now C is in just one of those two sets, which means only one of $\overline{AC}$ and $\overline{BC}$ have a point on l, where l exits from $\triangle ABC$, showing Pasch's axiom. ∎

The congruence axioms III-2 and III-3 make precise Euclid's first two common notions. Axioms III-1 and III-4 guarantee the existence and uniqueness of congruent segments and angles. These axioms replaced Euclid's use of circles and constructions. Axiom III-5 is Euclid's Proposition I-4, the familiar SAS property of high school geometry. Euclid's proof has a logical gap because he used motion without providing any axioms about movement.

Hilbert's fourth group contains only Playfair's axiom, which by Theorem 1.2.1 is logically equivalent to Euclid's fifth postulate. This axiom distinguishes Euclidean geometry from hyperbolic geometry, which we introduce in Chapter 3.

Although the earlier axioms ensure lines have infinitely many points, group V implies that the points on a line correspond to the real numbers. Axiom V-1, the Archimedean axiom, eliminates infinitely large and infinitely small line segments. [Archimedes (circa 287–212 B.C.) used this idea in some of his proofs.] Hilbert's complicated axiom of linear completeness, V-2, ensures that lines have no gaps while avoiding concepts from analysis.

DAVID HILBERT

> But in the present century, thanks in good part to the influence of Hilbert, we have come to see that the unproved postulates with which we start are purely arbitrary. They MUST be consistent; they HAD BETTER lead to something interesting.
> —*Coolidge*

The work of David Hilbert (1862–1943) symbolizes the modern abstract, axiomatic approach to mathematics. He contributed significant results in many fields, including algebraic invariants, number theory, partial and ordinary differential equations, integral equations, geometry, and the foundations of mathematics. He won international renown at age 26 when he published a result in algebraic invariants, a precursor of modern abstract algebra, that had seemed beyond possibility. His proof neatly avoided the laborious and limited constructive methods of previous mathematicians. One of these mathematicians, Paul Gordan, at first disdained Hilbert's radical nonconstructive proof saying, "This is not mathematics; it is theology." He later added, "I have convinced myself that theology also has its advantages." Hilbert's work in integral equations led to Hilbert space, an infinite dimensional analog to Euclidean space important in the study of quantum mechanics in physics.

Hilbert was very influential in the foundations of mathematics. Developments in nineteenth century analysis and geometry made clear the need for a careful scrutiny of axiomatics. To make hidden assumptions explicit, Hilbert realized that mathematicians must isolate formal axioms from their meaning. He devised ground-breaking axiom systems for both the real numbers and Euclidean geometry. He proved the relative consistency of geometry, assuming the consistency of the real numbers. Hilbert's program, which he hoped would prove the absolute consistency of all of mathematics, led others to a penetrating analysis of logic and the foundations of mathematics. Gödel's famous incompleteness theorems came out of that program and proved, among other things, the impossibility of Hilbert's goal. However, Gödel's theorems did show how deeply Hilbert's axiomatic approach enabled mathematicians to probe the foundations of mathematics.

Hilbert's formal axiomatics devoid of meaning was never intended to replace mathematics. Throughout his career, Hilbert stressed the dynamic interplay of concrete problems and general, abstract theories. His famous list of 23 problems in 1900 illustrates well this link. Each problem has led mathematicians to a more profound understanding of an area of mathematics. His own research revealed the power of particular problems to inspire deep mathematics and of abstract mathematics to elucidate particular problems.

1.3.2 Axiomatic systems

An axiomatic system provides an explicit foundation for a mathematical subject. Axiomatic systems include seven parts: the logical language, rules of proof, undefined terms, axioms, definitions, theorems, and proofs of theorems. We discuss the last five parts—the ones with geometric content.

Consider Euclid's definition of a point as "that which has no part." This definition is more a philosophical statement about the nature of a point than a way to prove statements. Euclid's definition of a straight line, "a line which lies evenly with the points on itself," is unclear as well as not useful. In essence, points and lines were so basic to Euclid's work that there is no good way to define them. Mathematicians realized centuries ago the need for undefined terms in order to establish an unambiguous beginning. Otherwise, each term would have to be defined with other terms, leading either to a cycle of terms or an infinite sequence of terms. Neither of these options is acceptable for carefully reasoned mathematics. Of course, we then define all other terms from these initial, undefined terms. However, undefined terms are, by their nature, unrestricted. How can we be sure that two people *mean* the same thing when they use undefined terms? In short, we can't. The axioms of a mathematical system become the key: They tell us how the undefined terms *behave*. Axioms are sometimes called *operational definitions* because they describe how to use terms and how they relate to one another, rather than telling us what terms "really mean." Indeed, mathematicians permit any interpretation of undefined terms, as long as all the axioms hold in that interpretation. In Section 1.4 we explore the interplay between axioms and their interpretations in models.

Unlike the Greek understanding of axioms as self-evident truths, we do not claim the truth of axioms. However, this does not mean that we consider axioms to be false. Rather, we are free to choose axioms to formulate the fundamental relationships we want to investigate. From a logical point of view, the choice of axioms is arbitrary; in actuality, though, mathematicians carefully pick axioms to focus on particular features. For example, in perspective drawing parallel lines intersect at a point on the horizon. Projective geometry is an axiomatic system in which any two lines intersect in a point. This system, discussed in Chapter 6, enables us to understand many consequences stemming from perspective. However, we don't need to decide whether "in truth" there are parallel lines or if all lines intersect. Indeed, in the concrete world of atoms and energy, there are no mathematical lines at all. Nevertheless, these axiomatic systems and many others have given us a profound understanding of the world. Axiomatic systems allow us to formulate and logically explore abstract relationships, freed from the specificity and imprecision of real situations.

Mathematicians build two basic types of axiomatic systems. One completely characterizes a particular mathematical system. For example, Hilbert's axioms characterize Euclidean geometry completely.[1] The second focuses on the common features of a family of structures, such as vector spaces. Although infinitely many different vector spaces exist, all satisfy certain essential properties. The general study of vector spaces greatly aids the development of theories in economics, physics, mathematics, and other fields. Such axiomatic systems unite a wide variety of examples within one powerful theoretical framework.

Mathematical definitions are built from undefined terms and previously defined terms. For example, Hilbert defines the angle $\angle ABC$ as a point B and two rays $\overrightarrow{BA}$ and $\overrightarrow{BC}$. In turn, the ray $\overrightarrow{BA}$ is the set of points X on the line $\overleftrightarrow{AB}$ such that X is between A and B, X is A, or A is between B and X. Thus Hilbert reduces the notion of an angle to the undefined terms *point*, *line*, *on*, and *between*. These same undefined terms are sufficient to define *convex*, an important concept in modern mathematics. The Greeks

---

[1] This statement may appear to contradict Gödel's incompleteness theorem. However, axiom V-2 is a "second-order axiom," a concept beyond the level of this text. (See Delong [6] for information on second-order logic.)

26 **Chapter 1** **Euclidean Geometry**

Figure 1.27 Nonconvex and convex sets.

never distinguished between nonconvex sets, which "bend," and convex sets (Fig. 1.27), perhaps because they never considered betweenness. Intuitively, in a convex set, every point can "see" every other point.

Definition 1.3.1 A set S is *convex* iff for distinct points P and Q in S, $\overline{PQ}$ is entirely in S.

Exercise 2 Rewrite the definition of convex, using only the undefined terms *between* and *point*.

Exercise 3 Which of the terms in Euclid's definitions (Appendix A) do you think should be undefined terms? Which of the remaining definitions fit our modern understanding of definitions?

Theorems and their proofs are the most distinctive parts of mathematics, whether in an axiomatic or some other system. In an axiomatic system, a theorem is a statement whose proof depends only on previously proven theorems, the axioms, the definitions, and the rules of logic. This condition ensures that the entire edifice of theorems rests securely on the explicit axioms of the system.

Proofs of theorems in an axiomatic system cannot depend on diagrams, even though diagrams have been part of geometry since the ancient Greeks drew figures in the sand. We need the powerful insight and understanding that such diagrams provide. However, a corresponding risk comes with the use of pictures: We are liable to accept as intuitive a step that does not follow from the given conditions. Euclid's first proof, discussed previously, shows how easy it is to include implicit assumptions. Euclid's minor "sins of omission" never led him to an erroneous result, but the potential remains for a diagram to mislead us, as in Example 1, used with permission from Dubnov [7, 15]. Even though diagrams are not permissible in a proof in an axiomatic system, they certainly can and

1.3 A Critique of Euclid–Modern Axiomatics

should be included to help us understand the ideas. They must be studied critically to ensure that the illustrated relationships are proved or legitimately assumed.

Example 1 *Claim.* A rectangle inscribed in a square is a square.

Verify that the claim is incorrect. Then try to find the error in the "proof."

Proof. Let rectangle $MNPQ$ be inscribed in square $ABCD$ (Fig. 1.28). Drop perpendiculars from P to $\overline{AB}$ and from Q to $\overline{BC}$ at R and S, respectively. Clearly, $\overline{PR} \cong \overline{QS}$ because these segments match the sides of the square $ABCD$. Furthermore, the rectangle's diagonals are congruent: $\overline{PM} \cong \overline{QN}$. So $\triangle PMR \cong \triangle QNS$, and hence $\angle PMR \cong \angle QNS$. Consider the quadrilateral $MBNO$, where O is the intersection of $\overline{QN}$ and $\overline{PM}$. Its exterior angle at vertex N is congruent to the interior angle at vertex M, so the two interior angles at vertices N and M are supplementary. Thus the interior angles at vertices B and O must be supplementary. But $\angle ABC$ is a right angle and hence $\angle NOM$ must also be a right angle. Therefore the diagonals of rectangle $MNPQ$ are perpendicular. Hence $MNPQ$ is a square.

The preceding argument is correct up to the conclusion $\angle PMR \cong \angle QNS$. Then the diagram shown in Fig. 1.28 misleads us to think that $\angle ONB$ is supplementary to $\angle OMR$. However, these angles can be congruent if we switch N and S in the diagram and correspondingly move Q down. Illustrate this second case. ●

How can we possibly make all assumptions explicit and eliminate all risk of incorrect proofs? Mathematical logic, developed by Hilbert and others, involves the use of a formal language so austere that a proof can be checked in a purely mechanical manner, free from human intuition. In principle, a computer could check such a proof to decide its validity. Consider, for example, the statement, "Two distinct points have a unique

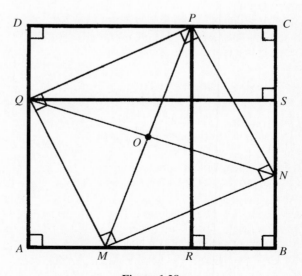

Figure 1.28

line on them." We can make the statement more explicit as follows: "For all points A_1 and A_2 if $A_1 \neq A_2$, then there is a unique line B_1 such that A_1 is on B_1 and A_2 is on B_1." Finally we could turn this more explicit statement into pure symbols:

$$\forall A_1 \forall A_2 (\neg(A_1 = A_2) \Rightarrow (\exists! B_1 : (R_1(A_1, B_1) \land R_1(A_2, B_1)))).$$

Clearly, we need to decode these symbols so that people not familiar with them can read the statement. (We don't use these logic symbols, but for your information, $\forall \neg \Rightarrow \exists ! \land$ represent *for all*, *not*, *implies*, *there exists*, *a unique*, *and*, respectively. The variable R_1 represents the relation *is on*.) If the entire axiomatic system, including the rules of proof, is encoded into such a formal language, we can mechanically determine whether a given string of symbols encodes a logical proof of the string of symbols encoding the theorem. There will be no risk of an inappropriate inference, but there will be an incredible barrier to human understanding. Indeed, finding a proof in such a language is a daunting task.

Axiomatic systems are a workable compromise between the austere formal languages of mathematical logic and Euclid's work, with its many implicit assumptions. Mathematicians need both the careful reasoning of proofs and the intuitive understanding of content. Axiomatic systems provide more than a way to give careful proofs. They enable us to understand the relationship of particular concepts, to explore the consequences of assumptions, to contrast different systems, and to unify seemingly disparate situations under one framework. In short, axiomatic systems are one important way in which mathematicians obtain insight. (Heath's edition of Euclid's *Elements* [13] provides detailed commentary on its logical shortcomings. Wilder [29] explores axiomatic systems in more detail.)

PROBLEMS FOR SECTION 1.3

1. **a)** From Group I of Hilbert's axioms, how many points and lines must exist? Prove your answer.
   b) Given two distinct lines prove from Group I of Hilbert's axioms that at most one point lies on both lines.
   c) Include axiom IV-1 and repeat part (a).
2. **a)** Compare Hilbert's axiom III-4 with Euclid I-7.
   b) Read in Euclid's *Elements* [13, vol. I, 247ff] the proofs of I-4 and I-7 and discuss the logical gaps in these proofs.
3. Discuss how well Euclid's definition of an angle (number 8) fits your intuition and how easy it would be to apply in a proof.
4. **a)** Use Hilbert's definition of an angle to define the interior of an angle. Define a triangle and its interior. How does the interior of a triangle relate to the interiors of its angles? Illustrate your definitions.
   b) Define a convex set from Hilbert's undefined terms. Are the interior of an angle and of a triangle, as you defined them, convex? Justify your answer.
5. **a)** List which of Hilbert's axioms can refer to points on just one line.
   b) Use the axioms in part (a) that are in groups I and II to find infinitely many points on any line. [*Hint:* Use induction.]
   c) Fix P_0 and P_1 on a line m. Use the axioms in part (a) to prove by induction that for all positive integers n there are points P_n on m such that P_n is between P_0 and P_{n+1} and such that $\overline{P_n P_{n+1}} \cong \overline{P_0 P_1}$.
   d) In the notation of part (c), what properties should P_{-n} satisfy? Prove that there are points that satisfy these properties.
   e) Which axiom(s) guarantee(s) that between two points on a line there is a third point?

6. Consider the axiomatic system with the undefined terms *point* and *adjacent* and the following axioms. Use ⋈ to denote "is adjacent to."
   i) There is at least one point.
   ii) If $P \bowtie Q$, then $Q \bowtie P$ and $P \neq Q$.
   iii) Every point has exactly three distinct points adjacent to it.
   iv) If $P \bowtie Q$ and $P \bowtie R$, then not $Q \bowtie R$.
   a) Prove that there are at least six points.
   b) Suppose that you omit axiom (iv). Can you still prove that there are at least six points? If so, prove it; otherwise find the largest number of points that must exist and prove your answer.
   c) Suppose that we change axiom (iii) to require exactly four distinct points adjacent to any point but leave the other axioms unchanged. How many points must exist? Prove your answer. Generalize.

7. Consider the axiomatic system with the undefined terms *point*, *line*, and *on* and the following axioms.
   i) There are a line and a point not on that line.
   ii) Every two distinct points have a unique line on them both.
   iii) Every two distinct lines have at least one point on them both.
   iv) Every line has at least three points on it.
   a) Given two distinct lines prove that they have exactly one point on them.
   b) Prove that there are at least seven points.
   c) Given any point prove that it has at least three lines on it. [*Hint:* First consider a point not on the line of axiom (i).]
   d) Prove that there are at least seven lines.

8. In the axiomatic system with the undefined terms *point* and *between*, use $P(Q)R$ to denote "Q is between P and R." Define a set S to be *convex* iff whenever P and R are in S and $P(Q)R$, then Q is in S. The axioms are:
   i) If $P(Q)R$, then $R(Q)P$.
   ii) If $P(Q)R$, then not $P(R)Q$ and $P \neq R$.
   a) Prove that, if $P(Q)R$, then not $Q(R)P$, not $R(P)Q$, and not $Q(P)R$.
   b) Prove that, if $P(Q)R$ then P, Q, and R are three distinct points.
   c) Compare the axioms and parts (a) and (b) of this problem with Hilbert's axioms of order II-1, II-2, and II-3.
   d) Prove that, if S and T are convex, then $S \cap T$ is also convex.
   e) If each S_i is convex, for i in a finite or infinite index set I, prove that $\bigcap_{i \in I} S_i$ is convex, where $\bigcap_{i \in I} S_i = \{P : \text{for all } i \in I, P \in S_i\}$.

1.4 AXIOMATIC SYSTEMS, MODELS, AND METAMATHEMATICS

In Section 1.3 we showed the logical need for undefined terms in an axiomatic system, which ignores what those terms mean. However, people depend on meaning and intuition to create and understand mathematics. A string stretched taut between its ends provides a strong intuition of a line, but such an image is too imprecise for mathematics. Mathematical models provide an explicit link between intuitions and undefined terms. The usual analytic model of Euclidean plane geometry is the set $\mathbf{R}^2 = \{(x, y) : x, y \in \mathbf{R}\}$, where a *point* is interpreted as an ordered pair of numbers (x, y) and a *line* is interpreted as the points that satisfy an appropriate first degree equation $ax + by + c = 0$. High school students spend considerable time learning how this algebraic model matches geometric intuition and axioms. In making a model, we are free to interpret the undefined terms in any way we want, provided that all the axioms hold under our interpretation. Note that the axioms are not by themselves true; a context is needed to give meaning to the axioms in order for them to be true or false.

Definition 1.4.1 A *model* of an axiomatic system is a set of objects together with interpretations of all the undefined terms of the axiomatic system such that all the axioms are true in the set using the interpretations.

30 **Chapter 1 Euclidean Geometry**

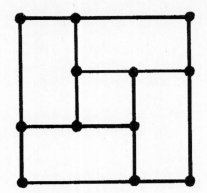

Figure 1.29 Interpret *point* by a dot and *line* by a line segment.

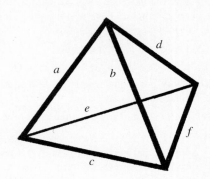

Figure 1.30 Interpret *point* by a line segment and *line* by a face of the pyramid.

Example 1 Figures 1.29 and 1.30 give different models of the axiomatic system with the undefined terms *point*, *line*, and *on* and the axioms:

  i) Every point is on exactly two lines.
  ii) Every line is on exactly three points.

In these and later models *on* has the obvious interpretation. ●

Exercise 1 In the model given by Fig. 1.30 reinterpret *lines* as the vertices (corners) of the pyramid (and *points* as line segments.) Is the result still a model of the axioms in Example 1.4.1?

Models do much more than provide concrete examples of axiomatic systems; they lead to important understandings about axiomatic systems. *Metamathematics* considers axiomatic systems and their models as a whole. The prefix *meta*, Greek for "beyond," distinguishes metamathematics from the mathematical properties proved within a system. For example, Metatheorem 1 implies that the axioms of a system are all that are needed to determine the models of the system. The proofs of metatheorems are beyond the level of this book. (See DeLong [6, Chapter 4] and Wilder [29, Chapter 2] for more on metamathematics.)

Metatheorem 1.4.1 If all the axioms of an axiomatic system are true in a model, then all the theorems of that system are also true in that model.

The most important property of an axiomatic system is consistency, which says that we cannot prove two statements that contradict each other. For example, a proof of both $1 = 2$ and $1 \neq 2$ would render proofs useless. However, consistency is difficult and often impossible to show directly. Fortunately Metatheorem 1.4.2 provides an easier method with models. In particular, the models for Example 1 show that the axiomatic system is consistent.

1.4 Axiomatic Systems, Models, and Metamathematics

Definition 1.4.2 An axiomatic system is *consistent* iff no contradictions can be proven from the axioms. An axiomatic system is *relatively consistent* iff its consistency can be proven assuming the consistency of another axiomatic system.

Metatheorem 1.4.2 (Gödel, 1930) An axiomatic system is consistent iff it has a model.

The analytic model of geometry shows the relative consistency of Euclidean geometry because analytic geometry depends on the real number system. Are the axioms for the real numbers consistent? The model of the real numbers we are likely to choose is a geometric one: a line in Euclidean geometry. We risk entering a vicious circle: The crucial property of a mathematical system, consistency, is beyond our reach for two central mathematical systems. Indeed, this problem and others like it inspired David Hilbert and others to investigate mathematical logic. Mathematical logic has given us profound insights into the nature of mathematics and proof, including their limitations. In particular, one of Gödel's incompleteness theorems, proven in 1930, shows that we cannot prove the absolute consistency of elementary arithmetic, and by extension, Euclidean geometry and the real number system. Thus relative consistency is the best we can do for such sophisticated systems. No one seriously doubts the consistency of these systems.

Other metamathematical properties, though not essential, give important insight into axiomatic systems and models.

Definition 1.4.3 A statement is *independent* of a set of axioms iff neither the statement nor its negation can be proved from the axioms. If, in an axiomatic system, each axiom is independent of the other axioms, the set of axioms is said to be *independent*. An axiomatic system is *complete* iff every statement based on the undefined terms can either be proved or disproved from the axioms.

We can use Metatheorem 1.4.1 to prove independence with models. Suppose that we have two models for a set of axioms and that a statement is true in the first model but not in the second model. Then Metatheorem 1.4.1 applied to the second model shows that the statement cannot be a theorem of these axioms. Similarly, the first model shows the negation of that statement cannot be a theorem.

Example 2 The models in Figs. 1.29 and 1.30 show the statement "there are exactly six points" to be independent of the axioms of Example 1. Because this statement is independent of the axioms, these axioms are not complete. ●

Example 3 Taxicab geometry alters the usual analytic model of Euclidean geometry by giving a different interpretation to distance. Rather than use the Pythagorean theorem to measure distance, we use the formula $d_T((x_1, y_1), (x_2, y_2)) = |x_2 - x_1| + |y_2 - y_1|$. This formula gives the distance a taxi goes if it travels only along north–south and east–west streets (Fig. 1.31). This model still satisfies Euclid's postulates, although circles look odd. However, Euclid I-4 (SAS) is no longer valid in this geometry (Fig. 1.32). This theorem holds in the usual analytic model, so it is independent of Euclid's postulates. That is, not only did Euclid's proof of I-4 have a flaw, but he could never actually prove that proposition from his axioms. (In fairness, Euclid's work shouldn't be faulted because of insights occurring 2000 years after his work.) ●

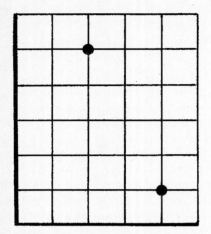
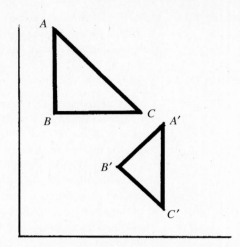

Figure 1.31 The taxicab distance between (2, 5) and (4, 1) is $|2-4| + |5-1| = 2 + 4 = 6$.

Figure 1.32 In taxicab geometry $\overline{AC}$ is longer than $\overline{A'C'}$, even though $\overline{AB} \cong \overline{A'B'}$, $\overline{BC} \cong \overline{B'C'}$, and $\angle ABC \cong \angle A'B'C'$.

Example 4 We can stretch a string taut on the surface of a sphere to construct a spherical line segment. Similarly, we can construct a spherical circle by fixing one end of a taut string and swinging the other end around it. Experiment informally on a ball or sphere. Confirm that, within reason, Euclid's first four postulates hold in this model. Now draw a spherical line segment with a length of between one-third and one-half the circumference of the sphere ($\overline{AB}$ in Fig. 1.33). Use this line segment to imitate Euclid's construction in Proposition I-1. (See Section 1.3.) Note that the two spherical circles with centers at A and B and radii of length AB do not intersect. This construction does not hold in this model but does in the usual model. Therefore it is independent of Euclid's first four postulates. ●

For more on taxicab geometry and spherical geometry, see Henderson [14] and Krause [19].

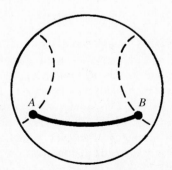

Figure 1.33

1.4 Axiomatic Systems, Models, and Metamathematics

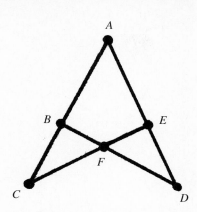

Figure 1.34 Interpret *point* by a dot and *line* by a line segment.

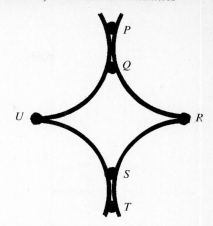

Figure 1.35 Interpret *point* by a dot and *line* by an arc.

Example 5 Hilbert's axiom system is both independent and complete, assuming that it is consistent. (The proof is beyond the level of this text. See the footnote in Section 1.3) ●

Hilbert's axioms are even stronger than being complete: All models of these axioms look exactly alike. In mathematical terminology, all are *isomorphic*. Two more models of the axiomatic system from Example 1 support this idea. The models in Figs. 1.34 and 1.35 both have six points and four lines, as the model in Fig. 1.30 did. They all look different on the surface, but the models in Figs. 1.30 and 1.34 are structurally the same (isomorphic), whereas the one in Fig. 1.35 is different. We can match the points and lines from Fig. 1.30 with those from Fig. 1.34 and have corresponding points on corresponding lines. Indeed, the labeling of these figures provides such a matching. However, no matter how we match the points in Fig. 1.35 with those of these two models, the relation *on* will never be the same. In particular, P and Q share two lines in common, something that never happens in the other models. The model shown in Fig. 1.29 has no chance of being isomorphic with any of these models because it has more points and lines.

Definition 1.4.4 Two models are *isomorphic* iff there is a one-to-one onto function that matches all the elements of one model with those of the other model and that transfers all the relations based on the undefined terms that hold in one model to relations that hold in the other model.

PROBLEMS FOR SECTION 1.4

1. Use graph paper to help you with this problem. (See Example 3.)

   a) Interpret $\overline{AB} \cong \overline{CD}$ in taxicab geometry to mean $d_T(A, B) = d_T(C, D)$. Which of Hilbert's congruence axioms hold in taxicab geometry?

   b) Define and describe "taxicab circles," using the distance d_T. What are the different types of intersection of two different taxicab circles?

   c) Determine whether SSS, ASA, and AAS hold in taxicab geometry.

2. Use a sphere or ball that you can draw on for this problem. (See Example 4.)

   a) Construct triangles of different sizes on a sphere and approximate their angle sum by using a copy of the circle in Fig. 1.36. (Place the center of the circle approximately over the vertex of an angle

34   **Chapter 1**   **Euclidean Geometry**

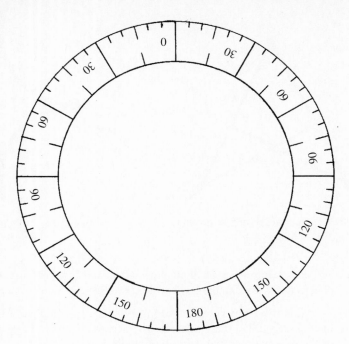

Figure 1.36   A protractor.

and the mark for 0° on one of the sides of the angle.) Does Theorem 1.1.1 hold in this model?

  b) Which of Hilbert's congruence axioms hold for the sphere?

  c) Determine whether SAS, SSS, ASA, and AAS hold for a sphere.

  d) Determine whether the Pythagorean theorem holds for a sphere.

3. Use the axiomatic system of Example 1.

  a) What is the smallest positive number of points of a model of this system?

Repeat for lines and explain your answers.

  b) Find a model with six points not isomorphic to any of the other models.

  c) Find two nonisomorphic models with nine points.

  d) Find and prove a theorem of this system.

4. Use the axiomatic system of Problem 6 of Section 1.3.

  a) Find several models of this system.

  b) Show that axiom (iii) is independent of the other axioms.

  c) Repeat part (b) for axiom (iv).

  d) Find two nonisomorphic models of this system, each with eight points.

  e) Find two nonisomorphic models of the system in Problem 6(c).

5. Use the axiomatic system of Problem 7 of Section 1.3.

  a) Find a model with seven points.

  b) Find models to show that all the axioms are independent.

6. Consider the axiomatic system with the undefined terms *corner*, *square*, and *on* and the following axioms.

  i) There is a square.

  ii) Each square is on four distinct corners.

  iii) For each square there are four distinct squares with exactly two corners on the given square.

  iv) Each corner is on four distinct squares.

  a) Show that these axioms are relatively consistent by finding an infinite model in the Euclidean plane.

b) Find a finite model for the first three axioms.

c) Find a finite model for all four axioms.

7. Interpret "Q is between P and R" by $d(P, Q) + d(Q, R) = d(P, R)$, where P, Q, and R are distinct and $d(X, Y)$ is the distance between X and Y.

a) Are all of Hilbert's axioms satisfied in the analytic model of Euclidean geometry with this interpretation?

b) On a graph, color the set of points in taxicab geometry between $(0, 0)$ and $(1, 2)$ using the given interpretation. Which of Hilbert's axioms of order are satisfied in taxicab geometry with this interpretation?

c) In Euclidean geometry, if Q is between P and R and R is between Q and S, then Q is between P and S. Does this property still hold in taxicab geometry with this interpretation of *between*? Explain.

d) If P and R are opposite points on a sphere, which points Q on the sphere would be between P and R under this interpretation?

1.5 SIMILAR FIGURES

The mathematical aspects of scale changes have been studied for thousands of years but continue to yield insights in various fields. The Euclidean tradition focuses on similar triangles and polygons. In Chapter 4 we use transformations to provide a deeper treatment of similar figures.

Definition 1.5.1   Nonzero numbers $a_1, a_2, \ldots$ are *proportional* to $b_1, b_2, \ldots$ iff there is a nonzero number k such that $k \cdot a_i = b_i$. The number k is the *ratio of proportionality*.

Exercise 1   Show that a_1, a_2 are proportional to b_1, b_2 only if $a_1 \cdot b_2 = a_2 \cdot b_1$.

Definition 1.5.2   Triangles $\triangle ABC$ and $\triangle A'B'C'$ are *similar* iff corresponding angles are congruent and the lengths of corresponding sides are proportional. We write $\triangle ABC \sim \triangle A'B'C'$.

Example 1   Show that the three triangles in Fig. 1.37 are similar.

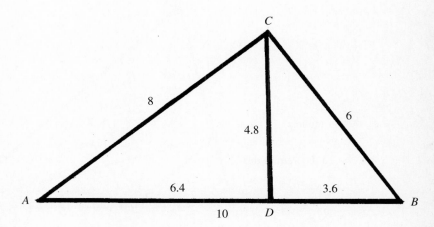

Figure 1.37

Solution. Each triangle is a right triangle by the converse of the Pythagorean theorem (Euclid I-48). Elementary trigonometry confirms that the other angles are congruent. Furthermore, the sides of the largest triangle (10, 6, and 8) and the smallest triangle (6, 3.6, and 4.8) are proportional, using $k = 0.6$. Similarly, the ratio $k = 0.8$ converts the sides of the largest triangle to the sides of the middle triangle (8, 4.8, and 6.4). ●

To show two triangles to be similar, we must in principle prove the congruence or proportionality for all their corresponding parts. Similarity theorems provide smaller sufficient sets of corresponding parts, analogous to the congruence theorems SAS, AAS, and SSS. We follow Euclid's approach, supplemented with our modern understanding of real numbers, to develop these theorems. Theorem 1.5.1 is Euclid's key to proving Theorems 1.5.2, 1.5.3, and 1.5.4.

Theorem 1.5.1 (Euclid VI-2) Let $\overrightarrow{AB}$ and $\overrightarrow{AC}$ be two rays with D between A and B and E between A and C. Then $\overline{BC} \| \overline{DE}$ iff $\triangle ABC \sim \triangle ADE$.

Proof. Construct segments $\overline{BE}$ and $\overline{CD}$, as in Fig. 1.38. First suppose that $\overline{BC} \| \overline{DE}$. Then the corresponding angles are equal by I-29. Triangles $\triangle BDE$ and $\triangle CED$ have the same area because they have the same base ($\overline{DE}$) and the same height (the distance between the parallel lines). Thus the areas of these triangles have the same ratio to the area of $\triangle ADE$, which also has base $\overline{DE}$. Now reconsider $\triangle ADE$ and $\triangle DBE$, using $\overline{AD}$ and $\overline{DB}$ as the bases so that these two triangles have the same height. Hence the lengths of their bases and their areas are proportional because of the formula Area = $\frac{1}{2}$(Base)(Height). That is, $\frac{1}{2}$(Height) is the ratio of proportionality. The same can be said about the triangles $\triangle ADE$ and $\triangle DCE$. Thus the bases $\overline{AD}$ and $\overline{DB}$ are in proportion to $\overline{AE}$ and $\overline{EC}$. By the addition properties of proportions, this result shows $\overline{AD}$ and $\overline{AB}$ are proportional to $\overline{AE}$ and $\overline{AC}$.

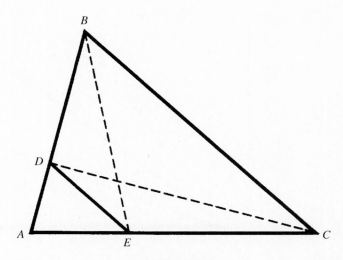

Figure 1.38

1.5 Similar Figures

Problem 7 asks you to prove that the sides $\overline{DE}$ and $\overline{BC}$ have the same proportion as the other two pairs of sides. Problem 8 asks you to show the other direction. ∎

Theorem 1.5.2 (Euclid VI-4) If two triangles have two pairs of corresponding angles congruent, then the triangles are similar.

Proof. Let $\triangle ABC$ and $\triangle A'B'C'$ have $\angle A \cong \angle A'$ and $\angle B \cong \angle B'$. By Theorem 1.1.1, $\angle C \cong \angle C'$. We must show that the three pairs of sides are proportional. If $\overline{AB} \cong \overline{A'B'}$, then the triangles are congruent by AAS. So, WLOG, suppose that $\overline{AB}$ is longer than $\overline{A'B'}$. Construct D on $\overline{AB}$ such that $\overline{AD} \cong \overline{A'B'}$, and construct $\overline{DE} \parallel \overline{BC}$ with E on $\overline{AC}$, as in Fig. 1.39. Then $\triangle ADE \cong \triangle A'B'C'$ by AAS. Furthermore, $\triangle ABC \sim \triangle ADE$ by Theorem 1.5.1. Then, by the definitions of congruent and similar triangles, $\triangle ABC \sim \triangle A'B'C'$. ∎

Theorem 1.5.3 (Euclid VI-5) If the three pairs of corresponding sides of two triangles are in proportion, then the two triangles are similar.

Proof. Problem 9. ∎

Theorem 1.5.4 (Euclid VI-6) If two pairs of corresponding sides of two triangles are proportional and their included angles are congruent, then the triangles are similar.

Proof. Problem 10. ∎

Definition 1.5.3 The *area* of a rectangle is the product of the lengths of two of its adjacent sides.

Exercise 2 Use Fig. 1.40 to derive the formula for the area of a triangle.

Theorem 1.5.5 If the lengths of the sides of $\triangle ABC$ are k times the lengths of the corresponding sides of $\triangle A'B'C'$, then the area of $\triangle ABC$ is k^2 times the area of $\triangle A'B'C'$.

Proof. Problem 11. ∎

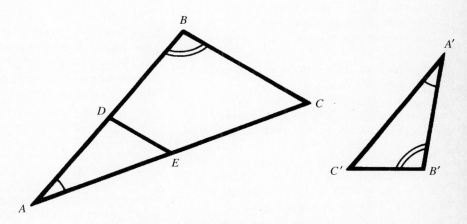

Figure 1.39

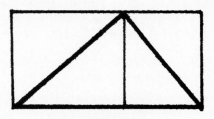

Figure 1.40

The generalization of Theorem 1.5.5 for areas and volumes of similar figures has important consequences for animals of different sizes. To state the generalization, let's suppose that there is a scaling ratio of k from figure A to B. Then the area (or surface area) of B is k^2 times the area of A, and the volume of B is k^3 times the volume of A. Consider increasing the size of an animal. The strength of bones and muscles grows roughly with the areas of their cross sections (k^2). However, the weight these bones and muscles must handle grows roughly with the volume of the animal (k^3). As animals get larger, the bones and muscles must get disproportionately thicker to support and move the greatly increased weight. Similarly, the lift of a bird's wings is roughly proportional to the surface area of the wings. The bird's weight is roughly proportional to the volume of the bird. To compensate for their much greater weights, large birds have evolved greater wingspreads than small birds. (McMahon and Bonner [22] provide more information on area and volume questions in nature.)

PROBLEMS FOR SECTION 1.5

1. Explain why in any triangle $\triangle ABC$ there is a unique point D on $\overrightarrow{AC}$ such that $\triangle ABC \sim \triangle ADB$. Describe AD in terms of the sides of $\triangle ABC$.

2. Consider a rectangle with length x and width y, where $x > y$.

   a) Find the ratio x/y so that you can halve the long side of this rectangle to get two smaller rectangles, both similar to the original rectangle. Explain.

   b) Redo part (a), but divide the rectangle into n smaller similar rectangles.

   c) Suppose that you remove a square with side y to leave a smaller rectangle. Find the ratio x/y so that the smaller rectangle is similar to the original rectangle. Explain.

3. Let A, B, C, and D be four points on a circle and $\overline{AC}$ and $\overline{BD}$ intersect at E, a point inside the circle. Prove that the product $AE \cdot EC$ equals $BE \cdot ED$. [*Hint:* Use Problem 9 of Section 1.2.]

4. Which of the following functions have graphs that are similar to the graph of $y = \sin(x)$? Explain your answer by using their graphs and your intuitive understanding of similarity. Generalize this problem.

   a) $y = 2\sin(x)$
   b) $y = \sin(2x)$
   c) $y = 2\sin(2x)$
   d) $y = \frac{1}{2}\sin(2x)$

5. a) Given any triangle $\triangle ABC$ and a square, explain why there are three points P, Q, and R on the square such that $\triangle PQR \sim \triangle ABC$.

   b) What shapes can you substitute for the square in part (a) and still have the result hold? Explain.

   c) Given any quadrilateral $ABCD$ and an equilateral triangle, can you find four points P, Q, R, and S on the triangle such that $PQRS$ is similar to $ABCD$? If this can work in general, explain why; if not, give a counterexample.

6. Let $\{A_i : i \in I\}$ and $\{B_i : i \in I\}$ be the vertices of two polygons A and B, respectively. Then define polygons A and B to be *similar* ($A \sim B$) iff both

i) there is a positive r such that for all $i, j \in I$, $A_i A_j = r B_i B_j$ and

ii) for all $i, j, k \in I$, $\angle A_i A_j A_k \cong \angle B_i B_j B_k$.

a) How does this definition compare with your intuition of similar polygons?

b) Prove that any two squares are similar.

c) Provide an example of two quadrilaterals whose sides are proportional but are not similar.

d) Provide an example of two quadrilaterals whose angles are congruent but are not similar.

e) Prove that any two regular polygons with the same number of sides are similar.

f) Prove that $A \sim B$ provided that for all $i, j, k \in I$, $\triangle A_i A_j A_k \sim \triangle B_i B_j B_k$.

g) If you allow the index set I to be infinite, does the given definition of *similar* still make sense? If so, give an example; if not, explain why not.

7. Fill in the proof of Theorem 1.5.1 by showing that the sides $\overline{DE}$ and $\overline{BC}$ are in the same proportion as the other two pairs of sides. [*Hint*: Construct $\overline{EF} \| \overline{AB}$ with F on $\overline{BC}$.]

8. Prove the remaining direction of Theorem 1.5.1.

9. Prove Theorem 1.5.3.

10. Prove Theorem 1.5.4.

11. Prove Theorem 1.5.5.

12. Generalize Theorem 1.5.5 to polygons with n sides, using induction. Assume that any polygon can be divided into triangles.

13. (Calculus) Recall that the area bounded by $y = f(x)$, the x-axis, $x = a$, and $x = b$ is $\int_a^b f(x)dx$ (Fig. 1.41).

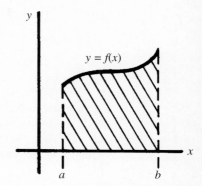

Figure 1.41

a) Explain why the region bounded by $y = kf(x/k)$, the x-axis, $x = ka$, and $x = kb$ is similar to the region shown in Fig. 1.41. In particular, explain why you need to use $f(x/k)$ instead of $f(x)$.

b) Use calculus to verify that the area of the region in part (a) is k^2 times the area of the region shown in Fig. 1.41.

c) Generalize this problem to volumes of similar regions in three dimensions.

1.6 THREE-DIMENSIONAL GEOMETRY

Although we live in a three-dimensional world, visualizing three-dimensional geometric figures is harder than visualizing two-dimensional figures. We explore nonaxiomatically the geometry of polyhedra (the plural of polyhedron) and the sphere to help you deepen your visual intuition. Working with physical models aids this understanding beyond what any textbook figures can do.

1.6.1 Polyhedra

Polyhedra continue to fascinate people, much as they did the ancient Greeks. We assume an intuitive understanding of a polyhedron because an exact definition is more complicated than our treatment warrants. (See Lakatos [20].) In brief, a *polyhedron* is composed of *vertices*, *edges*, and *faces*. The faces are polygons. At each vertex (or corner) at least three faces meet, and at each edge (line segment) two faces meet.

Pyramids are easy to visualize, and all polyhedra can be dissected into pyramids, much as all polygons can be dissected into triangles. A *pyramid* has a polygon with

40   **Chapter 1**   **Euclidean Geometry**

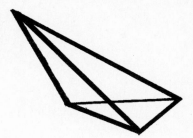

Figure 1.42

n edges for its *base*, one more vertex not in the plane of the base called the *apex*, and n triangular faces that are determined by the apex and the n edges of the base. The polyhedron with the fewest vertices, edges and faces is a triangular pyramid or *tetrahedron*, shown in Fig. 1.42.

Exercise 1   Verify that a pyramid with an n-gon for a base has $V = n + 1$ vertices, $E = 2n$ edges, and $F = n + 1$ faces.

Example 1   Use calculus to find a formula for the volume of a pyramid.

Solution.   WLOG select a coordinate axis so that the origin is at the apex of the pyramid and the x-axis is perpendicular to the base (Fig. 1.43). Suppose that the area of the base is B and that the x-coordinate of points on the base is $h > 0$, the height of the pyramid. Any cross section of the pyramid in a plane parallel to the base is a polygon similar to the base. If a cross section's x-coordinate is x, then Problem 12 of Section 1.5 shows its area to be $(x/h)^2 B$. Thus the volume of the pyramid is

$$\int_0^h \left(\frac{x}{h}\right)^2 B\,dx = \int_0^h \frac{x^2}{h^2} B\,dx = \left(\frac{x^3}{3h^2} B\right)\Bigg|_0^h = \frac{1}{3} hB.$$

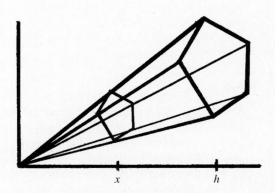

Figure 1.43

1.6 Three-Dimensional Geometry

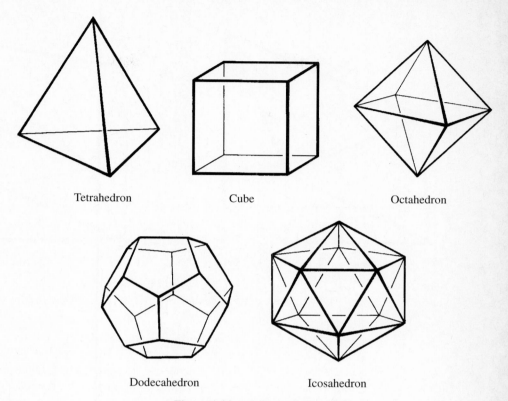

Figure 1.44 The regular polyhedra.

Exercise 2 Modify Example 1 to find the formula for the volume of a cone.

Exercise 3 Explain how, in principle, you can use pyramids to find the volume of any polyhedron.

The five polyhedra shown in Fig. 1.44 possess regular properties and a high degree of symmetry. These regular polyhedra are often called the Platonic solids because the Greek philosopher Plato was fascinated by them and his discussion of them is the oldest that survives. A convex polyhedron is *regular* provided that all its faces are the same regular polygon and the same number of polygons meet at each vertex. (See Coxeter [5, Chapter 10] for more information.)

The noted Swiss mathematician Leonhard Euler (1707–1783) developed a formula relating the number of vertices, edges, and faces for a large collection of polyhedra, including all convex ones. Because a careful proof, even for convex polyhedra, would require an overly long and technical development, we don't present one here. (See Beck et al [2] and Lakatos [20].) Euler's Formula has a wide variety of applications in geometry, graph theory, and topology.

Exercise 4 Find the number of vertices, V, edges, E, and faces, F, for the regular polyhedra in Fig. 1.44.

42 **Chapter 1** **Euclidean Geometry**

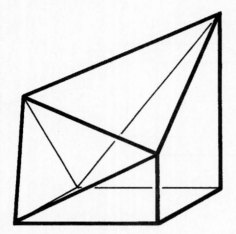

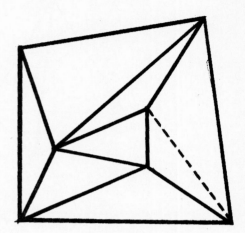

Figure 1.45 A polyhedron with 7 vertices, 13 edges, and 8 faces.   **Figure 1.46**

Theorem 1.6.1 **Euler's Formula** If a convex polyhedron has V vertices, E edges and F faces, then $V - E + F = 2$.

Outline of Reasoning. Given a convex polyhedron (Fig. 1.45) we can stretch the net of its vertices and edges to lay it out on a plane (Fig. 1.46). The number of vertices and edges remains the same, but the number of faces is one less than the original polyhedron. Next, if a face of this net is not a triangle, we divide it into triangles by adding edges. A technical argument shows that this division can always be made and that doing so increases the number of faces and edges the same amount. Another argument shows that we can carefully eliminate edges one at a time, each time eliminating either a face or a vertex. Thus we preserve the value $V - E + F$. (Try this process with Fig. 1.46.) In the end, we are left with a triangle, for which we have $V - E + F = 3 - 3 + 1 = 1$. Thus the original polyhedron must satisfy Euler's formula. ∎

Example 2 The angle sum of all the angles of a tetrahedron is $4 \times 180° = 720°$ because a tetrahedron has four triangles for faces. Similarly the six square faces of a cube give an angle sum of $6 \times 360° = 2160°$. A simple formula for the angle sum of all convex polyhedron depends on shifting our focus from the faces to the vertices, as Descartes discovered. •

Theorem 1.6.2 **Descartes's Formula** If a convex polyhedron has V vertices, then the angle sum of all of the angles around all the vertices is $360°(V - 2)$.

Proof. We derive this result from Euler's formula. We rewrite $V - E + F = 2$ as $V - 2 = E - F$. Note that the left side multiplied by 360 is Descartes's formula: $360°(V - 2) = 360°(E - F) = 180°(2E - 2F)$. Next we relate the right side to the angle sum.

Let F_i be the number of faces with i edges on them; for example, F_3 is the number of triangles. The number of faces is $F = F_3 + F_4 + F_5 + \cdots = \sum_{i=3}^{\infty} F_i$. (Of

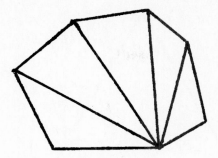

Figure 1.47

course, for a given polyhedron, the sum isn't infinite.) Figure 1.47 illustrates that a convex face with i edges has $i - 3$ diagonals from a vertex, dividing the face into $i - 2$ triangles. By Theorem 1.1.1 and induction, a face with i edges has an angle sum of $180°(i - 2)$. Thus the total angle sum is $\sum_{i=3}^{\infty} F_i 180°(i - 2)$. We can also count the number of edges in terms of the F_i. However, $\sum_{i=3}^{\infty} i F_i$ is too big, as each edge is on two faces and so is counted twice. Therefore $2E = \sum_{i=3}^{\infty} i F_i$. Thus the angle sum is $\sum_{i=3}^{\infty} F_i 180°(i - 2) = 180° \sum_{i=3}^{\infty} i F_i - 2 F_i = 180°(2E - 2F) = 360°(V - 2)$. ∎

Exercise 5 Verify Descartes's formula for the five regular polyhedra shown in Fig. 1.44.

The angle sum around each vertex of a convex polyhedron must be less than 360°. We call the difference between 360° and the angle sum at a vertex the *spherical excess* of that vertex. Descartes's formula tells us that the total spherical excess of all vertices is 720°.

1.6.2 Geodesic domes

Buckminster Fuller (1895–1983) related the concept of spherical excess to the strength of geodesic domes, which he invented. He was a prolific inventor who devoted years to designing economical, efficient buildings. Traditional buildings, designed as modified rectangular boxes, require the use of a lot of material to enclose a given volume and support a roof. In contrast, a sphere, the shape that minimizes the surface area for a given volume, is expensive to build because of its curved surface. Fuller avoided these drawbacks of the box and the sphere by starting with an icosahedron. The domed effect of the icosahedron distributes the weight of the structure evenly, as spherical domes do, without needing a curved surface. Furthermore, the triangular faces of the icosahedron are structurally stronger than the rectangular faces of traditional buildings. Fuller was able to increase building size by dividing each of the icosahedron's 20 faces into smaller triangles. To maximize the strength of the building, he found that he needed to arrange these smaller triangles so that all the vertices were on the surface of a sphere. He coined the name *geodesic dome* for a convex polyhedron whose faces are all triangles.

We consider only domes based on an icosahedron. The *frequency* of a dome is n, where each of the original triangles is divided into n^2 smaller triangles (Fig. 1.48). To determine a geodesic dome we need to know the measures of all the edges and angles.

44   **Chapter 1**   **Euclidean Geometry**

Figure 1.48   The division of one face of an icosahedron to make a three-frequency dome.

As Example 3 illustrates, such three-dimensional calculations rely extensively on two-dimensional geometry.

Example 3   Find the edge lengths and angle measures for the two-frequency dome shown in Fig. 1.49, if the radius OA of the sphere is 1.

Solution.   From Problem 6 of this section we know that the length $AB \approx 1.05146$. Because D' and E' are the midpoints of $\overline{AB}$ and $\overline{AC}$, $AD' = D'E' \approx 0.52573$. By the Pythagorean theorem, $OD' \approx 0.85065$. The ray $\overrightarrow{OD'}$ intersects the sphere at D and, again by the Pythagorean theorem, $AD \approx 0.54653$. By Theorem 1.5.4 the triangles $\triangle OD'E'$ and $\triangle ODE$ are similar. Hence $DE \approx 0.61803$. Next we can use the law of cosines, $c^2 = a^2 + b^2 - 2ab\cos(C)$, to determine that $m\angle DAE \approx 68.86°$. Theorem 1.1.1 gives us $55.57°$ for $m\angle ADE$ and $m\angle AED$. The angles of the other corner triangles are the same, and the angles of the center triangle $\triangle DEF$ are all $60°$ because it is equilateral. ●

The strength of an icosahedron relies on two facts. First, triangular faces are structurally stable on their own, a consequence of the congruence theorem SSS. Second, these triangular faces distribute forces well because of the angles where they meet. One good measure of these angles is the spherical excess at each vertex. For an icosahedron, the spherical excess at each vertex is $60°$. As the number of vertices increases with the frequency of the dome, Descartes's formula necessitates a decrease in the spherical excesses at the vertices. However, the proper design of the dome maximizes the smallest of these spherical excesses and so maximizes strength. Such a design will also make the triangles roughly equilateral, distributing the weight of the dome better and so strengthening the dome. High-precision technology and modern materials enable domes to be structurally stable with spherical excesses as small as $\frac{1}{2}°$.

Exercise 6   For the two-frequency icosahedron of Example 3, show that the spherical excesses of the two kinds of vertices are $15.7°$ and $17.72°$.

1.6 Three-Dimensional Geometry

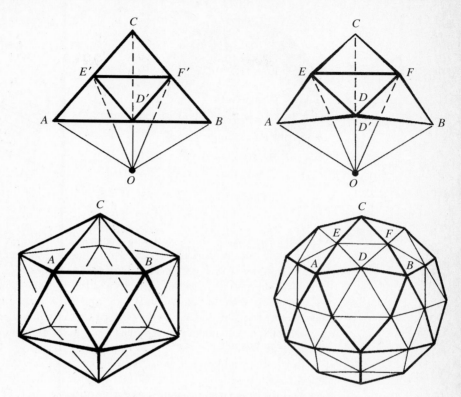

Figure 1.49 Each face ($\triangle ABC$) of the icosahedron becomes four faces on the two-frequency dome. Points D, E, and F are on OD', OE', and OF' and the same distance from O as A, B, and C are.

The strength of geodesic domes is extraordinary. The U.S. Air Force tested the strength of a 55-foot diameter fiber glass geodesic dome in 1955 before choosing domes to house the radar antenna of the Distant Early Warning system. They linked the vertices of the dome to a winch connected to a 17-ton concrete slab buried under the dome. They had intended to tighten the winch until the dome collapsed under the strain and then measure the breaking point. However, instead of collapsing, the dome withstood the stress and actually lifted the concrete slab. The great strength of geodesic domes enables them to enclose large spaces without interior supports. The largest geodesic dome, over 400 feet in diameter, far surpasses the largest space without interior support of any conventional building. (See Edmondson [8] and Kenner [17] for more on Fuller and geodesic domes.)

1.6.3 The geometry of the sphere

The needs of astronomy and global navigation have prompted the study of the geometry of the sphere for centuries. Astronomers use the inside surface of a sphere to describe the positions of the stars, planets, and other objects. For navigational purposes, the earth is

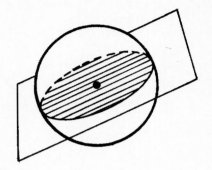

Figure 1.50 **Figure 1.51**

essentially a sphere. A ship's captain (or an airplane pilot) seeks the most direct route, following the curving surface of the earth rather than an Euclidean straight line. The shortest trip from Tokyo to San Francisco goes considerably north of either city (Fig. 1.50). The shortest path on the surface of a sphere connecting two points is a *great circle*—a circle with the same radius as the sphere. A great circle is the intersection of the sphere and a plane that passes through the center of the sphere (Fig. 1.51). Circles of longitude and the equator are great circles. The circles of latitude, except for the equator, are not great circles.

Great circles play much the same role on the sphere that straight lines do on the plane. In fact, a sphere satisfies the first four of Euclid's postulates in Appendix A. However, the geometric properties of lines and great circles differ in important ways. There are no parallel great circles because two distinct great circles always intersect in two diametrically opposed points. We use the same letter to denote them, placing a prime on one of them to distinguish one from the other. Two great circles divide the sphere into four regions called *lunes* (Fig. 1.52). The angle that these two great circles make is the angle at which the two planes meet (Fig. 1.53). Three great circles with no common point of intersection form eight *spherical triangles* (Fig. 1.54).

Exercise 7 Use Fig. 1.53 to explain why the area of a lune with an angle of $\alpha°$ is $(\alpha/90)\pi r^2$. Recall that the surface area of a sphere of radius r is $4\pi r^2$.

Exercise 8 Suppose that a spherical triangle $\triangle ABC$ has angles measuring $70°$, $80°$, and $90°$. Verify that the angle sums for the seven related spherical triangles, such as $\triangle ABC'$, also are more than $180°$.

In Theorem 1.6.3, we show not only that the angle sum of a spherical triangle is more than $180°$, but also that this angle sum is related to the triangle's area. The spherical triangle $\triangle ABC$ shown in Fig. 1.54 determines three overlapping lunes $BACA'$, $ABCB'$, and $ACBC'$. The opposite spherical triangle $\triangle A'B'C'$ determines three other lunes that do not intersect these three lunes for $\triangle ABC$. Together the six lunes cover the entire sphere.

1.6 Three-Dimensional Geometry 47

Figure 1.52

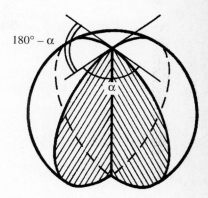

Figure 1.53

Theorem 1.6.3 The area of a spherical triangle is proportional to the excess of its angle sum over 180°. More precisely, on a sphere with radius r, the area of a spherical triangle with angle measures of $\alpha°$, $\beta°$, and $\gamma°$ is $[(\alpha + \beta + \gamma - 180)/180]\pi r^2$.

Proof. The area covered by the lunes $BACA'$, $ABCB'$, and $ACBC'$ is $2\pi r^2$, or half the sphere because the opposite lunes cover a symmetric region. Note that these three lunes each cover $\triangle ABC$. Then

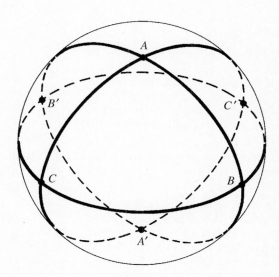

Figure 1.54

$$2\pi r^2 = \text{Area}(BACA') + \text{Area}(ABCB') + \text{Area}(ACBC') - 2 \cdot \text{Area}(\triangle ABC)$$
$$= \frac{\alpha}{90}\pi r^2 + \frac{\beta}{90}\pi r^2 + \frac{\gamma}{90}\pi r^2 - 2 \cdot \text{Area}(\triangle ABC).$$

Solving for $\text{Area}(\triangle ABC)$ gives $[(\alpha + \beta + \gamma - 180)/180]\pi r^2$. ∎

Exercise 9 Explain why, in spherical geometry, if two triangles on the same sphere are similar, they are congruent.

(For more information on spherical geometry, see McCleary [21, 3ff].)

PROBLEMS FOR SECTION 1.6

Physical models greatly aid in visualizing and solving these problems.

1. a) Construct models of the five regular polyhedra. (See Fig. 1.44 or Wenninger [28].)

   b) Prove that there are only five regular polyhedra, using spherical excess. [*Hint:* Why do regular polygons with more than five sides not need to be considered? Find the largest number of regular pentagons that can fit around a vertex. Repeat for squares and equilateral triangles.]

2. For a rectangular box with sides a, b, and c, explain why the diagonal d satisfies the "three-dimensional Pythagorean theorem" $a^2 + b^2 + c^2 = d^2$.

3. Suppose that the edge of a cube is 1 unit long.

   a) Find the distances from the center of the cube to the center of a face, to the midpoint of an edge, and to a vertex (Fig. 1.55).

   b) What percentage of a circumscribed sphere's volume does the cube occupy?

   c) Describe the polyhedron obtained by connecting four vertices of the cube, no two of which are adjacent. What is the volume of this polyhedron? [*Hint:* Find the volumes of the four pyramids cut away.]

4. Repeat Problem 3(a) and (b) for a regular octahedron and tetrahedron with edges 1 unit long (Figs. 1.56 and 1.57).

5. a) Construct the shape shown in Fig. 1.58 with three 3×5-in. note cards. Two of the cards need 3-in. slits in their centers. The third card needs this slit extended to one of the 3-in. sides. The 12 corners of these cards approximate the vertices of a regular icosahedron.

   b) Modify part (a) to find the exact coordinates of the vertices of a regular icosahedron as follows.

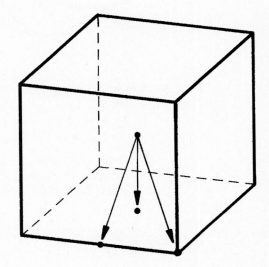

Figure 1.55

Let the slits be on the x-, y-, and z-axes and let the dimensions of the modified cards be $2 \times 2b$. Why are the 12 vertices at $(\pm 1, \pm b, 0)$, $(0, \pm 1, \pm b)$, and $(\pm b, 0, \pm 1)$? Find b.

6. Suppose that the edge $\overline{AB}$ of a regular icosahedron is 1 unit long (Fig. 1.59).

   a) Find the lengths of the "short diagonal" $\overline{AC}$ and the "long diagonal" $\overline{BC}$. (See Problem 5.)

   b) Find the length of the edge of a regular icosahedron inscribed in a sphere of radius 1.

   c) Repeat Problem 3(a) and (b) for the icosahedron.

7. Find eight vertices of a regular dodecahedron that are the vertices of a cube. (See Fig. 1.44.)

8. The *excess* of a spherical triangle is the difference between its angle sum and 180°. Without using

1.6 Three-Dimensional Geometry

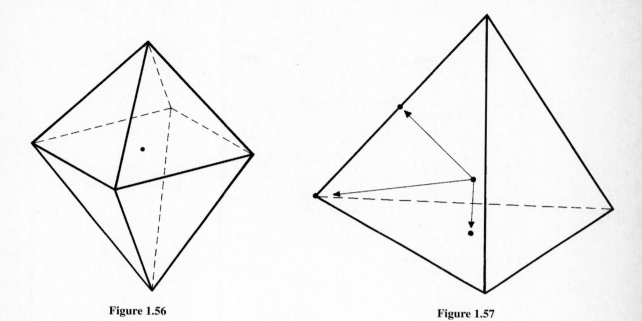

Figure 1.56

Figure 1.57

Theorem 1.6.3, show that excesses are additive: If D is on the short arc of the great circle through B and C, the excess of $\triangle ABC$ is the sum of the excesses of $\triangle ABD$ and $\triangle ADC$. Illustrate this situation.

9. On a sphere let N be the "north pole" and A and B be two points on the "equator" with $m\angle ANB = 90°$.

Let C, D, and E be the midpoints of $\widehat{AB}$, $\widehat{AN}$, and $\widehat{BN}$, respectively.

a) Explain and illustrate why $\widehat{CN}$, $\widehat{DB}$, and $\widehat{AE}$ intersect in a common point, F.

b) Find the angle sum of the spherical triangle $\triangle ACF$.

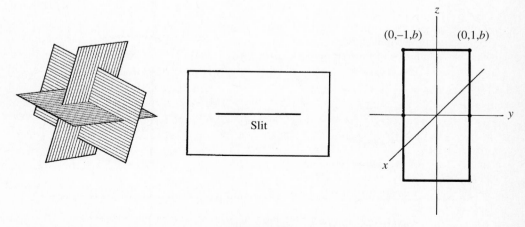

Figure 1.58

50 Chapter 1 Euclidean Geometry

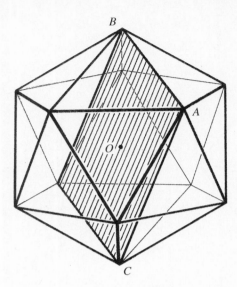

Figure 1.59

c) Verify that the area of $\triangle ACF$ agrees with Theorem 1.6.3.

10. Cavalieri's principle is often an axiom in high school texts because it provides an elementary way to prove results about volumes of curved shapes. Bonaventura Cavalieri (1598–1647), a pupil of Galileo, used his principle to find volumes (and areas) before the advent of calculus.

Cavalieri's Principle: Let A and B be two solids included between two parallel planes. If every plane P parallel to the given planes intersects A and B in sections with the same area, then A and B have the same volume.

a) Use Cavalieri's principle and Fig. 1.60 to show the following result of Archimedes. The volume of a cylinder whose height is twice its radius equals the volume of a sphere of the same radius plus twice the volume of a cone with the same radius and a height equal to the radius.

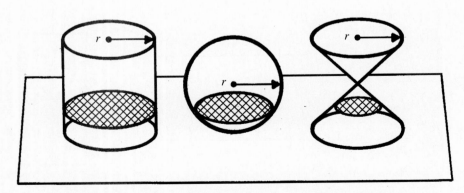

Figure 1.60 Compare the areas of the cross sections of a cylinder, a sphere, and a double cone at the same height.

b) Find two solids A and B with different volumes and a family of nonparallel planes so that A and B intersect each plane in sections of equal area.

11. (Calculus) Note that the derivative dA/dr of πr^2, the area of a circle of radius r, is the circumference of the circle. Explain geometrically why the area and circumference are related in this way. Explain why the same relationship holds for the volume and surface area of a sphere. [*Hint:* Consider adding a thin strip around a circle of radius r. Approximately how much area is in the thin strip?]

12. Figure 1.61 shows a design for a three-frequency dome. The nine triangles of the dome are "lifted" from the nine equilateral triangles dividing $\triangle ABC$ so that the vertices are all on a sphere of radius 1. Find the lengths and angle measures as follows, assuming that $DE = DI$.

a) Explain why there are just three different lengths, AD, DE, and DJ. Explain how to find all the angle measures from $m\angle DAI$ and $m\angle DJE$.

b) From Example 3, $AB \approx 1.05146$. Find $D'E'$, $D'M'$, OM', and OD'.

c) Use the law of cosines to find $m\angle D'OE'$, $m\angle AOD'$, AD, DE, OJ', $D'J'$, DJ, $m\angle DAI$, and $m\angle DJE$.

d) Find the spherical excess at A, D, and J.

13. a) Find the number of vertices on a two-frequency geodesic dome. (See Fig. 1.49.)

b) Find the number of vertices on a three-frequency dome (Fig. 1.61).

c) Show that the number of vertices on an n-frequency dome based on an icosahedron is $10n^2 + 2$. [*Hint:* Find formulas for the number of new vertices on one edge of the icosahedron and the number of new vertices on one face of the icosahedron. Then use the numbers of V, E, and F of an icosahedron.]

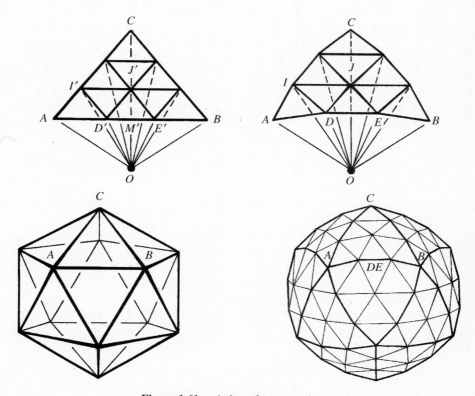

Figure 1.61 A three-frequency dome.

d) Use part (c) to find the average spherical excess at each vertex of an n-frequency dome. Compare your value when $n = 3$ with your answer in Problem 12(d). What is the largest practical value of n if the average spherical excess must be $\frac{1}{2}°$ for a dome?

14. Give an example on a sphere to show that Euclid I-16 is independent of Euclid's first four postulates.

15. A *spherical cap* is the set of all points on a sphere of radius R whose distance (measured along the sphere) is at most r from a point A on the sphere.
    a) Find the circumference of the spherical cap in terms of r and R.
    b) (Calculus) Verify that the area of the spherical cap is $2\pi R^2 - 2\pi R^2 \cos(r/R)$. [*Hint:* Use radians. Let x be as shown in Fig. 1.62. Recall that the area of the surface obtained by revolving $y = f(x)$ between $x = a$ and $x = b$ about the x-axis is $\int_a^b 2\pi f(x)\sqrt{1+(f'(x))^2}\,dx$. Calculate the integral for a spherical cap for general a and b. Then determine a and b in terms of r and R.]
    c) Archimedes found the surface area of the spherical cap to equal the area of a circle whose radius equals the (straight-line) distance from A to a point on the circumference of the cap. Verify Archimedes' theorem.

16. Curiously, the Euclidean plane is "triangle complete" but Euclidean space is not "tetrahedron complete." Triangle complete means that any three lengths that satisfy the triangle inequality ($a + b \geq c$) in any order can actually appear as the sides of a Euclidean triangle. However, there are sets of six lengths that satisfy the triangle inequality appropriately, but no tetrahedron in Euclidean space has its six edges with those lengths. More easily, there are four triangles whose corresponding sides are congruent that cannot be folded to make a tetrahedron. Find such an example. (Figure 1.63 shows four triangles that can be folded to make a tetrahedron.)

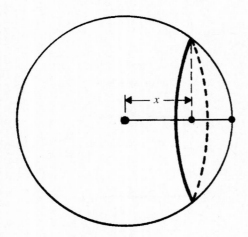

Figure 1.62

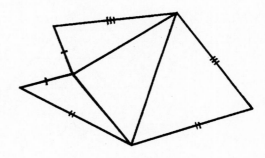

Figure 1.63

PROJECTS FOR CHAPTER 1

1. Archimedes and others estimated the ratio of the circumference of a circle to its diameter (π) by using regular polygons—the greater the number of sides, the more accurate the estimate of π.
    a) Consider polygons inscribed in a circle of radius 1 (Fig. 1.64). Find a formula for the length y of the side of a regular $2n$-gon in terms of the length x of the side of a regular n-gon.
    b) For a regular hexagon inscribed in a circle of radius 1, the side has a length of 1 and a perimeter of 6, an approximation of 2π. Use the formula from part (a) to find the perimeters of regular 12-gons, 24-gons, etc., to give better lower estimates of 2π.
    c) Write a computer program that will print out the approximations of π found by the formula in part (a) for polygons with 3×2^i sides, where $1 \leq i \leq 30$.

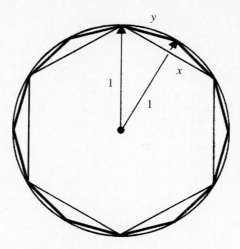

Figure 1.64 Find y in terms of x.

d) Find upper estimates for π by using circumscribed regular polygons in parts (a), (b), and (c). [*Hint:* The formula for $\tan(2\theta)$ will help.]

Remark The Arab mathematician Nasir-Eddin (1201–1274) used essentially this method (without modern notation or computers!) with a regular polygon of 3×2^{28} sides. He wanted to ensure that the error in his estimate of the circumference of the universe, based on Islamic ideas of its radius, was no more than the width of a horse's hair.

2. Tangram is a Chinese puzzle made with the seven shapes shown in Fig. 1.65. They can form a square, if properly arranged, and various interesting shapes.

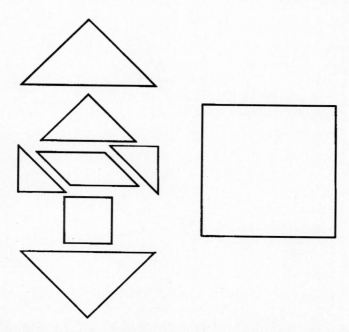

Figure 1.65 The seven pieces of a tangram can be arranged to form a square.

Any figure that can be made with these seven shapes is called a *tangram*. In this project, you are to construct all convex tangrams. Let the short sides of the smallest triangle have a length of 1. Find the lengths of the sides, the angles, and the area for each of the seven shapes. What angles can appear at the corners of a convex tangram? Why is there only one triangular tangram? Why can a convex tangram not have more than eight sides? After you have found the candidates for convex tangrams, construct them. (Read [24] gives all convex tangrams, some background, and an open mathematical question on tangrams.)

3. Figure 1.66 shows a pantograph, a device for enlarging a design.

   a) Explain why the design traced by the pencil must always be similar to the original design. [*Hint:* Why do the bars need to form a parallelogram?]

   b) Determine where the holes should be drilled so that the pencil will trace a figure whose dimensions are twice the original design and k times the original design.

   c) Make a pantograph and use it to enlarge several designs.

   d) How could you alter a pantograph to reduce a design?

4. Let's try to measure how close a nonconvex set is to being convex. Call a point P in a set a *guard point* provided that for every other point Q in the set, $\overline{PQ}$ is also in the set. (A guard at a guard point can see every spot in the set.) The ratio (area of guard points/area of set) is called the *inside convexity* of a set.

   a) Find the inside convexity for each shape depicted in Fig. 1.67. For each shape, find the "worst" place to put a guard, that is, the point from which the guard would see the smallest percentage of the total area. From that worst place, what percentage of the entire area does a guard see? Explain your answers.

   Another way to measure the convexity of a set is to measure how much must be added to the set to obtain a convex set. Define the *outside convexity* of a set S to be the ratio (area of S/area of the smallest convex set containing S).

   b) Find the outside convexity of the shapes shown in Fig. 1.67.

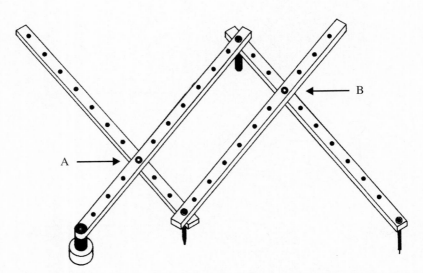

Figure 1.66 Fix the lower left base of the pantograph. As the stylus in the middle of the bottom traces a design, the pencil in the lower right draws an enlargement. Moving the connectors at A and B change the scale of the enlargement.

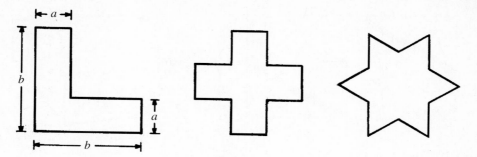

Figure 1.67

c) Explain why the inside (or outside) convexity of any convex set is 1.

d) Find a set with positive area whose inside (outside) convexity is 0.

e) Find a nonconvex set whose inside and outside convexity is 1. Explain.

f) Compare the definitions of *inside* and *outside convexity* with your intuition of how close a set is to convex. Look for a better definition.

5. Use Fig. 1.68 to give a plausible explanation of why the volume of a square pyramid is one-third the height times the base. Build a model of Fig. 1.68. Will this decomposition work if the cube is replaced with a rectangular box? Will this decomposition work if the cube is replaced with a triangular prism?

6. A ribbon wrapped around a box can be removed without cutting, stretching, or untying it (Fig. 1.69). Try to do so with an actual ribbon on a box. Then model this situation geometrically and explain why it works. It may be easier to explain first with a rectangle.

7. a) Construct the 13 Archimedean solids (Wenninger [28]). The Archimedean solids, together with an infinite family of prisms and another infinite family of antiprisms, are semiregular polyhedra. *Semiregular polyhedra* have two or more kinds of regular polygons for faces and the same number,

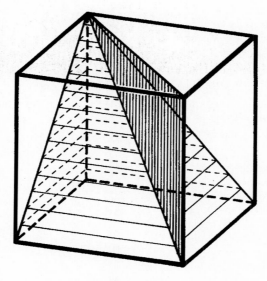

Figure 1.68

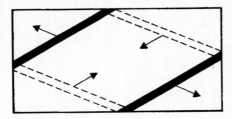 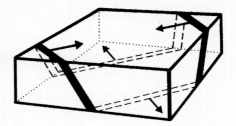

Figure 1.69

types, and arrangement of faces around each vertex.

b) Find the spherical excess at a vertex for each of these Archimedean solids.

c) Find the spherical excess at a vertex for semiregular prisms and antiprisms.

d) Explain why there can be no other Archimedean solids.

8. Define a *deltahedron* to be a strictly convex polyhedron all of whose faces are equilateral triangles. (A convex polyhedron is strictly convex if no two faces lie in the same plane. We use the name deltahedron because the capital Greek letter delta △ looks like a triangle.)

   a) Find an equation relating E and F in deltahedra. Justify your equation.

   b) Explain why there are at most five triangles at any vertex of a deltahedron. Use this to find an inequality relating V and E in a deltahedron.

   c) Use parts (a) and (b) to show that in a deltahedron $E \leq 30$ and E must be a multiple of 3. List all the possible candidates for values of V, E, and F of deltahedra.

   d) Build a complete set of deltahedra.

Remarks There is no deltahedron with $E = 27$. (See Beck et al. [2].)

9. Euler's formula applies to various polyhedra that are not convex. Imagine polyhedra as made of rubber and inflate them. Then the ones that can be inflated to look like a sphere will satisfy Euler's formula. The generalized Euler's formula applies to polyhedra with holes. Let H be the number of holes in the polyhedron. Find an equality relating V, E, F, and H. (The value $H = 0$ should give you Euler's formula.) You will probably need to draw or make several examples. You may find that you need to make the definition of a polyhedron explicit. Beck et al. [2] gives applications of this generalized formula. Generalize Descartes's formula to polyhedra with holes.

10. For a convex polyhedron, form its *dual* polyhedron as follows. The vertices of the dual are at the centers of the original polyhedron's faces. Two new vertices are connected with an edge provided that the corresponding original faces were joined at an edge.

    a) Describe the faces of the dual.

    b) Verify that the duals of regular polyhedra are again regular polyhedra.

    c) How are V, E, and F for a polyhedron and its dual related?

    d) Build duals of some of the Archimedean solids.

11. Build a geodesic dome. (See Kenner [17].)

12. Build some tensegrity figures, invented by Buckminster Fuller, such as that shown in Fig. 1.70. (See Kenner [17] and Pugh [23].)

13. Investigate the Geometer's Sketchpad, CABRI, or other computer programs designed to aid geometric exploration.

14. Investigate the geometry of four and more dimensions. (See Coxeter [5] and Rucker [25].)

15. Investigate the history of geometry. (See Aaboe [1], Eves [9], Kline [18], and Struik [27].)

16. Investigate axiomatic systems. (See Bryant [4] and Fishback [10].)

17. Develop an axiomatic system describing the geometry of pixels (points) and lines on a computer screen. Find and prove some theorems in this geometry.

18. Investigate the golden ratio and phyllotaxis. (See Coxeter [5] and Huntley [16].)

19. Investigate taxicab geometry. (See Krause [19].)

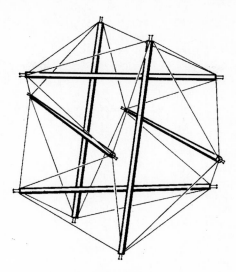

Figure 1.70

20. Investigate the geometry of the sphere. (See Henderson [14] and McCleary [21].)
21. Write an essay giving your understanding of why mathematics is certain and why it is applicable. Compare your ideas with those of Plato and Aristotle. (See also Grabiner [12].)
22. Write an essay on the roles of intuition and proofs in geometric understanding.
23. Write an essay on the different levels of proof demanded in mathematics, in a civil court case ("the preponderance of the evidence") and a criminal court case ("beyond a reasonable doubt"). What phrase would you offer to describe mathematical proofs? Compare the advantages and disadvantages of these differing levels of proof.
24. Write an essay explaining axiomatic systems and models to a high school geometry student. Use examples, preferably everyday ones.
25. Write an essay on the understanding of definition in mathematics, in other disciplines and in every day language. Discuss the advantages and disadvantages of these different notions of definition.

Suggested Readings

[1] Aaboe, A. *Episodes from the Early History of Mathematics.* Washington, D.C.: Mathematical Association of America, 1964.

[2] Beck, A., M. Bleicher, and D. Crowe. *Excursions into Mathematics.* New York: Worth, 1969.

[3] Boltyanskii, V. *Equivalent and Equidecomposable Figures.* Lexington, Mass.: D. C. Heath, 1963.

[4] Bryant, V. A fresh look at geometry. *Mathematics Magazine,* 1971, 44(1):180–182.

[5] Coxeter, H. *Introduction to Geometry.* New York: John Wiley & Sons, 1969.

[6] DeLong, H. *A Profile of Mathematical Logic.* Reading, Mass.: Addison-Wesley, 1971.

[7] Dubnov, I. *Mistakes in Geometric Proofs.* Lexington, Mass.: D. C. Heath, 1963.

[8] Edmondson, A. *A Fuller Explanation: The Synergetic Geometry of R. Buckminster Fuller.* New York: Birkhäuser, 1987.

[9] Eves, H. *An Introduction to the History of Mathematics.* New York: Holt, Rinehart and Winston, 1976.

[10] Fishback, W., *Projective and Euclidean Geometry.* New York: John Wiley & Sons, 1962.

[11] Gallian, J. *Contemporary Abstract Algebra*. Lexington, Mass.: D. C. Heath, 1990.
[12] Grabiner, J. The centrality of mathematics in the history of Western thought, *Mathematics Magazine*, 1988, 61(4):220–230.
[13] Heath, T. *The Thirteen Books of Euclid's Elements*, 3 vols. Mineola, NY: Dover, 1956.
[14] Henderson, D. *Experiencing Geometry on the Plane and Sphere*. Upper Saddle River, N.J.: Prentice-Hall, 1996.
[15] Hilbert, D. *The Foundations of Geometry*, 2d ed. (transl. E. Townsend). Peru, Ill.: Open Court, 1921.
[16] Huntley, H. *The Divine Proportion: A Study in Mathematical Beauty*. Mineola, N.Y.: Dover, 1970.
[17] Kenner, H. *Geodesic Math and How to Use It*. Berkeley: University of California Press, 1976.
[18] Kline, M. *Mathematical Thought from Ancient to Modern Times*. New York: Oxford University Press, 1972.
[19] Krause, E. *Taxicab Geometry: An Adventure in Non-Euclidean Geometry*. New York: Dover, 1986.
[20] Lakatos, I. *Proofs and Refutations: The Logic of Mathematical Discovery*. New York: Cambridge University Press, 1976.
[21] McCleary, J. *Geometry from a Differentiable Viewpoint*. Port Chester, N.Y.: Cambridge University Press, 1994.
[22] McMahon, T., and J. Bonner. *On Size and Life*. New York: Scientific American Library, Scientific American Books, 1983.
[23] Pugh, A. *An Introduction to Tensegrity*. Berkeley: University of California Press, 1976.
[24] Read, R. *Tangrams*. Mineola, N.Y.: Dover, 1965.
[25] Rucker, R. *Geometry, Relativity and the Fourth Dimension*. Mineola, N.Y.: Dover, 1977.
[26] Salmon, W. *Zeno's Paradoxes*. Indianapolis: Bobbs-Merrill, 1970.
[27] Struik, D. *A Concise History of Mathematics*. Mineola, N.Y.: Dover, 1967.
[28] Wenninger, M. *Polyhedron Models for the Classroom*. Reston, Va.: National Council of Teachers of Mathematics, 1966.
[29] Wilder, R. *The Foundations of Mathematics*. New York: John Wiley & Sons, 1965.

Suggested Media

1. "Curves of Constant Width," 16-minute film, International Film Bureau, Chicago, Ill., 1971.
2. "Equidecomposable Polygons," 10.5-minute film, International Film Bureau, Chicago, Ill., 1971.
3. "Geodesic Domes: Math Raises the Roof," 20-minute film, David Nulsen Enterprises, Santa Monica, Calif., 1979.
4. "Geometry: New Tools for New Technologies," 60-minute video, COMAP, Lexington, Mass., 1993.
5. "Mathematics of the Honeycomb," 13-minute film, Moody Institute of Science, Whittier, Calif., 1964.
6. "Music of the Spheres," 52-minute video from *The Ascent of Man* series, Time–Life Multimedia, New York, 1974.
7. "The Theorem of Pythagoras," 26-minute video, California Institute of Technology, Pasadena, Calif., 1988.
8. "The Tunnel of Samos," 30-minute video, California Institute of Technology, Pasadena, Calif., 1995.

2
Analytic Geometry

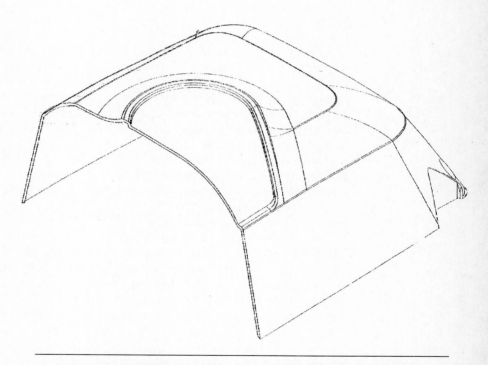

Computer modeling of a lawn mower shroud (housing). Computer graphics have greatly changed the designing of products. Eye-catching pictures and computing prowess easily attract attention, but the mathematics behind the pictures deserves fuller understanding. Analytic geometry helps represent the curves needed for a lawn mower part, an airplane, and other shapes. Transformational and projective geometries, the topics of Chapters 4 and 6, are the keys to enabling the computer show different views of a shape, including perspective.

> Though the idea behind it all is childishly simple, yet the method of analytic geometry is so powerful that very ordinary [youth] can use it to prove results which would have baffled the greatest of the Greek geometers—Euclid, Archimedes and Apollonius.
> —E. T. Bell

> How can it be that mathematics, being after all a product of human thought independent of experience, is so admirably adapted to the objects of reality?   —Albert Einstein

2.1 Overview and History

The fruitful union of algebra and geometry called analytic geometry has become an indispensable tool for mathematicians, scientists, and those in many other fields. Although René Descartes and Pierre Fermat deserve credit for creating analytic geometry, many others before and after shared in its development. (See Boyer [1].) After 150 B.C., Greek astronomers used coordinates to describe the positions of stars, and Greek and Roman geographers used coordinates to describe places on earth. In the fourteenth century Nicole Oresme used some examples of graphic representation. Arab and Renaissance mathematicians developed algebra into a powerful language. François Viète (1540–1603) used letters to represent unknown values and general situations. Viète called his approach *analysis*, following the ancient Greeks' meaning of the word. He started by assuming that the given problem had been solved and used a letter to represent this answer, which he then found algebraically in the modern sense. However, Viète retained the Greek limitation of adding only like quantities. Thus, in modern notation, Viète would be willing to add $x^3 + a^2x$, but not $x^3 + ax$, because the latter expression would represent a volume added to an area.

Pierre de Fermat (1601–1665) united the notational advances of Viète's algebra with traditional geometry. He realized that first- and second-degree equations correspond to lines and conics and investigated curves defined by higher degree equations. He solved some questions now considered part of calculus, such as finding maxima and minima. Although Fermat developed analytic geometry first, René Descartes (1596–1650) published sooner and was more influential. Descartes freed algebraic notation from Viète's restrictions of homogeneous dimensions. His famous book *Geometry*, published in 1637, showed the power of this new field, solving problems the Greeks had been unable to answer. Mathematicians began investigating the tremendous variety of curves suddenly described by algebraic equations. However, Descartes's book doesn't look like analytic geometry to us, for he didn't use coordinates and axes. Rather, he described the length of a line segment in terms of relationships of the lengths of various other line segments and translated these relationships into an algebraic equation. Nevertheless, we often call these coordinates Cartesian in honor of Descartes (and to distinguish them from other coordinate systems, such as polar coordinates).

During the first century of analytic geometry—and the early years of calculus—mathematicians didn't realize the power and simplicity of functions. Curves such as

RENÉ DESCARTES

René Descartes (1596–1650) won acclaim in philosophy, as well as in mathematics. In his most famous work, the *Discourse on the Method*, he tried to show how his rational approach could lead to new knowledge in any domain. He started by doubting everything that he did not know evidently to be true. His famous statement, "I think, therefore I am," was his first principle of philosophy, the idea that withstood all of his doubting. Next, he sought to reduce each difficulty he studied into many simpler parts and approach these parts systematically and exhaustively. This method may not seem revolutionary now, but Descartes was challenging traditional learning handed down for centuries. Descartes illustrated his method in three essays— *Optics*, *Geometry*, and *Meteorology*—that followed the *Method*. Of these, *Geometry* was by far the most influential.

In the first part of *Geometry*, Descartes showed how to solve a number of problems, including one that the Greeks had been unable to solve. He translated the geometric descriptions of curves into algebraic equations that he could then solve. In the second part, he provided a method of finding tangents to curves and attempted to classify curves. In the third part, he investigated the theory of equations, including his own rule of signs, giving a bound on the number of positive roots and what we call negative roots of a polynomial equation. (Descartes would have said that the equation $x + 5 = 0$ has 5 as a false root, rather than a root of -5.) (See Boyer [1] and Grabiner [7] for more on *Geometry*.)

Descartes intentionally wrote obscurely, making it difficult for others to extend his work. However, later editions of *Geometry* contained commentaries by others that explained his work, revealing the power of his method. The success of analytic geometry in posing and solving important problems has made it indispensable in mathematics. Of course, calculus depends crucially on analytic geometry, and it is no accident that so little time elapsed between Descartes's pioneering work in 1637 and the advent of calculus in the 1660s and 1670s.

---

the folium of Descartes, shown in Fig. 2.1, were tackled as were more familiar curves such as the one shown in Fig. 2.2. Leonhard Euler (1707–1783), the most prolific mathematician of all time, emphasized functions and recast analytic geometry and calculus in nearly modern form in his influential textbooks.

Mathematicians have continued to develop analytic geometry and extend it into new branches of mathematics. In the nineteenth century mathematicians used analytic geometry to overcome visual limitations and investigate four and more dimensions. Transformational geometry and differential geometry grew out of analytic geometry, as well as areas no longer thought of as geometry, such as linear algebra and calculus. The advent of computer graphics has renewed the interest in analytic geometry.

2.1.1 The analytic model

We make the connection between geometric concepts and their algebraic counterparts explicit by building a model of geometry in algebra. The familiar graphs of analytic geometry are not actually part of the model, but they make the model and its many applications understandable. Geometric axioms and theorems become algebraic facts to be verified. In turn, algebraic equations and relations can be visualized.

62 Chapter 2 Analytic Geometry

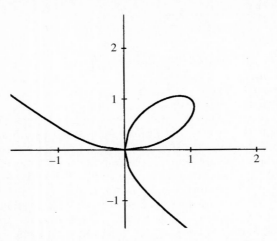

Figure 2.1 Folium of Descartes, $x^3 - 2xy + y^3 = 0$.

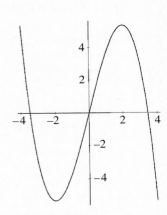

Figure 2.2 $y = \tfrac{1}{3}x^3 + 4x$.

Interpretation In the plane $\mathbf{R}^2$, by *point* we mean an ordered pair of real numbers (x, y). By *line* we mean a set of points of the form $\{(x, y) : ax + by + c = 0\}$, for $a, b, c \in \mathbf{R}$ with a and b not both 0. A point (u, v) is *on* the line $\{(x, y) : ax + by + c = 0\}$ iff $au + bv + c = 0$. The *distance* between two points $P = (x, y)$ and $Q = (u, v)$ is $d(P, Q) = \sqrt{(x - u)^2 + (y - v)^2}$.

Remarks As usual, we identify a line by its equation. Two lines $ax + by + c = 0$ and $mx + ny + p = 0$ are the same provided that there is a nonzero real number k such that $ak = m$, $bk = n$, and $ck = p$. The equation for the distance between two points given in the preceding interpretation is in essence the Pythagorean theorem—now no longer a theorem, but a definition.

Exercise 1 Find the slope and y-intercept of the line $ax + by + c = 0$, if $b \neq 0$. (When $b = 0$, the line is vertical and has no slope. If a and b are both 0, either no points or all points satisfy the equation $ax + by + c = 0$.)

Example 1 Verify that, for any two distinct points, there is only one line on both points. (This result shows that Hilbert's axioms I-1 and I-2 hold in this model.)

Verification. Let (x_1, y_1) and (x_2, y_2) be any two distinct points.

Case 1 $x_1 = x_2$. The line $x - x_1 = 0$ is on both points. Let $ax + by + c = 0$ be any line through these two points. Then $ax_1 + by_1 + c = 0$, and $ax_1 + by_2 + c = 0$. These equations reduce to $b(y_1 - y_2) = 0$. For the points to be distinct, $y_1 - y_2 \neq 0$. So $b = 0$, which forces $c = -ax_1$. Thus $ax + by + c = 0$ is a multiple of $x - x_1 = 0$, showing that only one line passes through these two points.

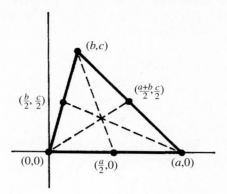

Figure 2.3

Case 2 $x_1 \neq x_2$. Verify $(y_2 - y_1/x_2 - x_1)x - y + y_1 - (y_2 - y_1/x_2 - x_1)x_1 = 0$ is a line on both points. As in case 1, verify that a line $ax + by + c = 0$ on both points is a multiple of the line given in this case. ●

Part of analytic geometry's power comes from our ability to solve geometric problems with algebra and to understand abstract algebraic expressions geometrically.

Example 2 Show that the medians of a triangle intersect in the point two-thirds the way from a vertex to the opposite midpoint.

Solution. WLOG, pick the axes so that the vertices of the triangle shown in Fig. 2.3 are $(0, 0)$, $(a, 0)$, and (b, c). Verify that the midpoints of the three sides are $(\frac{a}{2}, 0)$, $(\frac{b}{2}, \frac{c}{2})$, and $(\frac{a+b}{2}, \frac{c}{2})$. The *medians* connect midpoints to opposite vertices. Verify that the medians are $y = (c/(a + b))x$, $y = (c/(b - 2a))x - ac/(b - 2a)$ and $y = (2c/(2b - a))x - (ac/(2b - a))$. Verify that $(\frac{a+b}{3}, \frac{c}{3})$ is on all these lines and two-thirds the distance from each vertex to the opposite side. ●

(See Boyer [1] for more on the history of analytic geometry and Eves [6] for more on the model.)

PROBLEMS FOR SECTION 2.1

In these problems you may use familiar properties of geometry and analytic geometry such as: Nonvertical parallel lines have the same slope.

1. **a)** Guess what curve the midpoint of a ladder makes as the top of the ladder slips down a wall and the bottom of the ladder moves away from the wall. Draw a diagram.

   b) Model the situation in part (a) with a ruler and a corner of a sheet of paper, marking the various midpoints of the ruler on the paper.

   c) Use analytic geometry to find the set of all midpoints of segments of length 1 whose endpoints are on the x- and y-axes.

   d) Explore what happens in part (b) if the corner of the paper doesn't form a right angle or if you pick a point on the ruler other than the midpoint.

2. Use analytic geometry to show that the four midpoints of any quadrilateral always form a parallelogram.

3. Use analytic geometry to verify the law of cosines: $c^2 = a^2 + b^2 - 2ab\cos(C)$ (Fig. 2.4).

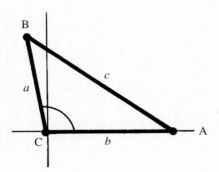

Figure 2.4 Law of cosines, $c^2 = a^2 + b^2 - 2ab\cos C$.

4. Verify Hilbert's axiom IV-1 (Appendix B): Through a given point P not on a given line k there passes at most one line that does not intersect k.

5. Define a circle in analytic geometry and verify Euclid's postulate 3 (Appendix A): To describe a circle with any center and radius.

6. For many applications, the use of different scales on the x- and y-axes is convenient (Fig. 2.5). Suppose that on the x-axis a unit represents a distance of k and on the y-axis a unit represents a distance of j. Explain why the equations of lines in this model have the same form as in the usual model. Develop a formula in this model for the distance between two points, $P = (x_1, y_1)$ and $Q = (x_2, y_2)$. Give the equation of a circle of radius r and center (a, b).

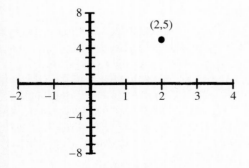

Figure 2.5

7. (Calculus) Before the advent of calculus, Fermat developed a method for finding the maximum or minimum of certain formulas, such as $bx^2 - x^3$. He reasoned first that if two values of x, say, u and v, give the same height, he would get $bu^2 - u^3 = bv^2 - v^3$.

   a) Verify that this equality reduces to $b(u + v) = u^2 + uv + v^2$, for $u \neq v$.

   b) Explain Fermat's reasoning that at a maximum (or minimum) for the formula the two x-values are equal. Replace v in part (a) with u and simplify to get $u = \frac{2}{3}b$. Verify, using calculus, that this value does indeed give a (relative) maximum for the function $y = bx^2 - x^3$. (Fermat didn't consider negative numbers or zero as answers.)

   c) Use Fermat's algebraic approach to find the (relative) maximum and minima for $x^4 - 2b^2x^2 + b^4$. Use calculus to verify that these values are correct.

   d) Explain any logical shortcomings of Fermat's approach.

   e) Explain the practical shortcomings of Fermat's method, by using $\cos(x) - \frac{1}{2}x$.

8. (Calculus) Most lines that intersect a curve do so in two (or more) places, say, a and b. When you solve the system of equations for the line and the curve, they reduce to $(x - a)(x - b)k = 0$, where k represents any other factors. However, tangents have a "double root" at the point of tangency: The system of a curve and its tangent line at a reduces to $(x - a)(x - a)k = 0$. You can use this idea to find tangents without calculus. First, find the tangent to $y = x^2 - x$ at $(2, 2)$.

   a) Verify that the equation of a general line through $(2, 2)$ is $y = mx - 2m + 2$.

   b) Substitute $mx - 2m + 2$ for y in $y = x^2 - x$ and find x in terms of m with the aid of the quadratic formula.

   c) Which value of m in part (b) gives one value of x (a double root)? [*Hint:* Consider what is under the $\sqrt{\ }$.] Use this m to find the tangent's equation.]

   d) Use calculus to verify your answer in part (c).

   e) Vary the preceding approach to find the two tangents to $y = x^2 - 2x + 4$ through the point $(0, 0)$. Graph this parabola and these tangents.

f) Vary this second approach to find the tangent to $y = 4x - x^2$ parallel to $y = 6x$.

g) Discuss any logical and practical shortcomings of this method.

9. a) Graph the following functions and decide which enclose a convex region of the plane: $y = x^2$, $y = x - x^2$, $y = x^3$, $y = x^4$, $y = e^x$, $y = \sin x$, and $y = \ln x$.

b) Functions f that satisfy $f((a+b)/2) \le (f(a) + f(b))/2$ for all a and b are called *convex functions*. Which of the functions in part (a) are convex functions? How do convex functions compare with functions that enclose a convex region of the plane? To explain the difference between these uses of *convex* for functions, define *concave* functions. Explain how concave functions relate to functions enclosing a convex region.

c) (Calculus) Find the second derivative of the functions in part (a). What is special about the second derivative of the convex functions? Use a graph to explain how the definition of a convex function fits with what you found out about the second derivatives of convex functions. What can you say about the second derivative of the concave functions you defined in part (c)?

10. The arithmetic of complex numbers (**C**) has a well-known geometric interpretation in $\mathbf{R}^2$. The complex number $a + bi$ can be represented as the point, or vector, (a, b) in the plane. Addition of complex numbers corresponds to vector addition: $(a + bi) + (c + di) = (a + c) + (b + d)i$.

a) Explain and illustrate on Cartesian axes why this addition satisfies the parallelogram law.

b) The *complex conjugate* of $a + bi$ is the number $a - bi$. Illustrate on Cartesian axes how these numbers are related geometrically.

c) The *modulus* of $a + bi$ is the real number $\sqrt{a^2 + b^2}$. What does the modulus tell you geometrically?

11. The formula for complex multiplication, $(a + bi) \times (c + di) = (ac - bd) + (ad + bc)i$, doesn't reveal the geometry.

a) How does the product of a complex number and its conjugate relate to the modulus? (See Problem 10.)

b) Illustrate with several examples on Cartesian axes the result of multiplying $a + bi$ by a real number $r + 0i$. What corresponds geometrically to multiplying by a real number?

c) Illustrate on Cartesian axes the result of multiplying $a + bi$ by i, a complex number on the unit circle. Also illustrate the result of multiplying $a + bi$ by $0.6 + 0.8i$ and by $-0.96 + 0.28i$, other points on the unit circle. What do you think multiplication by a point on the unit circle does geometrically to $a + bi$?

d) Explain why any complex number $c + di$ can be written as the product of its modulus with a complex number $x + yi$ on the unit circle (for which $x^2 + y^2 = 1$). Use parts (b) and (c) to describe what multiplication by a general complex number $c + di$ does geometrically to $a + bi$.

2.2 Conics and Locus Problems

The Greeks identified and studied the three types of conics: ellipses, parabolas, and hyperbolas. However, nearly two thousand years passed before the first of many applications of conics outside of mathematics appeared. We call these curves conics because they are the intersections of a (double-napped) cone with planes at various angles (Fig. 2.6). To find the familiar equations of these curves we use an easier characterization based on distance. The process of finding a set of points or its equation from a geometric characterization is called a *locus problem*. See Eves [6] for further information on these topics.

Example 1   Find the set (locus) of points P such that P is the center of a circle tangent to two given lines.

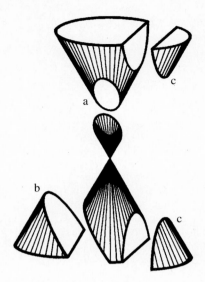

Figure 2.6 The intersection of a plane and a cone is (a) an ellipse, (b) a parabola, or (c) a hyperbola.

Solution. If the two lines are parallel, then a circle tangent to both must have its center on the line midway between these two lines. If the two lines intersect at a point Q, then the centers of circles tangent to both lines must be on one of the two angle bisectors of these two lines. Illustrate these two cases. ●

Definition 2.2.1 Given two points F and F', an *ellipse* is the set of points P in the plane such that the sum of the distances of P to F and F' is constant. Given two points F and F', a *hyperbola* is the set of points P in the plane such that the difference of the distances from P to F and F' is constant. Given a line m and a point F not on m, a *parabola* is the set of points P in the plane such that the distance of P from F equals the distance of P from m. The points F and F' are called *foci*. The line m is called the *directrix*. (Figs. 2.7, 2.8, and 2.9.)

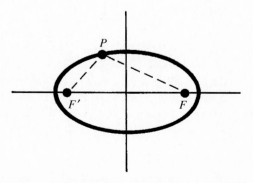

Figure 2.7 An ellipse.

PIERRE FERMAT

Although a lawyer and political councilor by profession, Pierre Fermat (1601?–1665) is best remembered for the mathematics he pursued in his spare time. He corresponded extensively with other mathematicians, rather than seeking publication. Indeed, his work in analytic geometry, even though done before Descartes, wasn't published until after Fermat died. He investigated the shapes of curves from their equations. This approach complemented Descartes's work, as Descartes had emphasized finding the equation of a curve defined by some geometric process. Fermat explicitly discussed first- and second-degree equations and explored many new curves with higher degree equations.

Fermat also contributed extensively and profoundly to number theory. He is best known for Fermat's last theorem: $a^n + b^n = c^n$ has no nontrivial integer solutions if $n > 2$. This statement holds the record for the greatest number of incorrect published "proofs," but Andrew Wiles finally proved it 1994. This simple-looking statement has led to extensive and profound investigations in number theory. Fermat's theorems in number theory concentrate on primes, divisibility, and powers. For example, Fermat's little theorem, an important tool in abstract algebra and coding theory, states that if a is not a multiple of a prime p, then p divides $a^{p-1} - 1$.

Other areas of mathematics also benefited from Fermat's creativity. The founding of probability as a mathematical subject grew out of letters between Fermat and Blaise Pascal. They corresponded at some length and eventually solved a problem on how to distribute fairly the wagers of an interrupted game of chance. Fermat was an important precursor of calculus with his method for finding maxima and minima. In addition, he understood the rules that we now describe as differentiating and integrating polynomials.

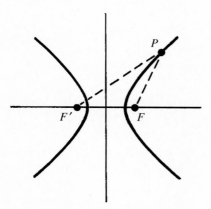

Figure 2.8 A hyperbola.

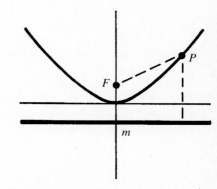

Figure 2.9 A parabola.

Exercise 1 Construct an ellipse as follows. Fix the two ends of a piece of string to different points on a piece of paper. Place a pencil on the paper so that it holds the string taut. Explain why the pencil sweeps out an ellipse as it moves.

Example 2 Find the equation of an ellipse.

Solution. WLOG, let the foci F and F' have coordinates $(f, 0)$ and $(-f, 0)$. Let $P = (x, y)$ be any point on the ellipse. The definition of an ellipse gives the following equation, which we adjust to a more familiar form.

$$\sqrt{(x-f)^2 + y^2} + \sqrt{(x+f)^2 + y^2} = k.$$

The sum of the distances is $k > 2f$.

$$\sqrt{(x-f)^2 + y^2} = k - \sqrt{(x+f)^2 + y^2}.$$

Square both sides and simplify.

$$x^2 - 2fx + f^2 + y^2 = k^2 - 2k\sqrt{(x+f)^2 + y^2} + x^2 + 2fx + f^2 + y^2.$$

$$-4fx - k^2 = -2k\sqrt{(x+f)^2 + y^2}.$$

Square again and simplify.

$$16f^2x^2 + 8fxk^2 + k^4 = 4k^2(x^2 + 2fx + f^2 + y^2).$$

$$x^2(16f^2 - 4k^2) - 4k^2y^2 = k^2(4f^2 - k^2).$$

Divide by the right side.

$$\frac{4x^2}{k^2} + \frac{4y^2}{(k^2 - 4f^2)} = 1.$$

We can rewrite this equation as

$$\frac{x^2}{a^2} + \frac{y^2}{b^2} = 1$$

because $k^2 - 4f^2 > 0$. The *diameters* of the ellipse along the x and y-axes are $2a$ and $2b$, respectively. ●

Exercise 2 Verify that a circle satisfies the definition of an ellipse.

Exercise 3 Outline a parabola, as indicated in Fig. 2.10. Fix a pin and a ruler on a piece of paper. Then place a right angled triangle or piece of paper with its right angle on the ruler and one leg touching the pin. Draw a line using the other leg. Move the triangle to draw different tangents to the parabola. Note that the ruler is not the directrix. (See Broman and Broman [2] for other constructions and an explanation.)

Example 3 The equation of a hyperbola is $x^2/a^2 - y^2/b^2 = 1$, and the equation of a parabola is $y = ax^2$.

Solution. See Problem 2. ●

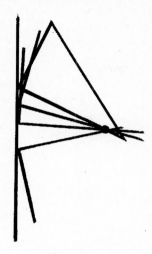

Figure 2.10

Examples 2 and 3 give the equations of the conics when we pick the easiest starting conditions. Less convenient conditions make the algebra more difficult and the final equation more complicated. The general equation of a conic is $ax^2 + 2bxy + cy^2 + 2dx + 2ey + f = 0$. The conic is an ellipse, parabola, or hyperbola, depending on whether $ac - b^2$ is positive, zero, or negative, respectively. (Circles are special ellipses with $a = c$ and $b = 0$.) However, some of these general second-degree equations are *degenerate*: that is, they are equations for one or two lines or a point or even the empty set. The locus is a conic provided that the determinant of

$$\begin{bmatrix} a & b & d \\ b & c & e \\ d & e & f \end{bmatrix}$$

is nonzero.

Example 4 If we pick $F = (\frac{\sqrt{2}}{2}, \frac{-\sqrt{2}}{2})$, $F' = (\frac{-\sqrt{2}}{2}, \frac{\sqrt{2}}{2})$ and $k = 2\sqrt{2}$, as shown in Fig. 2.11, the ellipse has the equation $\frac{3}{4}x^2 + \frac{1}{2}xy + \frac{3}{4}y^2 = 1$. Note that $ac - b^2 = (3/4)(3/4) - (1/4)(1/4) = 1/2 > 0$ and that the determinant is $-1/2$. ●

Example 5 Use calculus to show the reflection property of the parabola: All rays from the focus are reflected by the parabola in rays parallel to the axis of symmetry of the parabola.

Solution. Turn the parabola horizontally, as shown in Fig. 2.12, and use calculus and trigonometry. Verify that the parabola $y = \pm\sqrt{4kx}$ has focus $F = (k, 0)$. For any point $P = (x, \sqrt{4kx})$ on the upper half of the parabola, show that the slope of the line from F to P is $\sqrt{4kx}/(x - k)$ and that the slope of the tangent line at P is $\sqrt{k}/\sqrt{x}$. The reflection property says that the angle between these two lines equals the angle between the tangent line and a horizontal line. The slopes are tangents of the various angles. Recall that

$$\tan(\alpha - \beta) = \frac{\tan \alpha - \tan \beta}{1 + (\tan \alpha \tan \beta)}.$$

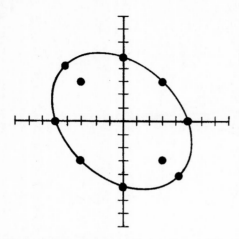

Figure 2.11 The ellipse $\frac{3}{4}x^2 + \frac{1}{2}xy + \frac{3}{4}y^2 = 1$.

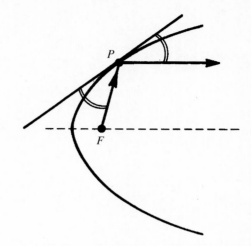

Figure 2.12 The reflection property of a parabola.

If $\tan \alpha = \sqrt{4kx}/(x - k)$ and $\tan(\beta) = \sqrt{k}/\sqrt{x}$, the right side of the formula for $\tan(\alpha - \beta)$ simplifies to $\sqrt{k}/\sqrt{x}$. It is the tangent for the angle between the tangent line and the horizontal line. Why is the lower half similar? The design of satellite disks makes use of this process in reverse: The disks take incoming parallel rays and focus them on one point. ●

PROBLEMS FOR SECTION 2.2

1. **a)** Find the locus of points in the plane equidistant from two points.
   b) Find the locus of points on a sphere equidistant from two points.
   c) Find the locus of points in space equidistant from two points. Relate this answer to your answers in parts (a) and (b).
   d) Repeat parts (a), (b), and (c) for three non-collinear points.

2. **a)** Derive the equation of a hyperbola.
   b) Derive the equation of a parabola. [*Hint:* Use $y = -k$ for the directrix and $(0, k)$ for the focus.]

3. **a)** Attach one end of a string to a ruler and fix the other end to a sheet of paper. Hold the ruler and a pencil as indicated in Fig. 2.13. As you slide the ruler back and forth, while keeping the string taut with the pencil, the pencil will trace out a curve. Identify which conic is created and explain why it is that conic.

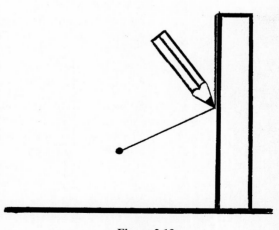

Figure 2.13

b) On wax paper or tracing paper draw a large circle and mark its center C and any other point A inside the circle. For various points P on the circle, fold and crease the paper so that points A and P coincide. The set of creases will outline an ellipse. Use the definition of an ellipse together with Fig. 2.14 to explain this result. Note that Q is a point on the ellipse.

c) Repeat part (b) with point A outside the circle. Modify Fig. 2.14 and explain this result.

4. a) Graph $y = \frac{1}{x}$, the most familiar equation for a hyperbola.

b) Rewrite this equation in the general form of a conic, verify that $ac - b^2 < 0$ and that the determinant is nonzero.

c) Verify that the foci of this hyperbola are at $(\sqrt{2}, \sqrt{2})$ and $(-\sqrt{2}, -\sqrt{2})$.

5. (Calculus) Every hyperbola has two *asymptotes,* or lines that the curve approaches but never intersects. A line $y = mx + b$ is an *oblique asymptote* of a function $y = f(x)$ provided that $\lim_{x \to \pm\infty} f(x) - (mx + b) = 0$.

a) Verify that the hyperbola $(x^2/a^2) - (y^2/b^2) = 1$ has asymptotes $y = (b/a)x$ and $y = -(b/a)x$. Show that the product of these two linear equations when rearranged gives a degenerate conic.

b) Use calculus to find the asymptotes of the hyperbola $y = x + (1/x)$. Verify that when you write this hyperbola's equation in the general form of a conic, $ac - b^2 < 0$ and that the determinant is nonzero.

c) Verify that all hyperbolas $y = x + (k/x)$, $k \neq 0$, have the same asymptotes as the hyperbola in part (b). Graph two hyperbolas, one with $k > 0$ and one with $k < 0$, on the same axes with the asymptotes. How does changing k change the graph? Show that every point not on the asymptotes is on just one of these hyperbolas.

6. Find and describe the locus of points P whose distance from a fixed point F is k times their distance from another fixed point F'. [*Hint:* Let $F = (k, 0)$ and $F' = (-1, 0)$. Consider the case $k = 1$ separately from other $k > 0$.]

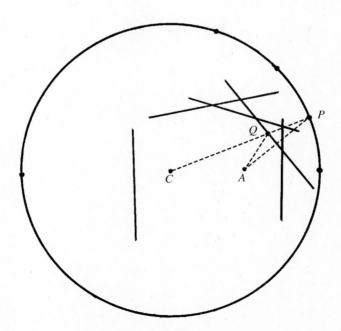

Figure 2.14

72 Chapter 2   Analytic Geometry

7. For each of the following equations, identify the type of conic it is, determine whether it is degenerate, and sketch its graph.
   a) $\frac{1}{2}y^2 + x - 2y = 0$.
   b) $\frac{1}{4}x^2 + y^2 - x - 6y + 9 = 0$.
   c) $\frac{1}{4}x^2 + y^2 - x - 6y + 10 = 0$.
   d) $2x^2 - 3xy + y^2 - 4 = 0$. [*Hint:* Factor $2x^2 - 3xy + y^2 = 0$ and then explain why the factors are the asymptotes.]
   e) $x^2 - xy + y^2 - 1 = 0$. [*Hint:* Pick x-values and find the y-values.]

2.3 Further Topics in Analytic Geometry

The great flexibility of algebra and vectors fosters many geometric and applied variations on the analytic geometry model. We present a few here.

2.3.1 Parametric equations

Leonhard Euler introduced parametric equations in his 1748 textbook on analytic geometry. Separate functions for the x- and y-coordinates allow graphs of complicated curves. These functions, $x(t)$ and $y(t)$, are in terms of a third variable, t, which we can think of as representing time. The point $(x(t), y(t))$ traces out a curve as t varies. As y does not depend on x, the curve can double back on itself or even cross itself. We can use more parametric equations to describe curves in three or more dimensions.

Example 1 (Calculus) For $x(t) = t^3 - t$, $y(t) = t^2 - t^4$, and $-1.1 < t < 1.1$, the point $(x(t), y(t))$ traces out the curve shown in Fig. 2.15. The function $y(t)$ controls the heights of points. Setting the derivative $y'(t) = 2t - 4t^3$ equal to 0 shows the curve to reach its maxima and local minimum when $t = \pm\sqrt{2}/2$ and $t = 0$. ●

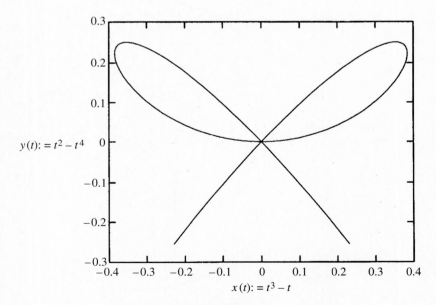

Figure 2.15

2.3 Further Topics in Analytic Geometry

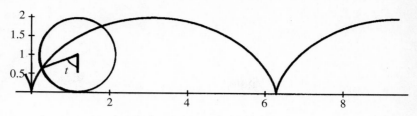

Figure 2.16 A cycloid.

Exercise 1 (Calculus) For which values of t does the curve cross itself at $(0, 0)$? What happens to the curve when $x'(t) = 0$?

Example 2 A *cycloid* is the curve traced by a point on the circumference of a circle as the circle rolls along a line. Find parametric equations for a cycloid for a circle of radius r.

Solution. Let the point start at $(0, 0)$ when $t = 0$ and let a circle of radius 1 roll along the x-axis. Figure 2.16 illustrates a general point on this cycloid. As the circle rolls, the center moves along the line $y = 1$. When the circle has turned an angle of t radians, the center is at $(t, 1)$. The point rotates around the center, so its coordinates will be $(t + f(t), 1 + g(t))$, for some functions f and g. From trigonometry, we get $(t - \sin t, 1 - \cos t)$. Note that $1 - \cos t = 0$ only when t is a multiple of 2π, or when the circle has rolled a whole number of turns. How is the equation altered if the radius is r? ●

2.3.2 Polar coordinates

Jakob Bernoulli (1645–1705) developed polar coordinates, an alternative analytic model of Euclidean geometry (Fig. 2.17a). The first coordinate of a point gives its distance from the origin (positive or negative), and its second coordinate gives the angle made with the x-axis. A *point* in polar coordinates is an ordered pair of real numbers (r, θ), but two ordered pairs (r, θ) and (r', θ') can represent the same point if (1) $r = r' = 0$, (2) $r = r'$ and θ and θ' differ by a multiple of $360°$, or (3) $r = -r'$ and θ and θ' differ by $180°$ or some odd multiple of $180°$. A *line* in polar coordinates is a set of one of two forms: Lines through the origin satisfy $\theta = \alpha$ for some constant α; or lines whose closest point to the origin is (a, α) satisfy $r = a \sec(\theta - \alpha)$ for constants $a \neq 0$ and α (Fig. 2.17b). The law of cosines is used to show the *distance* between two points (r, θ) and (a, α) is $\sqrt{r^2 + a^2 - 2ra \cos(\theta - \alpha)}$, provided r and a are both positive (Fig. 2.17c).

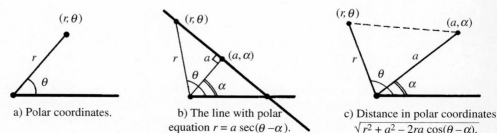

a) Polar coordinates.

b) The line with polar equation $r = a \sec(\theta - \alpha)$.

c) Distance in polar coordinates $\sqrt{r^2 + a^2 - 2ra \cos(\theta - \alpha)}$.

Figure 2.17

74   **Chapter 2**   **Analytic Geometry**

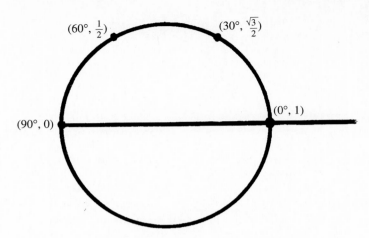

Figure 2.18 $r = \cos\theta$.

Example 3   The graph of $r = \cos\theta$ is the circle shown in Fig. 2.18. ●

2.3.3   Barycentric coordinates

Barycentric coordinates, devised by August Möbius in 1827, have a number of applications, including representing centers of gravity, which originally motivated Möbius. He later developed homogeneous coordinates in projective geometry from barycentric coordinates. (See Chapter 6.) Statisticians now use them in *trilinear plots*, as in Example 4.

Example 4   Figure 2.19 represents the election results of the 50 states and Washington, D.C. (the star), for the three-way presidential race in 1992. (We ignore other candidates for simplicity.) Three coordinates, the percentages, describe a state's outcome. The star for Washington, D.C., is at (86.3, 9.4, 4.3), the respective percentages for Clinton, Bush, and Perot. Two of the percentages determine the third, so the information is essentially two-dimensional. The dotted lines indicate where each candidate had a plurality. For example, the region for the winner, Clinton, is at the top with most of the dots. No dots are in Perot's region, indicating that he didn't win in any state. ●

Barycentric coordinates with two rather than three options are easier to use. Suppose that two points P and Q have weights w_p and w_q that add to 1. From Archimedes' law of the lever, the center of mass of these two points is at point R, between P and Q, which divides their distance in the same proportion as the weights with R closer to the heavier point. Möbius defined (w_p, w_q) to be the *barycentric coordinates* of the center of gravity of P and Q. ("Baro" is Greek for *weight* or *heavy*.) If P and Q have equal weights, R is the midpoint with coordinates $(0.5, 0.5)$. If P has all of the weight, $R = P = (1, 0)$. Example 5 shows that every point between P and Q can be represented with barycentric coordinates. Example 6 justifies extending barycentric coordinates relative to three points. This same reasoning extends to any number of points in any number of dimensions. If the weights at the original points are $\alpha, \beta, \gamma, \ldots$ and add to 1, the center of gravity is $(\alpha, \beta, \gamma, \ldots)$. We can extend barycentric coordinates to points outside

2.3 Further Topics in Analytic Geometry

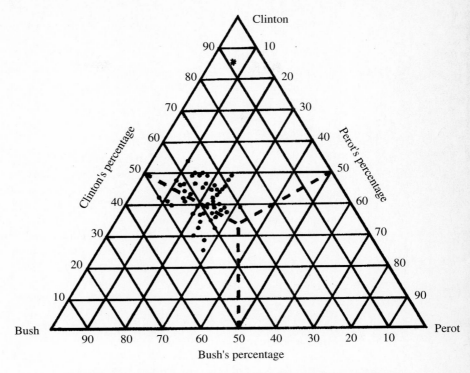

Figure 2.19 Trilinear plot of state and Washington, D.C., popular vote percentages in the 1992 presidential election.

the smallest convex set containing the original points, but some of the coordinates then will be negative.

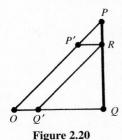

Figure 2.20

Example 5 Let $\overrightarrow{AB}$ denote the vector from A to B and let O be any point not on $\overleftrightarrow{PQ}$. Then any point R between P and Q satisfies $\overrightarrow{OR} = \alpha \overrightarrow{OP} + \beta \overrightarrow{OQ}$, for some positive α and β with $\alpha + \beta = 1$. Then (α, β) are the barycentric coordinates of R with respect to P and Q.

Solution. In Fig. 2.20 let $\overline{RP'} \| \overline{OQ}$ and $\overline{RQ'} \| \overline{OP}$. Then $\overrightarrow{OR} = \overrightarrow{OP'} + \overrightarrow{OQ'}$. As R is between P and Q, we can pick positive α and β such that $\overrightarrow{OR} = \overrightarrow{OP'} + \overrightarrow{OQ'} = \alpha \overrightarrow{OP} + \beta \overrightarrow{OQ}$. We need to show that $\alpha + \beta = 1$. Because $\overline{RP'} \| \overline{OQ}$ and $\overline{RQ'} \| \overline{OP}$, Theorem 1.5.1 shows $\triangle OPQ$ and $\triangle P'PR$ are similar. Thus the sides are proportional. Because $P'R = OQ' = \beta OQ$ and $OP' + P'P = OP$, we must have $\alpha + \beta = 1$. ●

Example 6 If S is in the interior of $\triangle PQR$ and O is a general point, $\overrightarrow{OS} = \alpha \overrightarrow{OP} + \beta \overrightarrow{OQ} + \gamma \overrightarrow{OR}$ for some positive α, β, and γ, with $\alpha + \beta + \gamma = 1$.

Solution. Draw a picture to illustrate this solution. Let T be the intersection of $\overleftrightarrow{QR}$ and $\overleftrightarrow{PS}$. By Example 5, $\overrightarrow{OS} = \alpha \overrightarrow{OP} + \delta \overrightarrow{OT}$, with $\alpha + \delta = 1$, and $\overrightarrow{OT} = \kappa \overrightarrow{OQ} + \lambda \overrightarrow{OR}$, with $\kappa + \lambda = 1$, and α, δ, κ, and λ are positive. Let $\beta = \delta \cdot \kappa$ and $\gamma = \delta \cdot \lambda$ so that both are positive. Then $\overrightarrow{OS} = \alpha\overrightarrow{OP} + \delta\overrightarrow{OT} = \alpha\overrightarrow{OP} + \delta(\kappa\overrightarrow{OQ} + \lambda\overrightarrow{OR}) = \alpha\overrightarrow{OP} + \beta\overrightarrow{OQ} + \gamma\overrightarrow{OR}$ and $\alpha + \beta + \gamma = \alpha + \delta(\kappa + \lambda) = 1$. ●

Exercise 2 If each vertex of a triangle has the same weight, what are the barycentric coordinates of the center of gravity? Explain why this center is where the medians of the triangle intersect.

2.3.4 Rational analytic geometry

We can form an analytic geometry with many algebraic structures besides the real numbers. These other models lead to different geometries because only the real plane $\mathbf{R}^2$ satisfies all of Hilbert's axioms in Appendix B. The model $\mathbf{Q}^2$ based on $\mathbf{Q}$, the rational numbers, loses the continuity of the real numbers. (Recall that a rational number is the quotient p/q of two integers.) A *rational point* is an ordered pair (x, y), where $x, y \in \mathbf{Q}$. A *rational line* is the set of all rational points $\{(x, y) : ax + by + c = 0\}$, where $a, b, c \in \mathbf{Q}$ and a and b are not both zero. Other interpretations remain the same as in the usual model, $\mathbf{R}^2$, and all of Euclid's postulates still hold. However, Euclid's construction of an equilateral triangle, proposition I-1, does not hold in $\mathbf{Q}^2$. Here, WLOG, let the given points be $(0, 0)$ and $(1, 0)$. Verify that the third vertex of the equilateral triangle must be $(\frac{1}{2}, \pm\frac{\sqrt{3}}{2})$. As $\sqrt{3}$ is irrational, this vertex is not a rational point. Hence there is no equilateral triangle in $\mathbf{Q}^2$ and proposition I-1, among others, is independent of Euclid's postulates.

Exercise 3 Explain why, if two lines with rational coefficients have a point of intersection, that intersection is a rational point in $\mathbf{Q}^2$.

(See a standard calculus text, such as Edwards and Penney [5, Chapters 10 and 12] for more on parametric equations and polar coordinates. Eves [6] has more varied information.)

PROBLEMS FOR SECTION 2.3

1. **a)** Explain why $C = \{(x, y) : x = \cos t, y = \sin t\}$ is a circle. Compare C and $C^* = \{(x, y) : x = \cos 2t, y = \sin 2t\}$.

    b) Graph $B = \{(x, y) : x = \cos t, y = \sin 2t\}$, a figure 8.

    c) Graph $S = \{(x, y) : x = t\cos t, y = t\sin t, t \geq 0\}$, a spiral.

    d) Compare $D = \{(x, y) : x = f(t), y = g(t), p \leq t \leq q\}$, and $D^* = \{(x, y) : x = f(kt), y = g(kt), p/k \leq t \leq q/k\}$, where $k \neq 0$. [*Hint:* See part (a).]

2. (Calculus) The *tangent vector* to $(x(t), y(t))$ is $(x'(t), y'(t))$. The tangent vector's length $\|(x', y')\| = \sqrt{(x')^2 + (y')^2}$ is the speed of the point along the curve at time t. Assume that t is in radians.

    a) Find the tangent vectors for C and C^* in Problem 1(a) and compare their lengths. Explain.

    b) Find the tangent vector for B in Problem 1(b). Compare the positions of the points and the directions of the tangent vectors when $t = \pi/2$ and $t = 3\pi/2$.

    c) Find the tangent vector for S in Problem 1(c). Compare the direction and length of the tangent vector when $t = 2\pi$ and when $t = 4\pi$. Describe the speed of the point along the spiral S as t increases.

    d) Find the points and the tangent vectors for D and D^* in Problem 1(d) for $t = c$ and $t = c/k$, respectively. Compare these tangent vectors and their lengths.

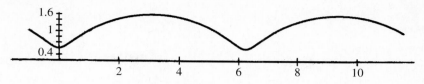

Figure 2.21 A modified cycloid.

3. (Calculus)
   a) Find the tangent vector (see Problem 2) of the cycloid of Example 2 and use it to find the velocity at $t = 2\pi$ and $t = \pi$. Use your answer to explain why, in a photograph of a moving train, the spokes of the wheels nearest the ground are always in focus but the ones at the top are often blurred.
   b) Find parametric equations for a modified cycloid similar to that in Example 2, but with the point inside the circle at $(0, k)$, $0 < k < 1$ (Fig. 2.21). Find the tangent vector for this curve. Compare the velocity at $t = 2\pi$ with that in part (a).
   c) Repeat part (b) with the point outside the circle ($k < 0$). Graph this curve.

4. Graph the curves having the following polar equations.
   a) $r = k\theta$ (spiral).
   b) $r^2 = \cos 2\theta$ (lemniscate of Bernoulli).
   c) $r = 2 \sin 3\theta$ (rose).
   d) $r = 1 + \cos \theta$ (cartioid).

5. a) Devise formulas to convert Cartesian coordinates (x, y) to polar coordinates (r, θ) and vice versa.
   b) Use part (a) to show how to write a function in polar coordinates $r = f(\theta)$ using parametric equations with $t = \theta$.
   c) Use part (b) to verify that $r = \cos \theta$ is a circle. (See Fig. 2.18.)
   d) Verify that $r = a \sec(\theta - \alpha)$, for $a \neq 0$, is the equation of a line. Find the Cartesian equation for the line with polar equation $r = \sqrt{2} \sec(\theta - 45°)$. Find the polar equation of the line having the Cartesian equation $\sqrt{3}x + y + 4 = 0$. [*Hint:* Use Fig. 2.17b), not part (a) of this problem.]
   e) Use the law of cosines (Problem 3, Section 2.1) to justify the distance formula in polar coordinates. Explain how to modify that distance formula when one or both of the points' first coordinates are negative.
   f) Explain why polar coordinates are a model of Euclidean geometry.

6. Draw a triangle $\triangle ABC$. Mark each of the following points on the triangle and give its barycentric coordinates with respect to A, B, and C: the midpoints of each side, the intersection of the medians, and the points A, B, and C. Mark the point in the triangle having the barycentric coordinates $(0.25, 0.25, 0.5)$.

7. Let $X = (1, 0)$, $Y = (0, 1)$, and $O = (0, 0)$. Explain why a point with Cartesian coordinates (α, β) has barycentric coordinates $(\alpha, \beta, 1 - \alpha - \beta)$ with respect to X, Y, and O.

8. Only a few of Hilbert's axioms (Appendix B) are false in the model $\mathbf{Q}^2$. Find them and explain why each fails in this model.

9. Consider the analytic geometry model $\mathbf{Z}^2$, where $\mathbf{Z}$ is the set of integers. Give suitable interpretations for Hilbert's undefined terms. Which of Hilbert's axioms (Appendix B) are true in the model $\mathbf{Z}^2$?

10. Assign coordinates to the points on a sphere, using their longitudes and latitudes (Fig. 2.22). Thus (x, y) represents a point, provided that $-90° \leq y \leq 90°$. Every point (x, y) has multiple representations, such as $(x + 360°, y)$.

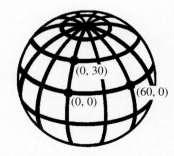

Figure 2.22 Coordinates on a sphere.

a) Describe all coordinates for the north and south poles.

b) Describe the curves on the sphere for $y = a$ and $x = b$.

c) Describe the curves on the sphere for $y = mx + b$.

d) If you interpret *point* as a point on the sphere and *line* as any of the curves of parts (b) and (c), which of Hilbert's axioms from group I (Appendix B) are true? Explain.

11. Assign coordinates to the points on the surface of a *torus* (a doughnut) using longitude and latitude measured in degrees (Fig. 2.23). Every point (x, y) has multiple representations, such as $(x + 360°, y + 720°)$. Interpret *point* as a point on the surface of a torus and *line* as the set of points satisfying an equation of the form $ax + by + c = 0$.

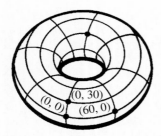

Figure 2.23 Coordinates on a torus.

a) Find values of a, b, and c so that the line $ax + by + c = 0$ is finitely long. Find other values so that that line is infinitely long.

b) Find a line that intersects n times with $y = 0$. Find a line that intersects $y = 0$ infinitely many times.

c) Which of Hilbert's axioms in group I and IV (Appendix B) hold in this model? Explain.

12. Taxicab geometry is an alternative analytic model with a distance function corresponding to the distances taxicabs travel on a rectangular grid of streets. The *taxicab distance* between $A = (a, b)$ and $S = (s, t)$ is $d_T(A, S) = |a - s| + |b - t|$. This geometry has quite different properties from Euclidean geometry. (See Section 1.4.)

a) How are two points that have the same Euclidean and taxicab distances related? If these distances are different, which is larger? If the smaller distance is 1, how large can the other distance be? Explain and illustrate.

b) A *taxicab circle* is the set of points at a fixed taxicab distance from a point. Describe taxicab circles. What are the possible types of intersection of two taxicab circles? Illustrate each.

c) A *taxicab midpoint* M of A and B satisfies $d_T(A, M) = d_T(M, B) = \frac{1}{2} d_T(A, B)$. Illustrate the different sets of midpoints two points can have.

d) The *taxicab perpendicular bisector* of two points comprises of all points equidistant from the two points. Describe and illustrate the different types of taxicab perpendicular bisectors.

e) In the plane find four points whose taxicab distances from each other are all the same.

13. Another distance formula on $\mathbf{R}^2$ uses the maximum of the x and y distances: For $A = (a, b)$ and $S = (s, t)$, $d_M(A, S) = Max\{|a - s|, |b - t|\}$. Redo Problem 12, using d_M.

2.4 CURVES IN COMPUTER-AIDED DESIGN

Computer graphics depend heavily on analytic geometry. The computer stores the coordinates of the various points and information about how they are connected. The easiest connections are line segments and arcs of circles, which are easily described with elementary analytic geometry. However, connecting a series of points smoothly requires calculus and more advanced techniques, which we discuss briefly in this section. Computer-aided design (CAD) uses polynomials—a flexible and easily computed family of curves. For a sequence of points, we need a curve or a sequence of curves that smoothly passes through the points in the specified order. A simplistic approach finds one polynomial through all the points, as in the first two examples.

2.4 Curves in Computer-Aided Design

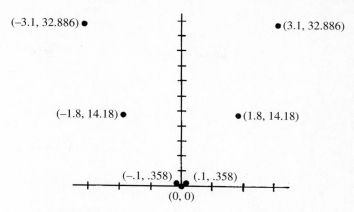

Figure 2.24 The given points for Examples 1, 2, and 3.

Example 1 Find the parabola $y = ax^2 + bx + c$ that goes through the three points $(0, 2)$, $(1, 3)$, and $(3, -7)$.

Solution. When we replace x and y by the values for each point we get three first-degree equations in a, b, and c: $2 = c$, $3 = a + b + c$, and $-7 = 9a + 3b + c$. Solving this system of three linear equations in three unknowns gives $y = -2x^2 + 3x + 2$. ●

Theorem 2.4.1 Given $n + 1$ points $P_j = (x_j, y_j)$, for $j = 0$ to n, with no two x_j equal, there is exactly one nth degree polynomial $y = a_0 + a_1 x + \cdots + a_n x^n$ such that for each j, $y_j = a_0 + a_1 x_j + \cdots + a_n x_j^n$.

Proof. See Strang [12, 80]. ■

Example 2 For the seven points shown in Fig. 2.24, Fig. 2.25 shows a freehand curve connecting these points with no unneeded bumps. Figure 2.26 shows that the graph of the

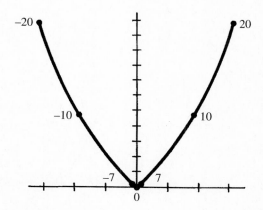

Figure 2.25 A freehand curve through the points in Fig. 2.24, with slopes as indicated.

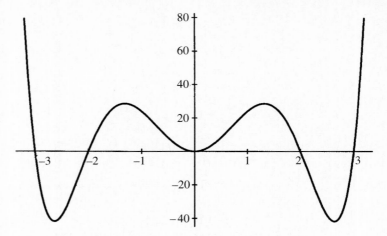

Figure 2.26 The curve $y = 35.93x^2 - 12.97x^4 + 0.998x^6$.

polynomial given by Theorem 2.4.1, $y = 35.93x^2 - 12.97x^4 + 0.9978x^6$, fits the freehand curve poorly. ●

The method of Theorem 2.4.1 has serious drawbacks, starting with the poor fit of Example 2. A second drawback is that the addition of another point or the shift of one point necessitates completely recomputing the polynomial, as Fig. 2.27 illustrates. Third, with a large number of points, the computations become time-consuming and are prone to roundoff errors. Fourth, functions of y in terms of x cannot curve back on themselves or represent space curves. Hermite curves solve all these problems. Instead of one polynomial going through all the points, we find a sequence of $n - 1$ parametric

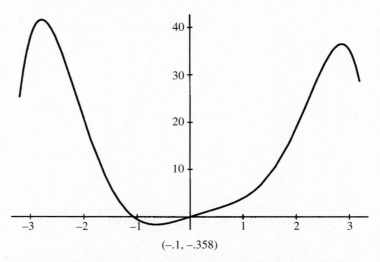

$(-.1, -.358)$

Figure 2.27 The curve resulting from shifting one point.

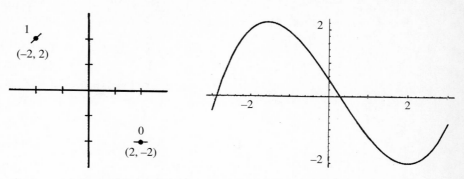

Figure 2.28 $y = \frac{1}{2} - \frac{7}{4}x - \frac{1}{8}x^2 + \frac{3}{16}x^3$ goes through $(-2, 2)$ and $(2, -2)$, with slopes of 1 and 0.

equations, using third-degree polynomials that connect neighboring points smoothly. That is, the derivative at the end of one curve is the derivative at the start of the next. The addition of a point in the middle or the repositioning of any point affects at most two cubics. Each cubic can be computed quickly, at least by a computer. Finally, the parametric form allows a curve to double back on itself, cross itself, or even twist into as many dimensions as are needed. Many applications use cubic spline curves, which build on the ideas presented here but do not need to be given the derivatives at the endpoints. (See Mortenson [9].) We motivate Hermite curves with the simpler, non-parametric situation of Theorem 2.4.2, illustrated in Fig. 2.28.

Theorem 2.4.2 Let two points be $P = (x_1, y_1)$ and $Q = (x_2, y_2)$ with $x_1 \neq x_2$ and the desired slopes m_1 and m_2 at P and Q, respectively. Then there is a unique third-degree polynomial $f(x) = a_0 + a_1 x + a_2 x^2 + a_3 x^3$ such that $f(x_1) = y_1$, $f(x_2) = y_2$, $f'(x_1) = m_1$, and $f'(x_2) = m_2$.

Proof. By elementary calculus, $f'(x) = a_1 + 2a_2 x + 3a_3 x^2$. The given points and slopes, P, Q, m_1, and m_2, determine four linear equations in the unknowns a_0, a_1, a_2, and a_3:

$$y_1 = a_0 + a_1 x_1 + a_2 x_1^2 + a_3 x_1^3;$$
$$y_2 = a_0 + a_1 x_2 + a_3 x_2^2 + a_3 x_2^3;$$
$$m_1 = a_1 + 2a_2 x_1 + 3a_3 x_1^2;$$
$$m_2 = a_1 + 2a_2 x_2 + 3a_3 x_2^2.$$

This system has a unique solution iff the determinant

$$\begin{vmatrix} 1 & x_1 & x_1^2 & x_1^3 \\ 1 & x_2 & x_2^2 & x_2^3 \\ 0 & 1 & 2x_1 & 3x_1^2 \\ 0 & 1 & 2x_2 & 3x_2^2 \end{vmatrix}$$

does not equal zero. This determinant equals $-x_1^4 + 4x_1^3 x_2 - 6x_1^2 x_2^2 + 4x_1 x_2^3 - x_2^4 = -(x_1 - x_2)^4$. Because $x_1 \neq x_2$, there is always a unique cubic. ∎

Example 3 Consider the points from Example 2, which we connect with a sequence of cubic polynomials. We use the slopes given previously in Fig. 2.25. For example, for the interval from $x = -3.1$ to $x = -1.8$, we have $f(-3.1) = 32.886$, $f(-1.8) = 14.18$, $f'(-3.1) = -20$ and $f'(1.8) = -10$. Theorem 2.4.2 gives the system:

$$a_0 - 3.1a_1 + 9.61a_2 - 29.791a_3 = 32.886;$$
$$a_0 - 1.8a_1 + 3.24a_2 - 5.832a_3 = 14.18;$$
$$a_1 - 6.2a_2 + 28.83a_3 = -20;$$
$$a_1 - 3.6a_2 + 9.72a_3 = -10.$$

This system gives the cubic $y = -0.141 - 8.254x - 1.466x^2 - 0.723x^3$. The other cubics, from left to right, have for their equations $y = -0.340 - 6.96x + 0.154x^2 - 0.256x^3$, $y = 37.4x^2 + 16x^3$, $y = 37.4x^2 - 16x^3$, $y = -0.340 + 6.96x + 0.154x^2 + 0.256x^3$, and $y = -0.14 + 8.25x - 1.466x^2 + 0.723x^3$. Figure 2.29 shows how well these polynomials fit the path of the curve determined intuitively. ●

Hermite curves, defined parametrically, allow even greater flexibility than the curves of Theorem 2.4.2. The variables $x = x_1$ and $y = x_2$ (and x_3, etc., for more dimensions) are separately determined by cubic polynomials in t. As in Theorem 2.4.2, we find the cubic from the coordinates of the endpoints and the desired derivatives $x'_i(t)$ at each endpoint.

Theorem 2.4.3 Let $P_0 = (u_1, u_2, \ldots, u_n)$ and $P_1 = (v_1, v_2, \ldots, v_n)$ be any two points in $\mathbf{R}^n$ and assume that the derivatives $dx_i/dt = x'_i(t)$ are given at $t = 0$ and $t = 1$. Then for each x_i, there is a unique cubic function $f_i(t) = a_{0i} + a_{1i}t + a_{2i}t^2 + a_{3i}t^3$ such that $P_0 = (f_1(0), f_2(0), \ldots, f_n(0))$, $P_1 = (f_1(1), f_2(1), \ldots, f_n(1))$, $f'_i(0) = x'_i(0)$, and $f'_i(1) = x'_i(1)$.

Proof. Apply Theorem 2.4.2 with $t = x$ and $x_i = y$. ■

Definition 2.4.1 The parametric curve formed by the n-component functions $f_i(t)$ in Theorem 2.4.3 is a *Hermite curve*.

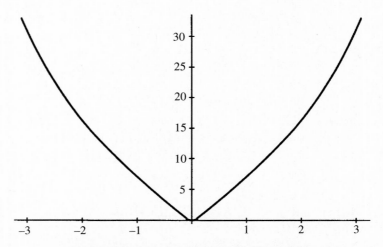

Figure 2.29 Six cubics fitting the original points.

2.4 Curves in Computer-Aided Design

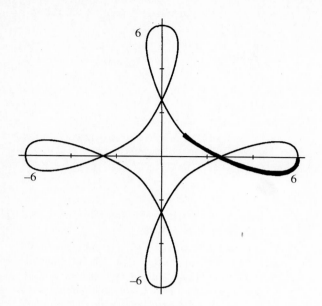

Figure 2.30

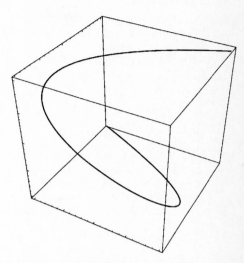

Figure 2.31 The graph of $x(t) = 16t^3 - 24t^2 + 9t$, $y(t) = 4t^2 - 4t$, and $z(t) = -2t^3 + 3t^2$.

Example 4 The shape shown in Fig. 2.30 contains eight Hermite curves. The thick portion of the curve starts at $(1, 1)$, with $x' = 3$ and $y' = -3$, and ends at $(6, 0)$, with $x' = 0$ and $y' = 1$. The parametric equations are $x(t) = 1 + 3t + 9t^2 - 7t^3$ and $y(t) = 1 - 3t - 2t^2 + 4t^3$. The parametric equations for the seven other portions of the curve shown in Fig. 2.30 use variations of these two functions. •

Example 5 Figure 2.31 shows the graph of the three-dimensional Hermite curve $x(t) = 16t^3 - 24t^2 + 9t$, $y(t) = 4t^2 - 4t$, and $z(t) = -2t^3 + 3t^2$. •

PROBLEMS FOR SECTION 2.4

Note: Computational devices are recommended for many of these problems.

1. Recompute the two cubics of Example 3 affected by the shift of the point $(-0.1, 0.358)$ to the new point $(-0.1, -0.358)$ with a slope of 1. Compare the graphs of these cubics with the one shown in Fig. 2.27.

2. Use Theorem 2.4.2 to find the following cubic polynomials, and then graph them.
   a) $f(-1) = -2$, $f(0) = 1$, $f'(-1) = 2$, and $f'(0) = 1$.
   b) $g(1) = -1$, $g(2) = 0$, $g'(1) = 0$, and $g'(2) = -1$.
   c) h defined so that it connects f and g smoothly.

3. Find and graph the following Hermite curves. Compare your result with that in Problem 2.
   a) $x(0) = -1$, $x(1) = 0$, $x'(0) = 1$, $x'(1) = 1$, $y(0) = -2$, $y(1) = 1$, $y'(0) = 2$, and $y'(1) = 1$.
   b) $x(0) = 1$, $x(1) = 2$, $x'(0) = 1$, $x'(1) = 1$, $y(0) = -1$, $y(1) = 0$, $y'(0) = 0$, and $y'(1) = -1$.
   c) $x(0) = 0$, $x(1) = 1$, $x'(0) = 1$, $x'(1) = 1$, $y(0) = 1$, $y(1) = -1$, $y'(0) = 1$, and $y'(1) = 0$.
   d) Give the equations for a new Hermite curve that is the mirror image in the x-axis of the one in part (a).

e) Give the equations for a new Hermite curve that is the result of rotating the one in part (a) 180° about the origin.

4. Design and graph Hermite curves to make a curve similar to the one shown in Fig. 2.32. [*Hint:* Make the first cubic go from $(1, 0)$ to $(2, 2)$. If the parametric equations for this cubic are $x(t) = f(t)$ and $y(t) = g(t)$, explain why the other curves are built from $\pm f(t)$ and $\pm g(t)$.]

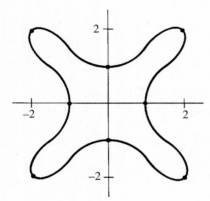

Figure 2.32

5. Repeat Problem 4 for the curve pictured in Fig. 2.33.

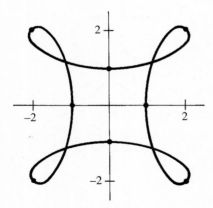

Figure 2.33

6. Suppose that you needed the curve in Theorem 2.4.2 to match given values for the second derivative at the start and endpoints, as well as the first derivative and heights there.

a) What degree polynomial would be needed to satisfy all these conditions?
b) Find a polynomial f that satisfies all the conditions $f(0) = 1$, $f(1) = 0$, $f'(0) = 1$, $f'(1) = 0$, $f''(0) = -1$, and $f''(1) = 2$.

7. Suppose that you improve Theorem 2.4.1 so that the polynomial not only goes through all the points, but also has a specified derivative at each point.

a) What would the degree of this polynomial need to be if there are three points?
b) Find a polynomial f that satisfies all the conditions $f(-1) = 0$, $f(0) = 1$, $f(1) = 2$, $f'(-1) = 0$, $f'(0) = 1$, and $f'(1) = 0$.
c) What would the degree of this polynomial need to be if there are n points?

8. The computational time for curve fitting is an important practical consideration. The methods of this section involve solving a system of n linear equations in n unknowns. Gaussian elimination, an efficient and widely used way to solve such systems, in general requires $(n^3 + 3n^2 - n)/3$ multiplications, which gives a good estimate of the computational time.

a) Use Theorem 2.4.1 to find the number of multiplications needed to identify one polynomial that goes through k points.
b) Repeat part (a) for the $k - 1$ cubics with designated slopes at these points, using Theorem 2.4.2.
c) Repeat part (b) for $k - 1$ two-dimensional Hermite curves.
d) Repeat part (c) for three-dimensional Hermite curves.
e) Explain why you need a polynomial of degree $2k - 1$ to find a curve that goes through k points and has a specified slope at each point. Repeat part (a) for this polynomial.
f) Compare the formulas you found in parts (a)–(e) and the ability of these curves to fit given conditions. How quickly does each formula grow as k increases? (Actual designs have dozens or hundreds of points.)

9. Compare Taylor polynomials and the polynomials of Theorems 2.4.1 and 2.4.2 as approximations for $y = \sin x$, as follows.

a) Find and graph from $-\pi$ to π the seventh-degree Taylor polynomial for $\sin x$ centered at 0.

b) Find and graph the fourth-degree polynomial described in Theorem 2.4.1 that goes through the points $(0, 0)$, $(\frac{\pi}{2}, 1)$, $(-\frac{\pi}{2}, -1)$, $(\pi, 0)$, and $(-\pi, 0)$.

c) Find and graph the cubic polynomials described in Theorem 2.4.2 that connect the points $(-\pi, 0)$, $(-\frac{\pi}{2}, -1)$, $(0, 0)$, $(\frac{\pi}{2}, 1)$ and $(\pi, 0)$ and have the same derivatives as $\sin x$ at those points.

d) Graph $y = \sin x$ and compare the advantages and disadvantages of each of the preceding approximations for $y = \sin x$.

2.5 HIGHER DIMENSIONAL ANALYTIC GEOMETRY

The formal, algebraic language of analytic geometry and vectors has enabled mathematicians to model higher dimensions as easily as two or three dimensions. The first investigations of geometry beyond three dimensions by Arthur Cayley and others starting in 1843 seemed puzzling and even nonsensical to most people, including many mathematicians. However, the variety and importance of applications in mathematics, physics, economics, and other fields in the twentieth century have provided convincing evidence of the significance and naturalness of higher dimensions in geometry. We briefly consider the analytic geometry of $\mathbf{R}^3$ before discussing polytopes, the higher dimensional analog of the polyhedra we considered in Section 1.6.

Definition 2.5.1 By a *point* in n-dimensional Euclidean geometry, we mean a vector $\vec{v} = (v_1, v_2, \ldots, v_n)$ in $\mathbf{R}^n$. By a *line* we mean a set of points $\{\alpha \vec{u} + \vec{v} : \alpha \in \mathbf{R}\}$, where $\vec{u}$ and $\vec{v}$ are fixed vectors and $\vec{u} \neq 0$. A point $\vec{w}$ is *on* the line above iff for some α, $\vec{w} = \alpha \vec{u} + \vec{v}$. Two lines $\alpha \vec{u} + \vec{v}$ and $\beta \vec{s} + \vec{t}$ are *parallel* iff $\vec{s}$ is a scalar multiple of $\vec{u}$. The *distance* between two points $\vec{u}$ and $\vec{v}$ is $d(\vec{u}, \vec{v}) = \| \vec{u} - \vec{v} \| = \sqrt{(u_1 - v_1)^2 + (u_2 - v_2)^2 + \cdots + (u_n - v_n)^2}$. A *plane* is the set of points $\{\alpha \vec{u} + \beta \vec{v} + \vec{w} : \alpha, \beta \in \mathbf{R}\}$, where $\vec{u}$ and $\vec{v}$ are independent vectors and $\vec{w}$ is any fixed vector.

Remarks Figures 2.34 and 2.35 illustrate lines and planes in $\mathbf{R}^3$. In the definition of a line, we shift the line through the origin (or one-dimensional subspace) $\{\alpha \vec{u} : \alpha \in \mathbf{R}\}$

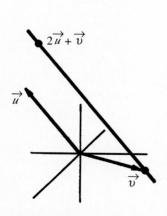

Figure 2.34 A line in $\mathbf{R}^3$.

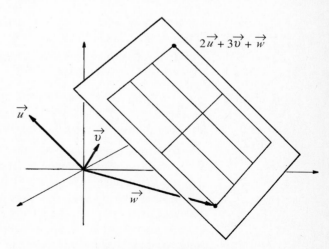

Figure 2.35 A plane in $\mathbf{R}^3$.

by adding the vector $\vec{v}$ so that we have a line through $\vec{v}$. Different vectors $\vec{v}$ give different parallel lines. The definition of a plane is similar to the definition of a line. The set $\{\alpha\vec{u} + \beta\vec{v} : \alpha, \beta \in \mathbf{R}\}$ is a two-dimensional subspace of $\mathbf{R}^n$. Adding the vector $\vec{w}$ translates this subspace to the parallel plane through $\vec{w}$. In $\mathbf{R}^3$, a plane can also be represented as the set of points (x, y, z) satisfying a linear equation $ax + by + cz + d = 0$. In general, a linear equation in n variables, such as $a_1x_1 + a_2x_2 + \cdots + a_nx_n + d = 0$, represents an $n-1$-dimensional *hyperplane* in $\mathbf{R}^n$. The formula for distance comes from the generalization of the Pythagorean theorem.

Exercise 1 Explain why the line through two points $\vec{s}$ and $\vec{t}$ consists of points of the form $\alpha(\vec{s} - \vec{t}) + \vec{t}$. [*Hint:* Consider $\alpha = 0$ and $\alpha = 1$.] Explain why the plane through three points $\vec{r}$, $\vec{s}$, and $\vec{t}$ consists of points of the form $\alpha(\vec{r} - \vec{t}) + \beta(\vec{s} - \vec{t}) + \vec{t}$.

Example 1 Use analytic geometry to verify that in three-dimensional geometry, if two distinct planes intersect, they intersect in a line.

Solution. Let's begin with two distinct planes $\alpha\vec{u} + \beta\vec{v} + \vec{w}$ and $\gamma\vec{r} + \delta\vec{s} + \vec{t}$. Then the coordinates of the six vectors are known, but the values of the four scalars can vary. If the point $\vec{a} = (a_1, a_2, a_3)$ is on both planes, the following system of three equations in the four unknowns α, β, γ, and δ has a solution.

$$\alpha u_1 + \beta v_1 + w_1 = \gamma r_1 + \delta s_1 + t_1 (= a_1).$$
$$\alpha u_2 + \beta v_2 + w_2 = \gamma r_2 + \delta s_2 + t_2 (= a_2).$$
$$\alpha u_3 + \beta v_3 + w_3 = \gamma r_3 + \delta s_3 + t_3 (= a_3).$$

However, as there are more unknowns than equations, elementary linear algebra assures us of infinitely many solutions. The planes are distinct, so these solutions must correspond to the points on a line, not a whole plane. ●

Conics (Section 2.2) generalize to conic surfaces in three dimensions. Some of these surfaces, such as the ellipsoid $(x^2/a^2) + (y^2/b^2) + (z^2/b^2) = 1$, the hyperboloid of two sheets $x^2/a^2 - y^2/b^2 - z^2/b^2 = 1$, and the paraboloid $z = ax^2 + ay^2$, can be defined using foci (and a plane as a directrix for the paraboloid) in the same way as conics are defined. More generally, we define a conic surface as the set of points satisfying a second-degree equation in x, y, and z. (Note that the terms xy, xz, and yz are considered second degree, as are x^2, y^2, and z^2.)

Exercise 2 Verify that, if $a^2 > b^2$, the ellipsoid must have foci at the points $(\sqrt{a^2 - b^2}, 0, 0)$ and $(-\sqrt{a^2 - b^2}, 0, 0)$ as follows. Verify that, for $x = 0$, a, or $-a$, the sum of the distances from any point (x, y, z) on the ellipsoid to these foci is constant.

A hyperboloid of one sheet (Fig. 2.36) has an unusual property: Lines lie entirely on the surface, even though the surface is curved.

Gaspard Monge (1746–1818) developed descriptive geometry, a method of representing three-dimensional geometric constructions by means of two (or more) two-dimensional projections. This method radically improved engineering design. Figure 2.37 illustrates how descriptive geometry enabled engineers to design on paper exact plans for their constructions. Computer-aided design now supersedes this hand-drawn approach but is based on the same analytic geometry.

GASPARD MONGE

At age 22, Gaspard Monge (1746–1818) became a professor of mathematics and at 25, a professor of physics. His brilliance and gift as a teacher were recognized throughout his life. He supported the French Revolution and made important contributions to it and to Napoleon's regime that followed. About the time the French Revolution erupted (1789), Monge developed descriptive geometry, leading to accurate two-dimensional engineering drawings of three-dimensional figures. The new government recognized the military importance of this method and classified it top secret. Monge served in various government positions, including a ministerial post. His greatest contribution was the founding in 1795 and the sustaining of the Ecole Polytechnique. Monge contributed greatly to our understanding of education with the organization of this renowned school and his own teaching. Monge and Napoleon became friends and, upon Napoleon's exile, Monge lost his official positions. No official notice was given of his death soon after, but his students mourned him.

In addition to his development of descriptive geometry, Monge was instrumental in championing geometric methods in mathematics following a period of dominance by algebra and calculus. He recognized the need for both analytic and synthetic geometric approaches and revived interest in projective geometry. He also made major contributions to what we now call differential geometry, making it into a separate branch of mathematics.

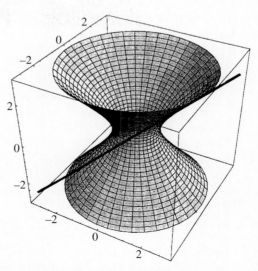

Figure 2.36 A hyperboloid of one sheet.

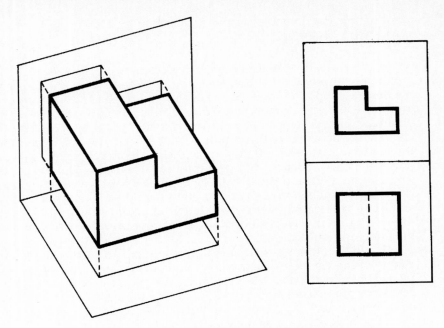

Figure 2.37 Two projections of a simple shape.

2.5.1 Regular polytopes

Mathematicians have studied polygons in two dimensions and polyhedra in three dimensions for millennia. By 1850 some of the earliest geometry of higher dimensions studied the analog of these shapes, called *polytopes*. A polyhedron is constructed by attaching polygons at their edges. Similarly, attaching polyhedra at their faces results in a four-dimensional polytope. As few of us can imagine such a construction, coordinates help us analyze polytopes. Here we consider only some of the regular polytopes. A four-dimensional convex polytope is *regular* provided that all the polyhedra in it are the same regular polyhedron and the same number of polyhedra meet at every vertex. Higher dimensional polytopes can be defined similarly. (See Coxeter [3, Chapter 22].)

Recall that there are five regular polyhedra in three dimensions. Several nineteenth century mathematicians discovered independently that there are a total of six regular polytopes in four dimensions, but for each dimension greater than four, there are only three regular polytopes. These three families of regular polytopes are the analogs of the cube, the tetrahedron, and the octahedron, which fortunately have simple Cartesian coordinates.

The Cube and Hypercubes.

Figure 2.38 illustrates one convenient way to place a square and a cube in Cartesian coordinates. Note that each coordinate of each point is either 1 or -1. Furthermore, for the cube, every possible combination of three such coordinates appears among the $8 = 2^3$ points. This suggests a way to generalize to four dimensions. Consider the $16 = 2^4$ points in $\mathbf{R}^4$ whose coordinates are all ± 1, some of which are labeled in

2.5 Higher Dimensional Analytic Geometry

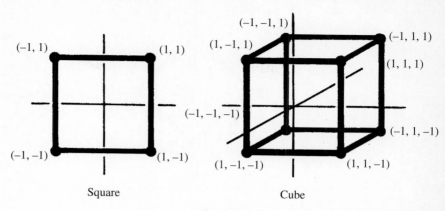

Figure 2.38

Fig. 2.39. These points are the vertices of a *hypercube*. The distance between two points joined by an edge in Fig. 2.39 is 2. Each point has four neighbors at this distance. The other possible distances between points on the hypercube are $2\sqrt{2}$, $2\sqrt{3}$, and 4.

Exercise 3 Find points on the hypercube at distances 2, $2\sqrt{2}$, $2\sqrt{3}$, and 4 from $(1, 1, 1, 1)$.

The Octahedron and Cross Polytopes.

The placement of the square and the regular octahedron in Fig. 2.40 suggests a way to generalize these figures. In $\mathbf{R}^4$, consider the eight points with 0 for three of their coordinates and 1 or -1 for the other coordinate. We designate these eight points as the vertices of the regular four-dimensional analog, called a *cross polytope*. Connect two of

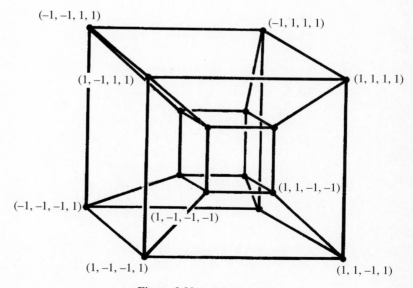

Figure 2.39 A hypercube.

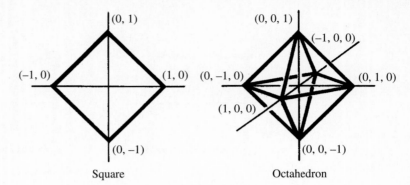

Square　　　　　　　　　　　　Octahedron

Figure 2.40

these vertices with an edge if their nonzero values appear in different coordinates. Thus $(0, 1, 0, 0)$ and $(0, 0, -1, 0)$ are connected with an edge, but $(0, 1, 0, 0)$ and $(0, -1, 0, 0)$ are not connected.

Exercise 4　Find the two possible distances between vertices for the four-dimensional analog of the octahedron. Verify that each vertex is on six edges.

The Tetrahedron and Regular Simplexes.

The rectangular coordinate system doesn't give easy three-dimensional coordinates for a regular tetrahedron. However, Fig. 2.40 suggests another approach. The octahedron has a triangular face with particularly simple coordinates: $(1, 0, 0)$, $(0, 1, 0)$, and $(0, 0, 1)$. Similarly, in four dimensions $(1, 0, 0, 0)$, $(0, 1, 0, 0)$, $(0, 0, 1, 0)$, and $(0, 0, 0, 1)$ are the vertices of a regular tetrahedron. Each vertex is a distance of $\sqrt{2}$ from the others. Now we can describe the regular *four-dimensional simplex*, as it is known. This shape (Fig. 2.41) has five vertices with coordinates $(1, 0, 0, 0, 0)$, $(0, 1, 0, 0, 0)$, $(0, 0, 1, 0, 0)$, $(0, 0, 0, 1, 0)$, and $(0, 0, 0, 0, 1)$. Each pair of vertices is connected for a total of 10 edges, all the same length. Similarly, every triple of vertices forms an equilateral triangle, and every quadruple of vertices forms a tetrahedron. Higher dimensional regular simplexes can be found in the same way.

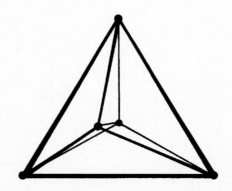

Figure 2.41　　A regular four-dimensional simplex.

2.5 Higher Dimensional Analytic Geometry

In this section, we discussed polytopes primarily to develop intuition for four-dimensional geometry. However, polytopes serve other purposes as well. For example, hypercubes provide a useful model in coding theory, where the vertices correspond to possible coded words. (See Section 7.3.) The simplex method in linear programming uses polytopes in many dimensions to find optimal solutions to many applied problems. Simplices form fundamental building blocks in topology.

PROBLEMS FOR SECTION 2.5.

1. **a)** Find the plane through $(0, 0, 0)$, $(1, 2, 4)$, and $(2, 0, 0)$ by using both vectors and an equation.

   b) Explain why the plane $2x + 3y - 4z = 0$ is perpendicular to the line through $(2, 3, -4)$ and the origin in three dimensions. [*Hint:* Use vectors.] Is $2x + 3y - 4z + 5 = 0$ also perpendicular to that line? Explain.

2. If two planes in $\mathbf{R}^4$ intersect in at least one point, must they intersect in more than that one point? (See Example 1.) If so, show why. If not, find an example of two planes in $\mathbf{R}^4$ intersecting in only one point.

3. (Descriptive geometry) In this problem you are to project onto the xy-plane and the yz-plane; that is, a point (x, y, z) will project to (x, y) and (y, z), and a line k has two projections, k_1 and k_2 (Fig. 2.42). The yz-projection directly above the xy-projection indicates that the y-coordinates match.

   a) Redraw Fig. 2.42 and include the line parallel to k through the point P. Explain why your line is parallel to k.

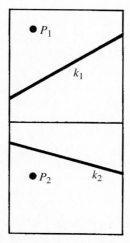

Figure 2.42 Projections of a point and a line.

   b) Consider Fig. 2.43. Explain whether lines k and m are parallel, intersect, or neither.

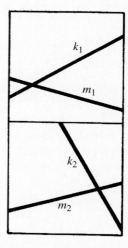

Figure 2.43 Projections of two lines.

   c) Find a convex solid other than a cube whose two projections are squares. Suppose that in addition the projection of this convex solid on the xz-plane is also a square. Must the solid be a cube? If so, explain why. If not, give an example.

4. **a)** Suppose that you know the distances of point (x, y) from points $(0, 0)$, $(1, 0)$, and $(0, 1)$. Explain why no other point $(\bar{x}, \bar{y})$ can be at the same respective distances from these points.

   b) Explain why part (a) holds if you replace $(0, 0)$, $(1, 0)$, and $(0, 1)$ with three points not on a line. What happens if the three points are on the same line?

   c) Generalize part (a) to three dimensions. How many points are needed?

   d) Redo part (c) in n dimensions.

5. a) Define an ellipsoid, a hyperboloid of two sheets, and a paraboloid, based on the corresponding definitions from Definition 2.2.1.

   b) Derive the text's equations for the conic surfaces of part (a).

6. a) Explain why the equation of the unit sphere is $x^2 + y^2 + z^2 = 1$.

   b) Explain why the equation of a great circle on the unit sphere is $ax + by + cz = 0$, where not all a, b, and c are zero. (See Section 1.6.)

   Represent a great circle by its triple $[a, b, c]$ and a point by (x, y, z). Then (x, y, z) is on $[a, b, c]$ provided that $ax + by + cz = 0$.

   c) Find the two points of intersection of the great circles $[2, 2, 1]$ and $[2, -2, 1]$. Find the great circle through the points $(\frac{2}{3}, \frac{2}{3}, \frac{1}{3})$ and $(\frac{2}{3}, \frac{-2}{3}, \frac{1}{3})$.

   d) Explain why the point (p, q, r) is perpendicular (orthogonal) as a vector to every point on the great circle $[p, q, r]$. [*Hint:* Use linear algebra.]

7. Plot, preferably with a computer graphics package, the following space curves and surfaces given parametrically. Describe the shapes in words.

   a) The curve $x(t) = \cos t$, $y(t) = \sin t$ and $z(t) = t$.

   b) The curve $x(t) = t \cos t$, $y(t) = t \sin t$ and $z(t) = t$, for $t \geq 0$.

   c) The surface $x(u, v) = \sqrt{1 - u^2} \cos v$, $y(u, v) = \sqrt{1 - u^2} \sin v$ and $z(u, v) = u$, for $-1 \leq u \leq 1$.

   d) The surface $x(u, v) = \sqrt{1 + u^2} \cos v$, $y(u, v) = \sqrt{1 + u^2} \sin v$ and $z(u, v) = u$.

8. The hyperboloid $x^2 + y^2 - z^2 = 1$ (See Fig. 2.36) has lines that lie on its surface. Start with a line through $(1, 0, 0)$ and a point on the hyperboloid at a height of $z = \sqrt{3}$.

   a) Explain why points on the hyperboloid with $z = \sqrt{3}$ can be written as $(2 \cos \theta, 2 \sin \theta, \sqrt{3})$. Explain why the line through this point and $(1, 0, 0)$ is given by $\alpha(2 \cos \theta - 1, 2 \sin \theta, \sqrt{3}) + (1, 0, 0)$.

   b) Explain why, if the line in part (a) is on the surface of the hyperboloid, it is also on $(2 \cos -\theta, 2 \sin -\theta, -\sqrt{3}) = (2 \cos \theta, -2 \sin \theta, -\sqrt{3})$. Show that θ satisfies $\cos \theta = \frac{1}{2}$. Find the two values of $\sin \theta$ and determine the two lines.

   c) Prove that every point on the lines in part (b) lies on the hyperboloid.

   d) Explain why an appropriate rotation of the two lines you found will give two lines through every point on the hyperboloid.

   e) Make a model of this hyperboloid by stretching strings between two circles. When one circle is rotated over the other, the strings lean to form a hyperboloid (Fig. 2.44).

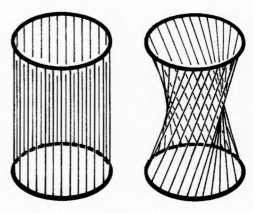

Figure 2.44

9. a) An edge, a square, a cube, and a hypercube are analogous shapes in one to four dimensions. For each, count the edges per vertex, faces per vertex, and so on, and the total number of vertices, edges, faces, and so on. Make a table of this information.

   b) Look for patterns, and explain any patterns you find. If possible, write formulas for these categories, based on only the dimension.

   c) Find the values corresponding to part (a) for the five-dimensional and six-dimensional hypercubes.

10. The octahedron is the *dual* of the cube: when you connect the centers of the cube's faces, you get an octahedron. Thus the octahedron has the same number of vertices as the cube has faces, and vice versa. How are their edges related?

   a) In what way is the four-dimensional cross polytope the dual of the four-dimensional hypercube? Use this duality to find the number of vertices, edges, faces, and octahedra of a four-dimensional cross polytope.

   b) Repeat part (a) for the five-dimensional cross polytope.

PROJECTS FOR CHAPTER 2

1. Define and investigate taxicab conics and other topics in taxicab geometry. (See Krause [8].) Note: Different positions of the foci and different slopes of the directrix lead to different-looking ellipses, hyperbolas, and parabolas.

2. Pick's theorem gives the area of many *lattice polygons*, that is, polygons whose vertices have integer coordinates, called *lattice points*.

   a) For each lattice polygon shown in Fig. 2.45, find its area and count the number B of lattice points on the boundary of the polygon and the number I of lattice points in the interior of the polygon. Find an equation relating the area with the numbers B and I. Does this equation hold for the polygons shown in Fig. 2.46?

   b) Prove your equation for lattice rectangles parallel to the axes.

   c) Do the lattice shapes shown in Fig. 2.47 satisfy your equation? Explain how these polygons differ from those in Figs. 2.45 and 2.46. State Pick's theorem, incorporating any needed hypotheses.

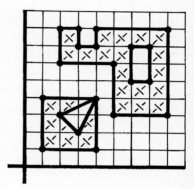

Figure 2.47

   d) Add an interior edge to divide a lattice polygon into two smaller lattice polygons (Fig. 2.48). How do B and I for the smaller polygons compare to those of the original polygon? How far can you carry this process of dividing? What can you say about the smallest lattice polygons? (See Coxeter [3, 209] for a proof of Pick's theorem.)

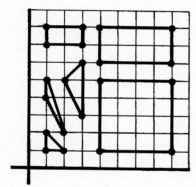

Figure 2.45

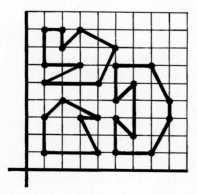

Figure 2.46

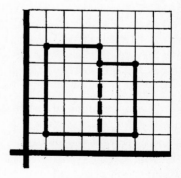

Figure 2.48

   e) Develop, state, and prove a restriction of Pick's theorem in three dimensions to rectangular boxes parallel to the axes (Fig. 2.49). Does this restriction work for pyramids? (Fig. 2.50.) Reeve [10] proves a general three-dimensional version.

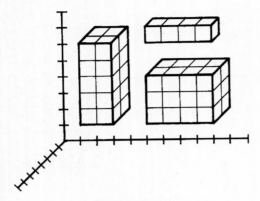

Figure 2.49

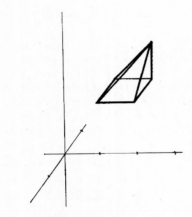

Figure 2.50 A pyramid with height 1.

3. Investigate computer-aided design. (See Mortenson [9].)

4. Investigate descriptive geometry. (See Douglass and Hoag [4].)

5. Investigate polytopes and four-dimensional geometry. (See Coxeter [3, Chapter 22] and Rucker [11].)

6. (Calculus) The folium of Descartes ($x^3 - 2xy + y^3 = 0$, shown in Fig. 2.1) is a challenge to graph by hand, even with the aid of calculus.

   a) Use implicit differentiation to find dy/dx.
   b) Find the values of x and y for which $dy/dx = 0$.
   c) When is dy/dx undefined? At these points the curve has a vertical tangent. How are these points related to those in part (b)?
   d) Find the points on the curve where $x = y$. Why is the curve symmetric with respect to this line?
   e) Graph this curve, using the previous information and Newton's method to plot points on the curve.
   f) Graph the folium of Descartes parametrically, using the following tricky substitution. In $x^3 - 2xy + y^3 = 0$, replace y by tx and solve for x to find $x(t)$. Then find $y(t) = tx(t)$. Verify that this graph matches the results in parts (a)–(e).

7. **Napoleon's theorem.** On each side of a triangle, construct an equilateral triangle lying outside (or inside) the triangle. The centers of these three triangles form an equilateral triangle.

   a) Show that in an equilateral triangle with side x, the distance from the center to any vertex is $x/\sqrt{3}$. Also, rewrite $\cos(A + 60°)$ by using the formula $\cos(A + B) = \cos A \cos B - \sin A \sin B$.

   WLOG let the vertices of the triangle be $S = (0, 0)$, $T = (p, 0)$, and $U = (q, r)$. Call the centers of the three constructed equilateral triangles V, W, and X (Fig. 2.51). Napoleon's theorem states that VW, VX, and WX are equal.

   b) Explain why showing that $VW = WX$ is sufficient.
   c) Find ST, SU, and TU and then SV, SW, TW, and TX.
   d) Explain why $\angle VSW$ is 60° wider than $\angle UST$ and similarly why $\angle XTW$ is 60° wider than $\angle UTS$.
   e) Use part (a) and the law of cosines to verify that both VW^2 and WX^2 equal $(p^2 + q^2 + r^2 - pq - \sqrt{3}pr)/3$.
   f) Write a proof of Napoleon's theorem.

8. Julius Plücker (1801–1868) extended the idea of a coordinate system enormously when he realized that entities other than points could have coordinates. For example, circles in the plane can have "circular coordinates" (a, b, r), where (a, b) is the center and $r > 0$ is the radius of the circle. Thus the set of circles in the plane is, in a sense, three dimensional.

a) Find the circular coordinates for the circle $x^2 + y^2 + 4x - 6y - 3 = 0$.

b) Find the equation of a circle having circular coordinates $(5, -4, 7)$.

c) Devise a set of "linear coordinates." How many "dimensions" does the set of lines in the plane have? Find, in your linear coordinates, the coordinates of the line whose equation is $y = 3x + 5$. Can two distinct pairs of your linear coordinates represent the same line? Are there any lines that do not have coordinates in your system? (Consider $x = 4$.) Are there any subsets of coordinates in your system that do not represent lines? (See Eves [6] for more information.)

9. Write an essay exploring the meaning of geometry in four or more dimensions. Suggestions: Compare multidimensional with plane geometry in a world that is strictly three dimensional. (See Rucker [11].)

10. Write an essay comparing analytic and synthetic geometry as ways to explain geometric concepts.

11. Write an essay comparing the certainty of algebraic derivations of geometric properties with the certainty of proving those properties in an axiomatic system. (See Grabiner [7].)

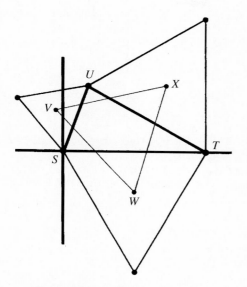

Figure 2.51   Napoleon's theorem.

Suggested Readings

[1] Boyer, C. *History of Analytic Geometry*, New York: Scripta Mathematica, 1956.

[2] Broman, A., and L. Broman. Museum exhibits for the conics. *Mathematics Magazine*, 1994, 67(3):206–209.

[3] Coxeter, H. *Introduction to Geometry*, 2d ed. New York: John Wiley & Sons, 1969.

[4] Douglass, C., and A. Hoag. *Descriptive Geometry*. New York: Holt, Rinehart and Winston, 1962.

[5] Edwards, C., and D. Penney. *Calculus with Analytic Geometry*, Englewood Cliffs, N.J.: Prentice Hall, 1994.

[6] Eves, H. *A Survey of Geometry*, vols. I and II. Boston: Allyn & Bacon, 1965.

[7] Grabiner, J. The centrality of mathematics in the history of Western thought. *Mathematics Magazine*, 1988, 61(4):220–230.

[8] Krause, E. *Taxicab Geometry*. Mineola, N.Y.: Dover, 1986.

[9] Mortenson, M. *Geometric Modeling*. New York: John Wiley & Sons, 1985.

[10] Reeve, J. On the volume of lattice polyhedra. *Proceedings of the London Mathematical Society*, 1957, 3(7):378–395.

[11] Rucker, R. *Geometry, Relativity and the Fourth Dimension*. Mineola, N.Y.: Dover, 1977.

[12] Strang, G. *Linear Algebra and its Applications*, 2d ed. New York: Academic Press, 1980.

Suggested Media
1. "Applications of Conic Sections," 10-minute video, Films for the Humanities and Sciences, Princeton, N.J., 1996.
2. "Curves from Parameters," 24-minute video, Films for the Humanities and Sciences, Princeton, N.J., 1996.
3. "Descartes and Problem Solving," 60-minute video, American Mathematical Society, Providence, R.I., 1992.
4. "Dimension," 13-minute film, Aims Instructional Media Services, Glendale, Calif., 1970.
5. "Flatland," 12-minute film, Contemporary/McGraw Hill Films, New York, 1965.
6. "The Hypercube: Projections and Slicing," 12-minute film, Banchoff/Strauss Productions, Providence, R.I., 1978.
7. "Locus of Points from Which Two Circles Are Seen Under the Same Angle," 4.5-minute film, Educational Solutions, New York, 1979.
8. "Mathematical Curves," 10-minute film, Churchill Films, Los Angeles, Calif., 1977.
9. "Reflecting on Conics," 24-minute video, Films for the Humanities and Sciences, Princeton, N.J., 1996.

3
Non-Euclidean Geometries

The repeated insects in this design clearly follow some rule, although the geometry behind this pattern may seem mysterious. Douglas Dunham, a geometer with an interest in computer graphics, has combined his knowledge of non-Euclidean geometries and computers to produce a large variety of such designs. The fundamentals of non-Euclidean geometries are basic to an understanding of the mathematics of these designs.

> The most suggestive and notable achievement of the last century is the discovery of non-Euclidean geometry. —*David Hilbert*
>
> To this interpretation of geometry I attach great importance for should I have not been acquainted with it, I never would have been able to develop the theory of relativity. —*Albert Einstein*

3.1 Overview and History

The classical understanding of axioms (postulates) as "self-evident truths" was shattered in mathematics by the introduction and development of non-Euclidean geometries during the nineteenth century. The story of non-Euclidean geometries plays an important role in the history of ideas, and mathematics students deserve to know it. In addition, non-Euclidean geometries illustrate the need to transcend the intuitive models of elementary mathematics, allowing us to think successfully about the much more abstract concepts of modern mathematics.

In this chapter we focus primarily on the specific non-Euclidean geometry now called *hyperbolic geometry*. This focus provides an in-depth look at an axiomatic system. In the final section of this chapter we consider two other non-Euclidean geometries: spherical and single elliptic geometries.

Until the nineteenth century, no one questioned the truth of Euclidean geometry, although many sought to remove a perceived blemish in Euclid's masterpiece. His fifth postulate about parallels was hardly a "self-evident truth." (See Section 1.2.) Mathematicians tried repeatedly to prove this postulate from the others. For the most part, though, they either explicitly or implicitly used equivalent assumptions, such as Playfair's, which is now used in geometry books. (Recall that Playfair's axiom states: "Through a given point not on a given line m there passes at most one line which does not intersect m.") Before 1800, the person who came closest to realizing that the fifth postulate could not be proved from the others was the Italian mathematician Girolamo Saccheri (1667–1733). His approach was to start from two negations of the fifth postulate and look for a contradiction. From one he found the desired contradiction. From the one leading to hyperbolic geometry he deduced increasingly bizarre consequences, such as the angle sum of a triangle is less than 180° and the existence of straight lines that approach each other but never cross. However, he found no explicit contradiction. Finally, he concluded that "the hypothesis . . . is absolutely false, because it is repugnant to the nature of the straight line." Bonola [1,43] Saccheri could receive credit for logically developing a non-Euclidean geometry, but he could as easily be seen as remaining inside the world view of asserting Euclidean geometry to be true and to be the geometry of the physical world.

That world view, accepted for centuries, found even more support in the 1700s. First, Immanuel Kant (1724–1804), the most influential philosopher of the time, held that the truths of mathematics differed fundamentally from incidental facts about the world, such as the earth has one moon. Mathematics, Kant taught, was *a priori* true— necessarily true, even before any particular experience we have. Geometry, as developed by Euclid, seemed a compelling example of how humans could obtain absolute knowledge about the world. In addition Kant argued that we needed an essentially inborn understanding of geometry and space before we could experience anything in space.[1] He thought that geometry had to be Euclidean.

In the eighteenth century, the Age of Enlightenment, mathematicians and philosophers built on the perceived absolute truth of mathematics in a second way. The astounding success of Newton's calculus and physics convinced his successors that mathematics wasn't just true in some metaphysical sense, but also in a tangible sense. The ideal world of mathematics, it seemed, was the real world. The physical meaning of much of the mathematics developed in the eighteenth century was so convincing that the rigorous deductive methods of Greek geometry seemed superfluous. The shock of the radically different mathematical results of the nineteenth century, starting with non-Euclidean geometries, forced mathematicians to reintroduce careful proofs.

The first person to break from the world view of Euclidean geometry, its unquestionable truth, and applicability to the physical world was Carl Friedrich Gauss (1777–1855). Despite his fame, Gauss never published anything on non-Euclidean geometries because he feared ridicule. Nicolai Lobachevsky (1793–1856) and János Bolyai (1802–1860), the two young mathematicians who did publish on non-Euclidean geometry, were greeted with silence for years after their publications in 1829 and 1832, respectively. Indeed, only with the publication of Gauss's notes after his death did the wider community of mathematicians start investigating non-Euclidean geometries.

3.1.1 The advent of hyperbolic geometry

Hyperbolic geometry, the non-Euclidean geometry that Gauss, Lobachevsky, and Bolyai developed, retains Euclid's first four postulates and changes the fifth postulate to the following axiom. In Sections 3.2, 3.3, and 3.4, we investigate hyperbolic geometry in much the same way that these mathematicians did. However, to make the process clearer, we make explicit certain logically necessary assumptions that had been overlooked until the end of the nineteenth century. (See Section 1.3.) Hyperbolic geometry is sometimes called Lobachevskian geometry to honor Lobachevsky's priority in publishing. Felix Klein called this geometry *hyperbolic* in his classification of geometries, which we discuss in Section 6.5.

Characteristic Axiom of Hyperbolic Geometry Given any line k and any point P not on k, there are at least two lines on P which do not intersect k (Fig. 3.1).

Various consequences follow from this change, including the many that Saccheri found. The most startling is the theorem that the measures of the angles of a triangle do not add up to 180°, as they do in Euclidean geometry. The goal in our study of hyperbolic geometry, Theorem 3.1.1, goes even further, relating the angle sum to the area of the triangle (Fig. 3.2). The greater the area of the triangle is, the smaller the angle sum is. Consequently, any triangle has a maximum area. As the sides of triangles can become indefinitely long, this consequence seems paradoxical.

---

[1] Studies of blind people and people with recovered sight indicate that Kant's argument is incorrect; sight is essential to developing the usual conception of space. (See Sacks [11, 124].)

NIKOLAI LOBACHEVSKY AND JÁNOS BOLYAI

Nikolai Lobachevsky (1793–1856) was a mathematics professor at the University of Kazan in Russia, where he first published on hyperbolic geometry in 1829. His extensive development included its trigonometry, corresponding to the trigonometry on a sphere of imaginary radius. He thought that this geometry might be pertinent to the study of astronomy because he realized that, as distances increased, the difference between Euclidean geometry and hyperbolic geometry became more noticeable. In his publications, Lobachevsky expounded on this geometry.

János Bolyai (1802–1860) was a Hungarian army officer with a reputation for dueling. One day he reportedly dueled several officers, playing his violin between duels. Bolyai found basically the same results as Lobachevsky. His publication of 1832, *The Science of Absolute Space*, also reported his investigation of properties common to both Euclidean and hyperbolic geometries.

The fame of Lobachevsky and Bolyai rests entirely on their work in hyperbolic geometry. Surprisingly the wider mathematical community neglected their publications, which deeply disappointed both men. Certainly, the original languages of Russian and Hungarian deterred readers. However, in 1840 Lobachevsky published a book on his research in German, then the foremost language of mathematics. Perhaps more significantly, geometric research focused then on projective geometry, which does not study parallel lines. Also, the trigonometric formulations may have impeded mathematicians interested in philosophical and geometric implications. Only after the posthumous publication of Gauss's notebooks in 1855 did an active study of non-Euclidean geometry begin. Because of the number of distinguished mathematicians who had failed to prove Euclid's fifth postulate, Lobachevsky and Bolyai should have found an interested, critical audience for their radical answer to this fundamental question in geometry.

Theorem 3.1.1   In hyperbolic geometry the difference, $180° - (m\angle A + m\angle B + m\angle C)$, between $180°$ and the angle sum of a triangle is proportional to the area of the triangle.

Exercise 1   Compare Theorem 3.1.1 with Theorem 1.6.3.

Theorem 3.1.1 suggests that we could decide "which geometry is true" by measuring real triangles. Since 1890, mathematicians and physicists have realized the futility

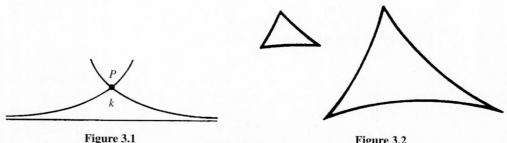

Figure 3.1   Figure 3.2

CARL FRIEDRICH GAUSS

Carl Friedrich Gauss (1777–1855) dominated mathematics during the first half of the nineteenth century, making fundamental contributions to virtually every area of mathematics. The prince of his small German state sponsored his education after learning of his prodigious abilities as a child. By age five he had found an arithmetic error in his father's accounts. In another story his grade school teacher made the class add the numbers from one to one hundred. After some thought Gauss found a formula and simply wrote down the correct answer, whereas his classmates toiled and made mistakes.

Gauss soon won acclaim as a mathematician. At 18 he developed the method of least squares, which is used extensively in statistics. A year later he constructed with straightedge and compass a regular 17-sided polygon and later characterized all constructible regular polygons, the first such advances since the ancient Greeks. He earned his Ph.D. at 22 by giving the first proof of the fundamental theorem of algebra, which says that every polynomial with real or complex coefficients has all of its roots in the complexes. Two years later, he published his first major treatise on number theory, rejuvenating that ancient area. That same year, 1801, Gauss astonished astronomers by determining the orbit of the first asteroid, Ceres, based on only a few observations and after it had been lost from view owing to weather conditions. To do so he had to generalize his method of least squares. Gauss subsequently became a professor of astronomy and director of the observatory in Göttingen.

By 1800 Gauss had become convinced that Euclid's parallel postulate could not be proven, and in installments he developed what he called non-Euclidean geometry. Although none of this work was published until after his death, he did correspond with a number of mathematicians about it. Gauss also made seminal contributions to differential geometry, including the curvature of a surface and geodesics. He showed that Gaussian curvature determined all the properties of a surface not related to how it is placed in a surrounding space.

Gauss's contributions on complex numbers persuaded mathematicians that complex numbers were essential in mathematics. He represented complex numbers geometrically, extended number theory to complex integers, and contributed fundamentally to what is now called complex analysis. He advanced knowledge in astronomy, magnetism, optics, and other applied fields.

Gauss's contemporaries revered him as the "Prince of Mathematicians," and he is often considered the greatest mathematician since Newton. He was also one of the last who could claim to know all the mathematics in existence at his time. Gauss's work can be seen both as crowning the great expansion of mathematics in the sixteenth and seventeenth centuries and as igniting the explosion of specialized, abstract mathematics since then.

---

of empirically deciding which geometry is correct because physical assumptions to test mathematical relations must be made. For example, we would have to assume that the path of a light ray is a straight line in order to measure even a moderately large triangle.

The first notable response to the advent of hyperbolic geometry came in 1854 in a lecture delivered by Georg Friedrich Bernhard Riemann (1826–1866) for his introductory lecture to the faculty at Göttingen University. Only Riemann's teacher, the aging

GEORG FRIEDRICH BERNHARD RIEMANN

Bernhard Riemann (1826–1866) had obvious mathematical ability early in his life. Nevertheless, he started studying theology at age 20 at the request of his father, a Lutheran minister. Within a year, though, he had turned to mathematics and finished his Ph.D. at age 25 under Gauss's direction. He became a professor at Göttingen University in Germany when he was 27 and remained there until his health deteriorated. Riemann suffered from tuberculosis for nearly the last four years of his life, working when he was well enough and dying before the age of 40.

Although Riemann is also remembered for Riemann sums and integrals that appear in calculus texts, his major work focused on physics and more advanced mathematics. Riemann made profound contributions to complex analysis and what later became topology and differential geometry. He blended deep physical and geometric intuition with insightful arguments. Some of his contemporaries criticized his proofs for a lack of rigor. However, his approaches, conjectures, and results have shaped all the areas he investigated.

Barely 10 years after mathematicians started exploring n-dimensional Euclidean geometry, Riemann's introductory lecture as a professor in 1854 developed a concept of space far more general. His vision of what we now call differential geometry included Euclidean, hyperbolic, spherical, and elliptic geometries, in any number of dimensions, as special cases. He showed how to base the concept of space on theoretical or physical measurements of distances. How those measurements of a space differed from the corresponding Euclidean measurements based on the Pythagorean theorem described how that space was curved. Sixty years after Riemann's death his work became the foundation for Einstein's general theory of relativity, which related those measurements to the effects of gravity.

Gauss, apparently caught the point of this lecture, entitled "On the Hypotheses which underlie Geometry." However, this talk, published after Riemann's death, focused geometric thought on a new field, differential geometry, and spurred an active debate on non-Euclidean geometries. Riemann had realized that the work of Gauss, Lobachevsky, and Bolyai was more than playing abstractly with a postulate. In essence, he recognized that the revised postulate implied that space had to be shaped differently than what Euclid's fifth postulate implied. He then articulated how infinitely many different geometries could be created, each with its own "shape."

Differential geometry, the field pioneered by Riemann, Gauss, and others, investigates geometries by looking at how they behave in small regions and, in particular, how they curve. Where Euclidean geometry is flat and spherical geometry is curved positively, hyperbolic geometry has a uniform negative curvature, as the model of the pseudosphere (Fig. 3.3) indicates. Riemann envisioned geometries in any number of dimensions with changing curvatures throughout. The general theory of relativity builds on Riemann's work on the curvature of space. Einstein used a four-dimensional geometry (three spatial dimensions and time) curved at each point by the gravitational forces acting there. Light waves travel along *geodesics*, paths of shortest length following the curvature of the surrounding space. In a sense, Einstein's theory settled the nineteenth

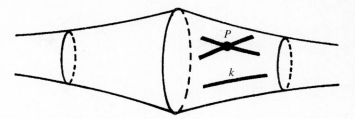

Figure 3.3 The effect of the characteristic axiom: A pseudosphere.

century question of the empirical truth of Euclidean and hyperbolic geometries in a surprising way: Both are false! (See do Carmo [3], Kline [6], McCleary [7], Rucker [10], and Weeks [13] for further information.)

The abstraction throughout mathematics and the strangeness of the new geometries led mathematicians to search again for absolute certainty in their mathematical arguments. No longer could mathematicians rely on an intuitive model to reveal the essential idea behind an argument. They examined the axioms for geometry, culminating in Hilbert's axioms for Euclidean geometry (see Section 1.3). Hilbert chose these axioms so that, by replacing the axiom for parallel lines with the characteristic axiom that we just presented, he would have axioms for hyperbolic geometry. (See Bonola [1] and Kline [6, Chapter 36] for further historical information.)

3.1.2 Models of hyperbolic geometry

In the latter part of the nineteenth century, various mathematicians developed models of hyperbolic geometry. We partially treat some of these models here. Each has some disadvantages, but they all help give a feel for this geometry. These models are based in part on a Euclidean plane or space, with suitable interpretations of the undefined terms. For the Poincaré model, we first need a definition: Two (Euclidean) circles are *orthogonal* iff the radii of these circles at the circles' points of intersection are perpendicular (Fig. 3.4). In all of these models the terms *on* and *between* have their usual meaning. In the language of Section 1.4, these models show the relative consistency of hyperbolic geometry based on Euclidean geometry.

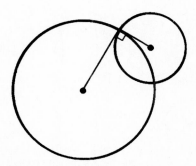

Figure 3.4 Orthogonal circles.

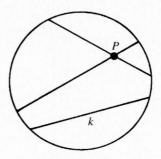

Figure 3.5 The effect of the characteristic axiom: The Klein model.

The Klein Model (1871). *Point* means a point in the interior of a particular Euclidean circle. *Line* means the portion of a Euclidean line in the interior of that circle. Figure 3.5 readily illustrates that the characteristic axiom holds in this model. This model's biggest advantage is the matrix representation of its transformations (see Chapter 6). However, both distances and angle measures are complicated in this model.

The Poincaré Model (1882). *Point* means a point in the interior of a particular Euclidean circle. *Line* means the portion interior to the given circle of any one of its diameters or of a Euclidean circle orthogonal to this circle. The chief advantage of this model is that angles are measured as they would be for curves in Euclidean geometry. Hence, as Fig. 3.6 illustrates, the angle sum of a triangle is less than 180°. The picture at the beginning of this chapter is based on the Poincaré model. Because angle measures are Euclidean, the repeated objects in this design are recognizably the same, although somewhat distorted to our eyes. The hyperbolic size of each of these creatures is the same in this model, which gives a sense of how distances are measured with the same complicated formula as the Klein model. In Section 4.6 we consider the transformations for this model and their connections with complex numbers.

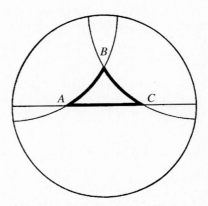

Figure 3.6 In the Poincaré model, the angle sum is less than 180°.

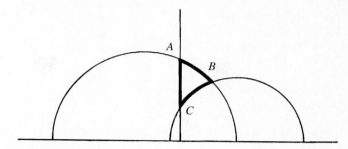

Figure 3.7 In the half-plane model, the angle sum is less than 180°.

The Half-Plane Model (1882). *Point* means a point in the upper half of the Euclidean plane [that is, points (x, y) with $y > 0$]. *Line* means the portion above the x-axis of either a vertical line or a circle with its center on the x-axis (Fig. 3.7). Poincaré developed this model and showed its close relation to the Poincaré model. This model also measures angles as Euclidean angles and, in addition, contains equations of lines that are easy to find. However, distances appear even more distorted than in the other models.

The Pseudosphere (1868). Eugenio Beltrami developed this first model of hyperbolic geometry. Figure 3.3 shows the surface of a pseudosphere. *Lines* are the paths of shortest distance along the surface (geodesics). Furthermore, both distances and angles are measured in a Euclidean manner. Thus this model is fairly natural. However, the pseudosphere doesn't entirely fit any Euclidean geometry. The "equator" of the pseudosphere shown in Fig. 3.3 is an *ad hoc* gluing together of two portions that on their own would simply curve out of Euclidean space.

Unless explicitly stated, the figures in the rest of this chapter are not based on any of these models. This procedure will encourage thinking of this geometry as an entire system, not just an isolated exercise applying only to an artificial model. Many of the lines drawn in the figures will be curved so that they will not appear to intersect. This corresponds to one of the findings of Saccheri, mentioned earlier, that lines can approach one another in this geometry without intersecting.

PROBLEMS FOR SECTION 3.1

1. Use the Klein model of hyperbolic geometry to investigate how many lines through a point P do not intersect a (hyperbolic) line k that is not on P.

2. Use the points inside the (Euclidean) unit circle $x^2 + y^2 = 1$ for the Poincaré model.
   a) Show that the circle $(x - 1.25)^2 + y^2 = (0.75)^2$ is orthogonal to the unit circle. Graph these two circles on the same axes.
   b) Find the intersections (inside the unit circle) of the circle from part (a) with $y = (9/13)x$ and $y = -(9/13)x$, two diameters of the unit circle. Include these two lines in the graph for part (a) to form a triangle.
   c) Verify visually that the angle sum of the triangle in part (b) is less than 180°.
   d) Find the measures of the three angles of the triangle in part (b) and verify that their sum is less than 180°. [*Hint:* The angle made by a line and a circle is the angle made by that line and the tangent to the circle at their intersection.]

The slope of a line equals tan α, where α is the angle the line makes with the x-axis. The formula $\tan(\alpha - \beta) = (\tan\alpha - \tan\beta)/(1 + \tan\alpha\tan\beta)$ converts slopes of two lines to the angle between them.]

3. Although three-dimensional Euclidean space satisfies the characteristic axiom, it doesn't satisfy all the axioms of plane geometry. Find which of Hilbert's axioms in Appendix B fail in three-dimensional Euclidean space.

4. Use the half-plane model of hyperbolic geometry.

   a) Find the equation of the line that is on the points (1, 1) and (4, 2).

   b) Explain why the equations of lines in this model are either $x = c$ or $y = \sqrt{r^2 - (x - c)^2}$, for appropriate constants c and r.

   c) Use a Euclidean argument to explain why two points in this model have one line which is on both of them.

   d) Verify that the point $(-4, 10)$ is not on the line $y = \sqrt{25 - (x + 4)^2}$. Find the equations of two lines on $(-4, 10)$ that do not intersect the line $y = \sqrt{25 - (x + 4)^2}$. Graph these three lines.

   e) Use a Euclidean argument to explain why the characteristic axiom always holds in this model. [Hint: Consider the two kinds of lines separately.]

   f) Verify that the two lines $y = \sqrt{25 - x^2}$ and $y = \sqrt{16 - (x - 1)^2}$ do not have a point of the model in common, although they do intersect in a Euclidean point as curves in the Euclidean plane. Find this Euclidean point. We call these lines *sensed parallels*.

   g) Verify that every point in the model has at least one sensed parallel to the line $y = \sqrt{25 - x^2}$ in the sense that the only Euclidean intersection of the sensed parallel with $y = \sqrt{25 - x^2}$ is the Euclidean point you found in part (e). Explain why there are exactly two sensed parallels to a line through a point not on that line.

5. (Calculus) In the half-plane model of hyperbolic geometry, consider the triangle formed by the lines $y = \sqrt{5 - (x - 3)^2}$, $y = \sqrt{10 - (x - 4)^2}$, and $x = 4$. Find the three vertices of this triangle and graph it. Find the measures of the three angles of this triangle. [*Hint:* See Problem 2(d). Use calculus and trigonometry. The angle sum is approximately 161.6°.]

6. A curious property of hyperbolic geometry is that any two lines that do not intersect and are not sensed parallels (described informally in Problem 5) have a common perpendicular. Use the half-plane model of hyperbolic geometry to illustrate this property with the two lines $y = \sqrt{1 - x^2}$ and $x = 2$. Graph these lines first. Then use relevant Euclidean concepts, including orthogonal circles, to find the hyperbolic line perpendicular to both of them.

3.2 PROPERTIES OF LINES AND OMEGA TRIANGLES

We develop hyperbolic geometry following the path of Saccheri, Gauss, Lobachevsky, and Bolyai, but adding the foundation of Hilbert and others. We can thus accentuate the geometric intuition of earlier work while having the logical basis required to avoid the flaws discovered later.

Our undefined terms are *point*, *line*, *on*, *between*, and *congruent*. Our axioms are Hilbert's axioms for plane geometry (see Appendix B), with the exception of axiom IV-1; Playfair's axiom, which we replace with the characteristic axiom of hyperbolic geometry. (See Section 3.1.) We also avail ourselves of Euclid's first 28 propositions, or those he proved without using the parallel postulate. (See Appendix A.) These propositions give us many of the key properties about lines and triangles, which we need in our development of hyperbolic geometry. They are provable in Hilbert's axiomatic system without our having to use either axiom IV-1 or the characteristic axiom.

Like the developers of hyperbolic geometry, we prove many theorems in this chapter, including Theorem 3.2.1, by contradiction. Until 1868, when Beltrami built the first model of hyperbolic geometry, mathematicians wondered whether the characteristic axiom itself was potentially a contradiction. A rigorous proof of even Theorem 3.2.1 requires considerable development from Hilbert's axioms. For example, in Fig. 3.8 we would need to show that B can always be chosen so that the line m enters $\triangle ABP$ at P. (By *enter* we mean that the line has P and another point inside of the triangle on it. (See Moise [8, Chapter 24] for a careful development of these basics, including the justification of right- and left-sensed parallels in Definition 3.2.1.)

Theorem 3.2.1   Given a point P not on a line k, there are infinitely many lines on P that have no points also on k.

Proof.   Let the two lines indicated by the characteristic axiom be l and m. Pick points A on k and B and C on l such that line m enters $\triangle ABC$ at P. By Pasch's axiom, m must also intersect the triangle on another side, WLOG at D on $\overline{AB}$. Now, for every point X on the segment $\overline{BD}$, we can draw $\overleftrightarrow{PX}$. (See Fig. 3.8.)

Claim. $\overleftrightarrow{PX}$ does not intersect k. Suppose, for a contradiction, that $\overleftrightarrow{PX}$ and k had the point Y in common. Then line m would enter triangle $\triangle XYA$ at D, and by Pasch's axiom m would have to intersect the triangle again on $\overline{XY}$ or on $\overline{YA}$. However, m already intersects $\overleftrightarrow{PX}$ at P, eliminating $\overline{XY}$. Furthermore $\overline{YA}$ is part of k, which has no point in common with m, which is a contradiction. Hence each of the infinitely many lines $\overleftrightarrow{PX}$ has no point in common with k. ∎

For the same k and P as in the preceding proof, the lines through P split into two categories: those that intersect k and those that do not. In Fig. 3.8, for Z on segment $\overline{AD}$, some of the lines $\overleftrightarrow{PZ}$, such as $\overleftrightarrow{PA}$, intersect k, but others such as $\overleftrightarrow{PD}$, don't. By the continuity of a line, some point W must separate these lines. Does $\overleftrightarrow{PW}$ intersect k? Draw a figure illustrating the following argument by contradiction showing that $\overleftrightarrow{PW}$ cannot intersect k. If T is the supposed intersection of $\overleftrightarrow{PW}$ and k, consider any point S on k with T between A and S. Now draw $\overleftrightarrow{PS}$, which clearly intersects k at S. Therefore $\overleftrightarrow{PW}$ would not be the last line, contradicting the assumption about W. Explain why a similar situation occurs on the left side of Fig. 3.8.

Definition 3.2.1   Given a point P not on line k, the first line on P in each direction that does not intersect k is the (*right-* or *left-*) *sensed parallel* to k at P. Other lines on P that do not intersect k are called *ultraparallel* to k. Let A be on k with $\overline{AP} \perp k$. Call the smaller of the angles

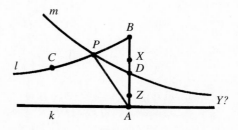

Figure 3.8

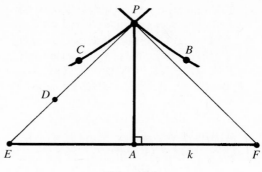

Figure 3.9

a sensed parallel makes with $\overline{AP}$ the *angle of parallelism* at P. (If the angles are equal, either is the angle of parallelism.)

One goal of this section is to show that the angle of parallelism depends only on the length of segment $\overline{AP}$.

Theorem 3.2.2 If l and m are the two sensed parallels to k at P, they have the same angle of parallelism.

Proof. As shown in Fig. 3.9, let $\overline{AP} \perp k$ and $\overleftrightarrow{PB}$ and $\overleftrightarrow{PC}$ be the sensed parallels to k at P. For a contradiction, let the angles of parallelism differ, say, $m\angle APB < m\angle APC$. Construct inside $\angle APC$ a new angle $\angle APD \cong \angle APB$. Because $\overleftrightarrow{PC}$ is a sensed parallel, $\overrightarrow{PD}$ intersects k, say, at E, giving a triangle $\triangle APE$. Let F be on k with $\overline{AF} \cong \overline{AE}$. Then $\triangle APF \cong \triangle APE$ by SAS (Euclid I-4). But $\angle APF \cong \angle APB$, which means that $\overleftrightarrow{PF}$ and $\overleftrightarrow{PB}$ are the same line by Hilbert's axiom III-4. However, $\overleftrightarrow{PB}$ is a sensed parallel to k, so it cannot intersect k at F, which is a contradiction. Hence the angles of parallelism must be the same. ∎

Corollary 3.2.1 All angles of parallelism are less than a right angle. Two lines with a common perpendicular are ultraparallel.

Proof. See Problem 2. ∎

Theorem 3.2.3 Let P be a point not on k and let m be a sensed parallel to k at P. If S is any other point on m, then m also is a sensed parallel to k at S.

Proof. WLOG, let m be a right-sensed parallel to k at P.

Case 1 Suppose that S is on m to the left of P. Because m and k do not intersect, m is either the right-sensed parallel to k at S or m is ultraparallel to k at S. For a contradiction, suppose that l, not m, is the right sensed parallel to k at S. As shown in Fig. 3.10, pick T on l and on the opposite side of m from k. Then $\overleftrightarrow{TP}$ crosses m at P. As m is the right-sensed parallel to k at P, $\overleftrightarrow{TP}$ must intersect k, say, at A. Then $\overline{SA}$ lies below l because l is a right-sensed parallel. Let U be on m with S between P and U. Then $\overline{UA}$ is below $\overline{SA}$. Now l enters $\triangle UPA$ at S and, by Pasch's axiom, l would need to intersect

3.2 Properties of Lines and Omega Triangles

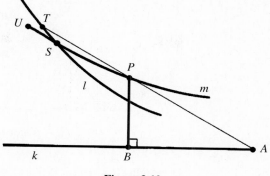

Figure 3.10

either $\overline{UA}$ or $\overline{SA}$, both of which are below it, which is a contradiction. Hence m is the right-sensed parallel to k at S.

Case 2 Let S be on m to the right of P. (See Problem 3.) ∎

3.2.1 Omega triangles

In both the Klein and Poincaré models of hyperbolic geometry (Fig. 3.11) sensed parallels "meet" on the circular boundary. Following the historical development, we say that sensed parallels meet at imaginary points called *omega points*. Although the originators of hyperbolic geometry didn't have any models, they benefited greatly from thinking of sensed parallels "meeting at infinity." In particular, they used *omega triangles*, which have one omega point, to prove theorems about ordinary triangles. They first showed, as do we here, that omega triangles share some key properties with regular triangles. In the Klein and Poincaré models, omega triangles (Definition 3.2.2) are triangles with one vertex on the boundary. These models also suggest Theorem 3.2.4, which we accept without proof. (For a proof, see Moise [8, 321–322].)

Definition 3.2.2 All lines right- (left-) sensed parallel to a given line are said to have the same *right (left) omega point*. An *omega triangle* $\triangle AB\Omega$ consists of two (ordinary) points A and B, the segment $\overline{AB}$, and the sensed parallel rays $\overrightarrow{A\Omega}$ and $\overrightarrow{B\Omega}$.

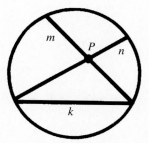 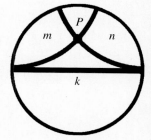

Figure 3.11 Sensed parallels in the Klein and Poincaré models.

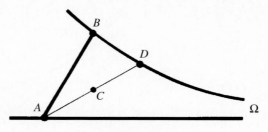

Figure 3.12

Theorem 3.2.4 If m is sensed parallel to k, then k is sensed parallel to m. If m is sensed parallel to l and l is sensed parallel to k with the same omega point, then m is sensed parallel to k with the same omega point.

Exercise 1 Draw an omega triangle for the Klein model.

Theorem 3.2.5 **Modified Pasch's Axiom for Omega Triangles** If a line k contains a point interior to $\triangle AB\Omega$ and k is on one of the vertices, then k intersects the opposite side of $\triangle AB\Omega$.

Proof. Let C be in $\triangle AB\Omega$, and draw line $\overleftrightarrow{AC}$ (Fig. 3.12). Because $\overleftrightarrow{A\Omega}$ is the sensed parallel to $\overrightarrow{B\Omega}$ at A, for $k = \overrightarrow{AC}$, k intersects $\overrightarrow{B\Omega}$, say, at D. The line $\overleftrightarrow{BC}$ is handled similarly. We can extend this theorem by treating Ω as a vertex. The line $\overleftrightarrow{C\Omega}$ enters the ordinary triangle $\triangle ABD$ at C, and by Pasch's axiom $\overleftrightarrow{C\Omega}$ must intersect $\overline{AB}$ or $\overline{BD}$. However, if $\overleftrightarrow{C\Omega}$ intersected $\overline{BD}$, there would be two sensed parallels to $\overleftrightarrow{A\Omega}$ at that point: $\overleftrightarrow{B\Omega}$ and $\overleftrightarrow{C\Omega}$. So $\overleftrightarrow{C\Omega}$ intersects $\overline{AB}$. ∎

Theorem 3.2.6 **Euclid I-16 for Omega Triangles** The measure of an exterior angle of an omega triangle is greater than the measure of the opposite interior angle.

Proof. Let $\triangle AB\Omega$ be an omega triangle and extend $\overline{AB}$ (Fig. 3.13). We prove $\angle CA\Omega$ to be greater than $\angle AB\Omega$ by showing that the other two possibilities lead to contradictions. For case 1, let $m\angle CA\Omega < m\angle AB\Omega$. Construct $\angle ABZ$ inside $\angle AB\Omega$ with

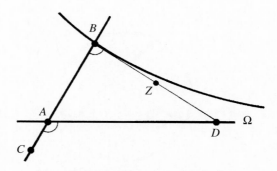

Figure 3.13 Case 1 of Theorem 3.2.6.

3.2 Properties of Lines and Omega Triangles

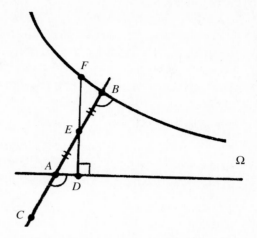

Figure 3.14 Case 2 of Theorem 3.2.6.

$\angle ABZ \cong \angle CA\Omega$. Then $\overleftrightarrow{BZ}$ intersects $\overrightarrow{A\Omega}$, say, at D because $\overleftrightarrow{B\Omega}$ is the sensed parallel to $\overrightarrow{A\Omega}$ at B. But then $\angle CA\Omega$ is an exterior angle to an ordinary triangle, $\triangle ABD$, with an opposite interior angle $\angle ABD \cong \angle CA\Omega$. This result contradicts Euclid I-16.

For case 2 (Fig. 3.14), we try to set $\angle CA\Omega \cong \angle AB\Omega$. If we let E be the midpoint of $\overline{AB}$, we can draw the perpendicular $\overline{DE}$ to $\overrightarrow{A\Omega}$. (The angle of parallelism is acute, so D is not A.) Construct F on $\overleftrightarrow{B\Omega}$ so that $\overline{FB} \cong \overline{AD}$ and F and D are on opposite sides of $\overleftrightarrow{AB}$. Note that $\angle FBE \cong \angle DAE$ by our extra assumption in this case. So by SAS $\triangle FBE \cong \triangle DAE$. Then the angles at E are vertical angles, ensuring that D, E, and F are on the same line, by Euclid I-14. Because $\angle ADE$ is a right angle, $\angle BFE$ is also. But then the angle of parallelism for $\overrightarrow{F\Omega}$ to $\overrightarrow{D\Omega}$ would be a right angle, which is a contradiction. Hence the assumption that $\angle CA\Omega \cong \angle AB\Omega$ must be wrong. Both alternatives are impossible, proving the theorem. ■

We can extend Euclid's concept of congruent triangles to omega triangles. However, the lengths of two of the "sides" of an omega triangle are infinite, and we can hardly measure the imaginary angle "at" the omega point. Hence there are only two angles and the included side to consider in each omega triangle. We show that if two omega triangles have two of these three parts congruent, they have their third parts congruent. In this case we define the omega triangles to be *congruent*.

Theorem 3.2.7 In omega triangles $\triangle AB\Omega$ and $\triangle CD\Lambda$, if $\overline{AB} \cong \overline{CD}$ and $\angle AB\Omega \cong \angle CD\Lambda$, then $\triangle AB\Omega \cong \triangle CD\Lambda$.

Exercise 2 Illustrate the proof of Theorem 3.2.7.

Proof. For a contradiction, assume that $\angle BA\Omega$ is not congruent to $\angle DC\Lambda$, and so WLOG $\angle DC\Lambda$ has a smaller measure. Construct $\angle BAP$ inside $\angle BA\Omega$ with $\angle BAP \cong \angle DC\Lambda$. Then $\overrightarrow{AP}$ intersects $\overrightarrow{B\Omega}$, say, at E. From the ordinary triangle $\triangle ABE$ we construct a congruent one in the other omega triangle. Let F be the point on $\overrightarrow{D\Lambda}$

with $\overline{DF} \cong \overline{BE}$. Then $\triangle DFC \cong \triangle BEA$ by SAS. However, this outcome means that $\angle DCF \cong \angle BAE \cong \angle DC\Lambda$, which implies that $\overleftrightarrow{C\Lambda}$ would intersect $\overleftrightarrow{D\Lambda}$, which is a contradiction. Hence the two angles, and so the two omega triangles, are congruent. ∎

Theorem 3.2.8 If corresponding angles of omega triangles are congruent, then the omega triangles are congruent.

Proof. See Problem 8. ∎

If $\angle AB\Omega$ is a right angle in omega triangle $\triangle AB\Omega$, then $\angle BA\Omega$ is the angle of parallelism. Theorem 3.2.7 implies that the size of this angle depends only on the length AB. Conversely, if we know the angle of parallelism, Theorem 3.2.8 says that only one length has that angle of parallelism. A comparison with Euclidean geometry reveals the oddity of this situation. In Euclidean geometry the angles of a shape tell us nothing about the size of the shape. In particular, similar shapes have the same angles but are of different sizes. In hyperbolic geometry, the angles of shapes determine their size. In both geometries we can measure angles absolutely by comparing them with a right angle. However, Theorem 3.2.8 says that in hyperbolic geometry we can, in principle, measure lengths in an absolute way by using the angle of parallelism.

PROBLEMS FOR SECTION 3.2

1. Suppose that m is a left-sensed parallel to k and l is a right-sensed parallel to k. Use the Klein model to determine whether m and l can (a) intersect, (b) be sensed parallel, or (c) be ultraparallel.

2. Prove Corollary 3.2.1.

3. Prove case 2 of Theorem 3.2.3 [*Hint:* Pick T on l between m and k.]

4. Relate Euclid I-27 and I-28 to Corollary 3.2.1.

5. Prove Pasch's axiom (Hilbert II-4) for omega triangles.

6. Suppose that $\overleftrightarrow{AC} \perp \overleftrightarrow{C\Omega}$, B is between A and C and that $\overleftrightarrow{A\Omega}$, $\overleftrightarrow{B\Omega}$, and $\overleftrightarrow{C\Omega}$ are all right-sensed parallels. Theorem 3.2.8 implies that the angles of parallelism for $\triangle AC\Omega$ and $\triangle BC\Omega$ cannot be equal. Which is larger? Prove your answer.

7. Let $\overleftrightarrow{AC}$ and $\overleftrightarrow{BD}$ be sensed parallels, $\overline{AB} \perp \overleftrightarrow{BD}$ and $\overline{CD} \perp \overleftrightarrow{BD}$. Suppose, as Fig. 3.15 suggests, that $\overline{CD}$ is shorter than $\overline{AB}$. Use your answer to Problem 6 to compare 360° with the angle sum of the quadrilateral $ABCD$.

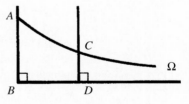

Figure 3.15

8. Prove Theorem 3.2.8.

9. If M is the midpoint of $\overline{AB}$ in $\triangle AB\Omega$, prove that $\angle A \cong \angle B$ iff $\angle AM\Omega$ is a right angle.

10. In omega triangles $\triangle AB\Omega$ and $\triangle CD\Lambda$, if $\angle A \cong \angle B$, $\angle C \cong \angle D$, and $\overline{AB} \cong \overline{CD}$, prove that $\triangle AB\Omega \cong \triangle CD\Lambda$.

3.3 SACCHERI QUADRILATERALS AND TRIANGLES

We now turn from unbounded sensed parallels and omega triangles to bounded regions, including the quadrilaterals that Saccheri used in his investigations. The Arab mathematician Omar Khayyam (circa 1050–1130) developed this quadrilateral and, in

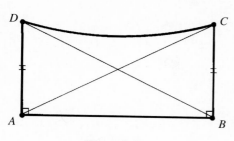

Figure 3.16   **Figure 3.17**

essence, Theorem 3.3.1, in his attempt to prove Euclid's fifth postulate. Saccheri is believed to have built on Khayyam's work.

Definition 3.3.1 A *Saccheri quadrilateral* has two opposite congruent sides perpendicular to one of the other sides. The common perpendicular is the *base*, and its opposite side is the *summit*.

Theorem 3.3.1 The summit angles of a Saccheri quadrilateral are congruent. The base and summit are perpendicular to the line on their midpoints.

Proof. Let $\overline{AB}$ be the base of Saccheri quadrilateral $ABCD$ and $\overline{AC}$ and $\overline{BD}$ the diagonals (Fig. 3.16). Then $\triangle ABC \cong \triangle BAD$ by SAS, and by SSS $\triangle ADC \cong \triangle BCD$. Then the summit angles, $\angle ADC$ and $\angle BCD$, are congruent. For the second statement of the theorem, let E and F be the midpoints of $\overline{AB}$ and $\overline{CD}$ (Fig. 3.17). Draw $\overline{DE}$ and $\overline{CE}$. Then $\triangle DAE \cong \triangle CBE$ by SAS, and $\triangle DEF \cong \triangle CEF$ by SSS. Hence $\angle DFE \cong \angle CFE$, so $\overline{EF} \perp \overline{CD}$. Similarly, $\overline{EF} \perp \overline{AB}$. ∎

Theorem 3.3.2 The summit angles of a Saccheri quadrilateral are acute.

Proof. On Saccheri quadrilateral $ABCD$ with base $\overline{AB}$, let $\overleftrightarrow{C\Omega}$ and $\overleftrightarrow{D\Omega}$ be left-sensed parallels to $\overleftrightarrow{AB}$ (Fig. 3.18). By Theorem 3.2.6, the exterior angle $\angle ED\Omega$ of omega triangle $\triangle DC\Omega$ is larger than $\angle DC\Omega$. Theorem 3.2.7 implies that the angles of parallelism $\angle AD\Omega$ and $\angle BC\Omega$ are congruent. Hence $\angle EDA$ is larger than $\angle DCB$. The summit angles are congruent, so $\angle EDA$ is bigger than $\angle CDA$. These two angles form a straight line, so the smaller, a summit angle, must be acute. ∎

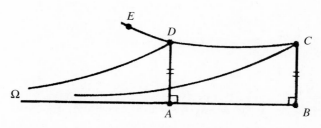

Figure 3.18

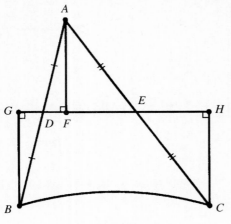

Figure 3.19

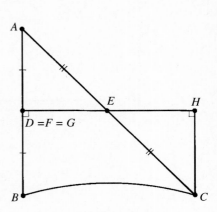

Figure 3.20

A direct consequence of Theorem 3.3.2 is that the angle sum of a Saccheri quadrilateral is always less than 360°. We convert this fact to the corresponding fact about triangles in Theorem 3.3.3 by dissecting a triangle to form a Saccheri quadrilateral (Fig. 3.19). Then we show that the sum of the measures of the summit angles of the Saccheri quadrilateral equals the angle sum of the triangle.

Theorem 3.3.3   The angle sum of any triangle is less than 180°.

Proof.   In any triangle $\triangle ABC$, let D be the midpoint of $\overline{AB}$ and E be the midpoint of $\overline{AC}$. Construct the perpendiculars $\overline{AF}$, $\overline{BG}$, and $\overline{CH}$ to $\overline{DE}$.

Claim.   Either D is between F and G or $D = F = G$. (See Problem 3.) This claim implies that the construction looks like that shown in Fig. 3.19, 3.20, or 3.21.

Cases 1 and 2   These cases correspond to Figs. 3.19 and 3.20. (See Problem 4.)

Case 3   Let D lie between F and the other points G, E, and H on this line (Fig. 3.21). In $\triangle AFD$ and $\triangle BGD$, we know that $\overline{AD} \cong \overline{BD}$, $\angle AFD \cong \angle BGD$, and $\angle ADF \cong \angle BDG$. Hence, by AAS, $\triangle AFD \cong \triangle BGD$. Similarly, $\triangle AFE \cong \triangle CHE$. Then $\overline{BG} \cong \overline{AF} \cong \overline{CH}$, which shows that $GHCB$ is a Saccheri quadrilateral. Thus both summit angles, $\angle CBG$ and $\angle BCH$, are acute. Finally we need to show that the angle sum of the triangle equals the sum of the measures of these two summit angles. Following the labeling in Fig. 3.21, we have

$m\angle ABC + m\angle BAC + m\angle ACB$
$= m\angle 2 + m\angle 1 + m\angle 4 + m\angle 5 = m\angle 2 + m\angle 3 + m\angle 4 + m\angle 5 = m\angle 2 + m\angle 6 + m\angle 5$
$= m\angle GBC + m\angle HCB.$ ∎

Exercise 1   If we dissect a Euclidean triangle as in Theorem 3.3.3, what type of quadrilateral results? Is it a Saccheri quadrilateral?

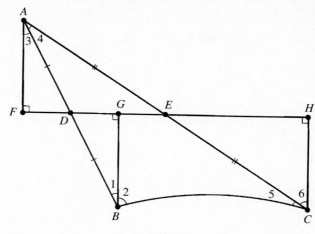

Figure 3.21

Exercise 2 Explain why the argument in the preceding proof wouldn't show that for Euclidean triangles the sum of the measures of their angles is less than 180°.

Corollary 3.3.1 The angle sum of a quadrilateral is less than 360°.

Proof. In any quadrilateral $ABCD$ draw $\overline{BD}$. If $ABCD$ is the union of $\triangle ABD$ and $\triangle CBD$, we can use Theorem 3.3.3 twice. Otherwise, A and C are on the same side of $\overline{BD}$. The definition of a polygon (Problem 4 of Section 1.2) implies that A is in the interior of $\triangle CBD$ or that C is in the interior of $\triangle ABD$. Either way $ABCD$ is the union of $\triangle ABC$ and $\triangle ADC$. Now use Theorem 3.3.3 twice. In Problem 5 you are asked to fill out this argument. ∎

The next theorem shows that the powerful Euclidean notion of similar triangles plays no role in hyperbolic geometry. In hyperbolic geometry similar triangles are always congruent. Indeed, as Saccheri noted, the existence of noncongruent but similar triangles is equivalent to Playfair's axiom.

Theorem 3.3.4 If two triangles have corresponding angles congruent, then the triangles are congruent.

Proof. Let $\triangle ABC$ and $\triangle A'B'C'$ have corresponding angles congruent (Fig. 3.22). If one pair of corresponding sides is congruent, then by ASA the triangles would be congruent. For a contradiction, assume WLOG that $\overline{AB}$ is longer than $\overline{A'B'}$. Construct D between A and B such that $\overline{AD} \cong \overline{A'B'}$ and construct E on the ray $\overrightarrow{AC}$ so that $\overline{AE} \cong \overline{A'C'}$. Then by SAS, $\triangle ADE \cong \triangle A'B'C'$, which gives the two cases shown in Fig. 3.23 to consider. In Problem 6 you are asked to derive a contradiction in each case. These contradictions show that $\overline{AB}$ cannot be longer than $\overline{A'B'}$, and so the triangles are congruent. ∎

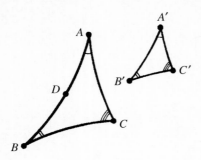

Figure 3.22

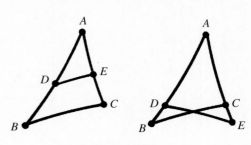

Figure 3.23

PROBLEMS FOR SECTION 3.3

1. Use the unit circle for the Poincaré model and construct a Saccheri quadrilateral as follows. Let the base be on $y = 0$, the two sides be on $[x \pm (5/3)]^2 + y^2 = (4/3)^2$ and the summit be on $x^2 + [y - (5/3)]^2 = (4/3)^2$. Graph these four hyperbolic lines. Explain why the two sides are perpendicular to the base. Why can you reasonably expect the sides to be congruent? Verify visually that the summit angles are acute.

2. a) Prove that the supposition in Problem 7 of Section 3.2 is correct. [*Hint:* Consider X on $\overrightarrow{DC}$ such that $ABDX$ is a Saccheri quadrilateral. Use Theorem 3.3.2.]

   b) Explain why the supposition you showed in part (a) implies, as Saccheri proved, that sensed parallels approach one another.

3. (Proof of the claim in Theorem 3.3.3.) If D, F, and G are the same point or D is between F and G, the claim holds. Suppose, for a contradiction, some other situation occurs. That is, either D coincides with just one of the other points or that F and G are both on the same side of D. Find a contradiction for each of these options. [*Hint:* Use Euclid I-16 for the second option.]

4. Prove the first two cases of Theorem 3.3.3, based on Figs. 3.19 and 3.20.

5. Illustrate and complete the proof of Corollary 3.3.1.

6. a) Finish the proof of Theorem 3.3.4. [*Hint:* Use Corollary 3.3.1 and Euclid I-16.]

   b) Explain why part (a) doesn't apply in Euclidean geometry.

7. As shown in Fig. 3.24, $\triangle ABC$ and $\triangle ABD$ have angle sums less than 180°. Which triangle, if any, has the larger sum? Prove your answer.

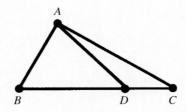

Figure 3.24

8. Generalize Corollary 3.3.1 to convex polygons with n sides. Prove your generalization by using induction. Does this generalization hold if the polygon isn't convex? Explain.

9. In Section 3.4 we will define the *defect* of a triangle to be the difference between 180° and the angle sum of the triangle. Prove that the defects of triangles are additive in the following sense. If S lies between Q and R in $\triangle PQR$, show that the defect of $\triangle PQR$ equals the sum of the defects of $\triangle PQS$ and $\triangle PSR$.

3.4 AREA AND HYPERBOLIC DESIGNS

Areas in hyperbolic geometry do not have easily remembered formulas, such as $A = \frac{1}{2}bh$ for a Euclidean triangle. In place of that, Theorem 3.4.5 asserts that the area of a triangle is proportional to the *defect* of the triangle, or the amount by which its angle sum is less than 180°. Without formulas for areas, we need to make explicit what we mean by area. Whether in Euclidean, hyperbolic or another geometry, area satisfies four axioms.

  i)   Area is a nonnegative real number.
  ii) Congruent sets have the same area.
  iii) The area of a disjoint union of sets is the sum of their areas.
  iv) The area of any point or line segment is zero.

Example 1 Show that a triangle and the associated Saccheri quadrilateral from Theorem 3.3.3 have the same area.

Solution. From case 1 of Theorem 3.3.3, $\triangle AFD \cong \triangle BGD$ and $\triangle AEF \cong \triangle CEH$, and quadrilateral $BDEC$ is congruent to itself (Fig. 3.19). By axiom (ii), the pieces of $\triangle ABC$ have the same areas as the corresponding pieces of the Saccheri quadrilateral. But we cannot yet use axiom (iii), for these pieces are not disjoint. However, the intersections of the pieces are just line segments and so have zero area by axiom (iv). Thus the areas of the triangle and its associated Saccheri quadrilateral are equal. Cases 2 and 3 are left as exercises. ●

Theorem 3.4.1 generalizes Example 1 and justifies the definition of equivalent polygons. The notion of equivalent polygons applies in many geometries. For example, W. Bolyai and P. Gerwien showed that if two Euclidean polygons have the same area, they are equivalent. Theorem 3.4.2 enables us to show that two triangles have the same area by comparing their Saccheri quadrilaterals.

Theorem 3.4.1 If a set S is the union of a finite number of sets $A_1, A_2, \ldots, A_n$ and the intersection of any two sets A_i and A_j is a finite number of line segments, then the area of S is the sum of the areas of $A_1, A_2, \ldots, A_n$.

Proof. Problem 1. ∎

Definition 3.4.1 Two polygons A and B are *equivalent* iff there are finitely many triangles $A_1, A_2, \ldots, A_n$ and $B_1, B_2, \ldots, B_n$ such that A is the union of the A_i; B is the union of the B_i; for each i, $A_i \cong B_i$; and the intersection of any A_i and A_j (or B_i and B_j) is at most a line segment.

Theorem 3.4.2 If polygons A and B are equivalent and polygons B and C are equivalent, then A is equivalent to C.

Proof. Divide A and B into families of congruent triangles, A_i and B_i. Again, divide B and C into families of congruent triangles, B'_j and C'_j, where B_i and B'_j can be different (Fig. 3.25). We subdivide the triangles B_i and B'_j into smaller triangles B_{ijk} so that we can reassemble them into either A or C. The intersection B_{ij} of the convex sets B_i and B'_j is convex by Problem 8(e) of Section 1.3. Because B_{ij} is a convex polygon, it can be subdivided into triangles B_{ijk}. Then the union of all these small triangles B_{ijk} must give B and, by the assumptions of equivalence, can be rearranged to form both A

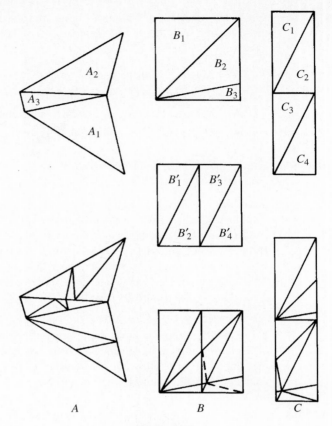

Figure 3.25

and C. This result shows that A and C can be divided into triangles A_{ijk} and C_{ijk} to make A and C equivalent. ∎

Theorem 3.4.3 Two Saccheri quadrilaterals with congruent summits and congruent summit angles are congruent and so have congruent sides and bases.

Proof. Let $ABCD$ and $EFGH$ be two Saccheri quadrilaterals, with $\overline{CD} \cong \overline{GH}$, $\angle ADC \cong \angle EHG$, and bases $\overline{AB}$ and $\overline{EF}$. Showing that $\overline{BC} \cong \overline{FG}$ and $\overline{AB} \cong \overline{EF}$ is sufficient to show that the quadrilaterals are congruent, as their corresponding angles are already congruent. Suppose, for a contradiction, that their sides are not congruent. WLOG, say, $BC > FG$ (Fig. 3.26). Hence there is a point B' on $\overline{BC}$ such that $\overline{B'C} \cong \overline{FG}$. Similarly, there is an A' on $\overline{AD}$ such that $\overline{A'D} \cong \overline{EH}$. From the midpoints P and Q of the summits $\overline{CD}$ and $\overline{GH}$ draw the line segments $\overline{A'P}$, $\overline{PB'}$, $\overline{EQ}$, and $\overline{QF}$. In Problem 2 you are asked to complete the proof of this theorem. ∎

Definition 3.4.2 The *defect* of a triangle is the difference between 180° and the angle sum of the triangle.

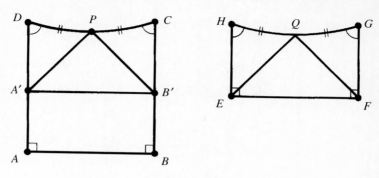

Figure 3.26

Theorem 3.4.4 Triangles with the same defect have the same area.

Proof. *Case 1* Suppose that the two triangles $\triangle ABC$ and $\triangle A'B'C'$ have the same defect and, further, that they have a pair of corresponding sides, $\overline{BC} \cong \overline{B'C'}$, congruent. We construct the associated Saccheri quadrilaterals for each triangle, $GHCB$ and $G'H'C'B'$ (Fig. 3.27). Recall that each triangle is equivalent to its Saccheri quadrilateral (Example 1) and that the angle sums of the summit angles of these two Saccheri quadrilaterals equal the angle sum of the corresponding triangles (the proof of Theorem 3.3.3). These sums are equal, so the summit angles of the Saccheri quadrilaterals must be congruent. By Theorem 3.4.3 the Saccheri quadrilaterals are congruent and so

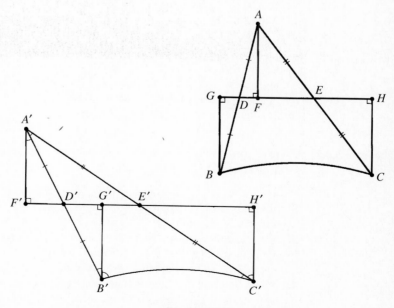

Figure 3.27

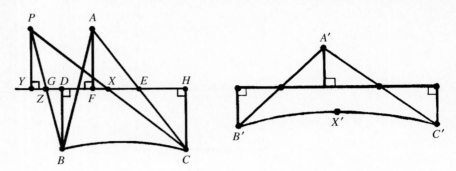

Figure 3.28

are equivalent. Then Theorem 3.4.2 shows the two triangles are equivalent and so have the same area.

Case 2 For the general case, we suppose only that $\triangle ABC$ and $\triangle A'B'C'$ have the same defect. We construct a third triangle with one side congruent to one side of the first triangle and another side congruent to one side of the other triangle and with the same defect as the given triangles. Then we use case 1 twice to conclude that $\triangle ABC$ and $\triangle A'B'C'$ are both equivalent to this third triangle. Hence we need only to construct this triangle and prove that it satisfies the needed properties (Fig. 3.28). Let $\overline{B'C'}$ be the longest of the six sides of the two triangles and let X' be the midpoint of $\overline{B'C'}$. There is a point X on line $\overleftrightarrow{GH}$, which includes the base of the Saccheri quadrilateral $\overline{GH}$, such that $\overline{XC} \cong \overline{X'C'}$. (We can find such an X because $\overline{EC}$ is shorter than $\overline{X'C'}$.) Construct P on $\overleftrightarrow{XC}$ with X the midpoint of P and C. Then $\triangle PBC$ is the candidate for the third triangle. Next, show that the intersection Z of $\overline{PB}$ with $\overleftrightarrow{GH}$ is the midpoint of $\overline{PB}$ so that $\triangle PBC$ has $GHCB$ for its Saccheri quadrilateral. Let Y be the point where the perpendicular from P meets $\overleftrightarrow{GH}$. Then $\triangle PYX \cong \triangle CHX$ by AAS, implying that $\overline{PY} \cong \overline{CH}$, which is congruent to $\overline{BG}$. Thus $\triangle PYZ \cong \triangle BGZ$, by AAS, and Z is indeed the midpoint of $\overline{PB}$. Finally, as $\triangle PBC$ has the same Saccheri quadrilateral as $\triangle ABC$, it has the same defect. We can now use case 1 twice. ∎

Theorem 3.4.5 There is a real number k such that, for every triangle $\triangle ABC$, the area of $\triangle ABC$ is $k \cdot$ (Defect of $\triangle ABC$).

Proof. The defect and the area are functions of the triangle. First, Problem 9 of Section 3.3 shows the defect function to be additive: For P between B and C, the defects of $\triangle ABP$ and $\triangle ACP$ add to the defect of $\triangle ABC$. Area axiom (iii) shows that the area function also is additive. Moreover, both functions are continuous. A result from calculus states that a continuous function f that is additive [that is, it satisfies $f(x + y) = f(x) + f(y)$] must be of the form $f(x) = cx$, for some constant c. Both the defect function and the area function for triangles have this form, so one must be a multiple of the other: The area of $\triangle ABC$ is $k \cdot$ (Defect of $\triangle ABC$). ∎

3.4 Area and Hyperbolic Designs

Figure 3.29

Figure 3.30

Theorem 3.4.6 The area of a convex polygon is proportional to the defect of that polygon.

Proof. See Problem 5(b). ∎

The tie between the area of a polygon and its defect leads to curious possibilities. In Euclidean geometry, all equilateral triangles must have 60° angles. In hyperbolic geometry, however, the angles of an equilateral triangle, though congruent, must be less than 60°. Furthermore, as the sides of the triangle lengthen continuously, this angle measure must decrease continuously, by Theorem 3.4.5 (Fig. 3.29). Thus some unique length for this angle measures exactly 45°, so we can fit eight equilateral triangles around a point. We can extend this pattern to cover the entire plane, as partially illustrated in Fig. 3.30 for the Poincaré model. We can create a corresponding pattern with seven or more equilateral triangles around a point. Similarly, we can create patterns with five or more "squares," where a "square" is a figure with four congruent sides and four congruent angles. With the help of the geometer H. S. M. Coxeter, the Dutch artist M. C. Escher laboriously created several imaginative patterns, building on hyperbolic designs like that shown in Fig. 3.30. More recently, Douglas Dunham [4] has used geometry and computer graphics to create more varied patterns relatively quickly (Figs. 3.31 and 3.32). From one copy of the repeating motif the computer constructs congruent copies to fill out the plane. Of course, the computer needs to be programmed to "draw" in hyperbolic geometry instead of Euclidean geometry. Inversions, which we discuss in Section 4.6, provide one key to drawing congruent shapes for the Poincaré model. We also need to compute distances between points in this model. Let A and B be any two hyperbolic points. The hyperbolic line on these two points intersects the boundary of the Poincaré model in the two omega points of that line, say, Ω and Λ. Poincaré found the hyperbolic distance between A and B in terms of the Euclidean distances among the four points A, B, Ω, and Λ (Fig. 3.33). (The formula in Definition 3.4.3 involves the cross-ratio, which we discuss in Chapter 6.)

Definition 3.4.3 In the Poincaré model, the *hyperbolic distance* between A and B is

$$d_H(A, B) = c \cdot \left| \log \left(\frac{A\Omega / B\Omega}{A\Lambda / B\Lambda} \right) \right|,$$

122 **Chapter 3** **Non-Euclidean Geometries**

Figure 3.31   **Figure 3.32**

where XY is the Euclidean distance, c is some constant, and Ω and Λ are the two omega points of line $\overleftrightarrow{AB}$.

Example 2 Verify that neighboring points P_i in Fig. 3.34 are equally spaced. The x-coordinates of the points are $P_0 = 0$, $P_1 = \frac{1}{3}$, $P_2 = \frac{3}{5}$, $P_3 = \frac{7}{9}$, $P_4 = \frac{15}{17}$, $P_5 = \frac{31}{33}$, $P_{-i} = -P_i$, $\Omega = -1$, and $\Lambda = 1$.

Solution. The Euclidean distances between these points are simply the differences of their x-coordinates. Then $(P_0\Omega/P_1\Omega) \div (P_0\Lambda/P_1\Lambda) = (1/(4/3)) \div (1/(2/3)) = 1/2$. Similarly, $(P_1\Omega/P_2\Omega) \div (P_1\Lambda/P_2\Lambda) = ((4/3)/(8/5)) \div ((2/3)/(2/5)) = 1/2$. All of the corresponding quotients equal 1/2 or 2. In turn, the absolute values of their logarithms are all the same. Hence, whatever the constant c is, the distances will all be equal. ●

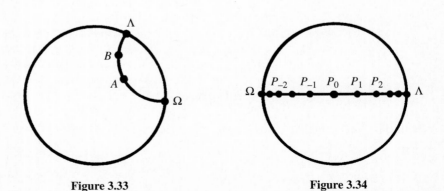

Figure 3.33   **Figure 3.34**

PROBLEMS FOR SECTION 3.4

1. Prove Theorem 3.4.1.
2. Complete the proof of Theorem 3.4.3 as follows.
   a) Prove that $\triangle A'DP \cong \triangle EHQ$.
   b) Prove that $\triangle A'PB' \cong \triangle EQF$.
   c) Prove that quadrilateral $AA'B'B$ would have to have four right angles, which is a contradiction. This contradiction forces $\overline{BC} \cong \overline{FG}$ and $\overline{AD} \cong \overline{EH}$.
   d) Prove that the bases, $\overline{AB}$ and $\overline{EF}$, must be congruent.
3. Prove that there is some real number K such that the area of every triangle is less than K. (The smallest such K is the area of a "triangle" that has three omega points for its vertices.)
4. Use Fig. 3.35 and the theorems of this section to prove that hyperbolic triangles having the same height and congruent bases don't necessarily have the same area. What happens to the area of $\triangle AB_iB_{i+1}$ as $i \to \infty$?

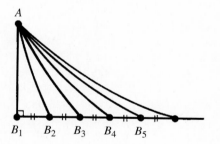

Figure 3.35

5. Recall that the angle sum of a convex Euclidean polygon with n sides is $(n-2)180°$.
   a) Prove by induction that the angle sum of a convex hyperbolic polygon with n sides is less than $(n-2)180°$.
   b) The defect of a convex hyperbolic polygon with n sides is the difference between $(n-2)180°$ and its angle sum. Prove that the area of the polygon is proportional to its defect.
6. Generalize Problem 3 to show polygons with n sides have a largest area. Find the relationship between the least upper bound K_n of the areas of polygons with n sides and $K = K_3$, the least upper bound for the areas of triangles.
7. We can construct the inner circle of equilateral triangles shown in Fig. 3.30 with the help of Fig. 3.36.

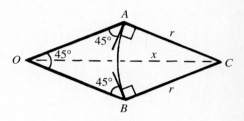

Figure 3.36

   a) Construct the unit circle, the x-axis, the y-axis, and $y = \pm x$.
   b) The remaining side of each of the eight triangles is an arc of a circle orthogonal to the unit circle. Explain why these circles all have their centers on the lines forming angles of $22\frac{1}{2}° + k \cdot 45°$ with the x-axis. Explain why these circles all have their centers the same distance, $x = OC$, from the origin and have the same, as yet unknown, radius r.
   c) Explain why, in Fig. 3.36, OA and OB equal r.
   d) Use the Poincaré model to explain why, in Fig. 3.36, $x^2 = 1 + r^2$.
   e) The law of cosines gives a second equation in x and r, namely, $x^2 = r^2 + r^2 - 2r^2 \cos 135°$. Find x and r.
   f) Finish constructing the eight equilateral triangles from part (a).
8. a) Verify that $C : (x - \frac{5}{3})^2 + y^2 = (\frac{4}{3})^2$ is orthogonal to the unit circle.
   b) Find the intersections Ω and Λ of circle C with the unit circle. Verify that $P = (\frac{1}{3}, 0)$, $Q = (\frac{1}{2}, \frac{\sqrt{15}}{6})$, and $R = (\frac{2}{5}, \frac{\sqrt{39}}{15})$ are on C. Assume $c = 1$ and find $d_H(P, Q)$, $d_H(Q, R)$, and $d_H(P, R)$. Why would we expect the sum of the two smaller distances to equal the larger distance? Verify that they do.
9. Show the following properties for d_H in the Poincaré model.

a) $d_H(A, B) = d_H(B, A) \geq 0$ and $d_H(A, B) = 0$ iff $A = B$.

b) If B is between A and C on the diameter AC, then $d_H(A, B) + d_H(B, C) = d_H(A, C)$.

10. a) Find the general pattern for the x-coordinates of the points P_i in Example 2.

b) Use part (a) to show that $d_H(P_i, P_{i+1}) = d_H(P_0, P_1)$.

3.5 SPHERICAL AND SINGLE ELLIPTIC GEOMETRIES

In one sense, mathematicians have studied the geometry of the sphere for millennia. However, before Bernhard Riemann in 1854 no one had thought of spherical geometry as a separate geometry, but only as properties of a Euclidean figure. The characteristic axiom of spherical geometry is that every two lines (great circles) always intersect in two points. (See Section 1.6.)

To retain the familiar notion of Euclidean and hyperbolic geometries that two points determine a line, Felix Klein in 1874 saw the need to modify spherical geometry. The usual way to do so was to identify opposite points on the sphere as the same point and study this "collapsed" geometry, which Klein called *single elliptic geometry*. Thus the characteristic axiom of single elliptic geometry is that every two distinct lines intersect in only one point. (Klein called spherical geometry *double elliptic geometry* because lines intersect in two points.) Spherical and single elliptic geometries share many theorems in common, such as the angle sum of a triangle is greater than 180°. In addition, single elliptic geometry possesses some unusual features worth noting. We can represent single elliptic geometry as the half of a sphere facing us (Fig. 3.37) so long as we remember that a line (or curve) that leaves the part facing us immediately reappears directly opposite because opposite points are identified.

A line in either of these geometries has many of the same properties as a circle in Euclidean geometry. First, we can't determine which points are "between" two points because there are two ways to go along a line from one point to another point. Note that we can use two points to "separate" two other points (Fig. 3.38). Second, the total length of a line is finite. A single elliptic line has another, more unusual property: It doesn't separate the whole geometry into two parts, unlike lines in Euclidean, hyperbolic, and spherical geometries. Figure 3.37 indicates how to draw a path connecting any two points not on a given line so that the path does not cross that line.

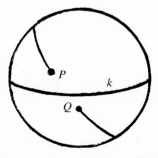

Figure 3.37 In the single elliptic geometry there is a path from P to Q that does not intersect k.

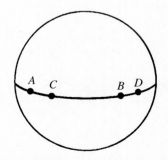

Figure 3.38 A and B separate C and D.

3.5 Spherical and Single Elliptic Geometries

In certain ways, Euclidean geometry is intermediate between spherical and single elliptic geometries on the one hand and hyperbolic geometry on the other hand. For example, in Euclidean geometry, the angle sum of a triangle always adds to 180°. As we know in hyperbolic geometry, the corresponding sum falls short of 180° in proportion to the area of the triangle. In spherical and single elliptic geometries, this sum is always more than 180° and the excess is proportional to the area of a triangle. (Theorem 1.6.3 shows this condition for Euclidean spheres.) Indeed, in these geometries triangles can have three obtuse angles, so the sum can approach 540°.

In our development of hyperbolic geometry we assumed that Euclid's first 28 propositions hold, for they used only Euclid's first four postulates, but not the fifth postulate. Many of these propositions, including two of the triangle congruence theorems (SAS and SSS), continue to hold in spherical and single elliptic geometries. However, most of the propositions after I-15, including AAS, do not hold in these geometries, even though they do not depend on the fifth postulate.

Figure 3.39 illustrates Euclid's approach to showing, as I-16 states, that in any triangle an exterior angle, such as $\angle BCD$, is larger than either of the other two interior angles, $\angle ABC$ and $\angle BAC$. From the midpoint E of $\overline{BC}$, Euclid extended AE to F so that $\overline{EF} \cong \overline{EA}$. Then by SAS $\triangle ECF \cong \triangle EBA$. He then concluded that $\angle BCD$ is larger than $\angle ECF$, which is congruent to the interior angle $\angle EBA$. Figure 3.39 supports this conclusion, but the similar situation shown in Fig. 3.40 for single elliptic geometry reveals that the conclusion depends on implicit assumptions. In Fig. 3.40, the part of $\overleftrightarrow{AE}$ that looks like segment $\overline{AE}$ covers more than half the length of the line. Hence the corresponding part of $\overleftrightarrow{EF}$ overlaps this apparent segment. Euclid implicitly assumed that lines extend infinitely in each direction. Postulate 2 only says, "to produce a finite straight line continuously in a straight line." The overlapping "segments" $\overline{AE}$ and $\overline{EF}$ in Fig. 3.40 satisfy the letter and, within reason, the spirit of postulate 2. Nevertheless, I-16 is false here because $\angle BCD$ can be smaller than $\angle ECF$.

Exercise 1 Draw the figure in spherical geometry corresponding to the situation depicted in Fig. 3.40.

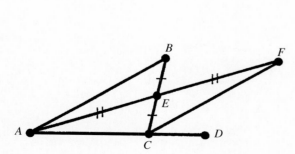

Figure 3.39 Euclid's diagram for proposition I-16.

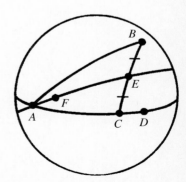

Figure 3.40 The diagram from Fig. 3.39 in single elliptic geometry.

Chapter 3 Non-Euclidean Geometries

We partially develop single elliptic geometry as we did hyperbolic geometry. (For a more thorough development, see Gans [5].) Theorems 3.2.1–3.2.8 do not relate to geometries where all lines intersect. Therefore we start with Theorem 3.3.1 concerning Saccheri quadrilaterals, which we repeat here.

Theorem 3.3.1 The summit angles of a Saccheri quadrilateral are congruent. The base and summit are perpendicular to the line on their midpoints.

Exercise 2 Verify that the proof of Theorem 3.3.1 holds in spherical and single elliptic geometry.

In hyperbolic geometry, Theorem 3.3.2 showed that the summit angles of a Saccheri quadrilateral were acute. Theorem 3.5.2 shows they are now obtuse. Theorem 3.5.1 provides a key step to this end.

Exercise 3 In Euclidean geometry what can you say about the summit angles of a Saccheri quadrilateral?

Theorem 3.5.1 In single elliptic geometry, all lines perpendicular to a given line intersect in one point.

Proof. From Theorem 3.3.1 we know that, in Saccheri quadrilateral $ABCD$, $\overline{EF}$ is perpendicular to both $\overline{AB}$ and $\overline{CD}$. By the characteristic property of single elliptic geometry, $\overleftrightarrow{AB}$ and $\overleftrightarrow{CD}$ intersect in a unique point, say, P.

Claim. $d(A, P) = d(B, P)$ and $d(C, P) = d(D, P)$. (In a model, the distance is along the shortest path. As shown in Fig. 3.41, the shortest path from B to P goes "around behind" because we can "jump" from the right edge to the left edge.)

For a contradiction, construct P' on $\overleftrightarrow{AB}$ so that $\overline{AP} \cong \overline{BP'}$ and P and B separate A and P'. (Intuitively P' is on the "other" side of B, as depicted in Fig. 3.41.) Then $\triangle ADP \cong \triangle BCP'$ by SAS. Thus $\angle BCP'$, which is congruent to $\angle ADP$, is supplementary to $\angle BCD$. Then by Euclid I-14, $\overleftrightarrow{CP'}$ is the same line as $\overleftrightarrow{CD}$. This result would give two points of intersection of $\overleftrightarrow{AB}$ with $\overleftrightarrow{CD}$, which is a contradiction. So $P' = P$,

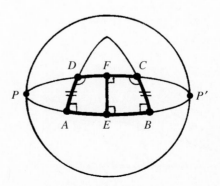

Figure 3.41

showing that $d(A, P) = d(B, P)$ and $d(C, P) = d(D, P)$. Furthermore, the distance from E to P is the same by way of A as by way of B, which is the maximum distance two points can be separated. The same holds for the distance from F to P, and $d(E, P) = d(F, P)$ by Euclid I-6, for isosceles triangles. Moreover, note that $d(E, P)$ and $d(F, P)$ do not depend on the length of $\overline{EF}$. That is, common perpendiculars are always the same length and intersect in the same point. ∎

Theorem 3.5.2 In single elliptic geometry, the summit angles of a Saccheri quadrilateral are obtuse.

Proof. In Saccheri quadrilateral $ABCD$ construct $\overleftrightarrow{DG}$ perpendicular to $\overline{AD}$ and let Q be the intersection of $\overleftrightarrow{DG}$ and $\overleftrightarrow{AB}$ (Fig. 3.42). Then distance $d(A, P)$ is less than or equal to $d(A, Q)$ because Q is the point furthest from A by Theorem 3.5.1. If these distances were equal, $P = Q$, implying that $A = E$ and giving a trivial Saccheri quadrilateral. Hence $d(A, P) < d(A, Q)$. Thus $\overleftrightarrow{DP}$ enters $\triangle QDA$ at D, and so $\angle PDA$ must be smaller than $\angle QDA$, which is a right angle. Hence $\angle ADC$, a summit angle and supplementary to $\angle PDA$, is obtuse. ∎

Once we have Theorem 3.5.2, the development of single elliptic geometry follows the lines of hyperbolic geometry. (See Project 5.)

Single elliptic geometry has another property worth discussing. Consider what happens to an asymmetrical object in single elliptic geometry as it moves along the path of a line (Fig. 3.43). This object actually has two opposite representations, as the "front" one slides out of view, it is "replaced" by the "rear" one, which is fundamentally different. The rear representation won't match the front representation because it has an opposite orientation. That is, these two representations are mirror images. No matter how you twist and turn two objects with opposite orientations in Euclidean geometry, as long as you stay in the plane, you cannot make them coincide. In single elliptic geometry these representations are the same, so we have the curious property that simply moving an object around in this geometry can switch its orientation. Single elliptic geometry is topologically a *nonorientable* surface, whereas spherical, Euclidean, and hyperbolic geometries are *orientable*. On a nonorientable surface there is no consistent way to define a clockwise direction.

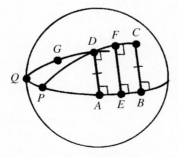

Figure 3.42

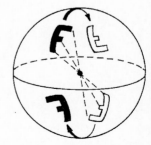

Figure 3.43 Single elliptic geometry is not orientable.

PROBLEMS FOR SECTION 3.5

1. Call a triangle with two right angles a *doubly right triangle*.
   a) Prove in single elliptic geometry that two doubly right triangles are congruent if their sides between the right angles are congruent.
   b) Prove in single elliptic geometry that two doubly right triangles are congruent if their third angles are congruent.

2. Euclid I-26 has two congruence theorems: AAS and ASA.
   a) Explain why Theorem 3.5.1 shows that AAS is not a triangle congruence theorem in single elliptic and spherical geometries.
   b) Prove that ASA is a triangle congruence theorem in single elliptic and spherical geometries, using SAS and a proof by contradiction.

3. Reformulate and prove Theorem 3.5.1 for spherical geometry.

4. Prove Theorem 3.5.2 in spherical geometry.

5. Do Problem 1 in spherical geometry.

6. Assume the analog to Theorem 3.4.5 in single elliptic and spherical geometries: The area of a triangle is proportional to the *excess* of the angle sum of a triangle over 180°. Each of the three angles of a triangle must have a measure of less than 180°.
   a) What would be the "triangle" if all three angles had a measure of 180°? What would be the area of this "triangle" if the sphere has radius r?
   b) Explain why the proof of Theorem 1.6.3 isn't sufficient to establish the analog to Theorem 3.4.5.

PROJECTS FOR CHAPTER 3

1. Construct an empirical model of hyperbolic geometry, using two regular mirrors and one convex mirror, as illustrated in Fig. 3.44. To make this curved mirror, lay some flexible reflective surface, such as mylar, on the outside of a cylindrical frame. Place some design inside the mirrors. You should see replicas of your design as though it were a part of hyperbolic geometry. As the angles at A, B, and C change, you will get different numbers of copies of your design around these points.

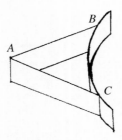

Figure 3.44

2. Use appropriate software (for example, Geometer's Sketchpad or CABRI) to create the Poincaré model of hyperbolic geometry. Investigate how the angle sum of a hyperbolic triangle changes as the size of the triangle changes. Construct a quadrilateral that has four interior angles of 60°.

3. In Euclidean geometry develop and prove the analogs, if any, to Theorems 3.3.3–3.4.6 using any of Euclid's propositions in Appendix A. How do they differ from the non-Euclidean cases?

4. In spherical geometry develop and prove the analogs to Theorems 3.3.3–3.4.6 using Problem 4 of Section 3.5.

5. In single elliptic geometry develop and prove the analogs to Theorems 3.3.3–3.4.6 using Theorem 3.5.2.

6. Investigate patterns in spherical geometry. Find all "regular patterns," or those where the same number of the same regular spherical polygons fit around every point. How do these patterns compare with polyhedra and regular patterns in Euclidean and hyperbolic geometries?

7. Investigate ultraparallel lines in hyperbolic geometry. (Cederberg [2] proves that any two ultraparallel lines have a common perpendicular.)

8. Investigate *limiting curves* or *horocycles*, the points at a constant distance from a hyperbolic line. These sets act like circles of infinite radius. (See Smart [12].)

9. Investigate the historical development of non-Euclidean geometry. (See Bonola [1], Kline [6, Chapter 36] and Richards [9].)
10. Investigate other non-Euclidean geometries. (See Yaglom [14].)
11. Investigate the concept of space. (See, for example, Rucker [10] and Weeks [13].)
12. Investigate curvature and differential geometry. (See McCleary [7].)
13. Write an essay comparing Euclidean and hyperbolic geometry. Address your essay to someone at the level of calculus.
14. Write an essay comparing how models and synthetic development have helped you understand hyperbolic geometry. Discuss how convincing the proofs in hyperbolic geometry are to you.
15. Write an essay discussing the difference between spherical geometry as a separate geometry and the geometry of the sphere as a part of Euclidean geometry.
16. Write an essay discussing whether you think that mathematics was discovered or invented. Use your understanding of non-Euclidean geometries as an example. (See, for example, Kline [6, 1032–1038].)

Suggested Readings

[1] Bonola, R. *Non-Euclidean Geometry: A Critical and Historical Study of its Development*. Mineola, N.Y.: Dover, 1955.
[2] Cederberg, J. *A Course in Modern Geometries*. New York: Springer-Verlag, 1989.
[3] do Carmo, M. *Differential Geometry of Curves and Surfaces*. Englewood Cliffs, N.J.: Prentice-Hall, 1976.
[4] Dunham, D. Hyperbolic symmetry. *Comp. and Maths. with Appl.,* 1986, 12(1 and 2):139–153.
[5] Gans, D. *An Introduction to Non-Euclidean Geometry*. New York: Academic Press, 1973.
[6] Kline, M. *Mathematical Thought from Ancient to Modern Times*. New York: Oxford University Press, 1972.
[7] McCleary, J. *Geometry from a Differentiable Viewpoint*. Port Chester, N.Y.: Cambridge University Press, 1994.
[8] Moise, E. *Elementary Geometry from an Advanced Standpoint*. Reading, Mass.: Addison-Wesley, 1974.
[9] Richards, J. *Mathematical Visions: The Pursuit of Geometry in Victorian England*. San Diego: Academic Press, 1988.
[10] Rucker, R. *Geometry, Relativity and the Fourth Dimension*. Mineola, N.Y.: Dover, 1977.
[11] Sacks, O. *An Anthropologist on Mars*. New York: Knopf, 1995.
[12] Smart, J. *Modern Geometries*. Monterey, Calif.: Brooks/Cole, 1988.
[13] Weeks, J. *The Shape of Space*. New York: Dekker, 1985.
[14] Yaglom, I. *A Simple Non-Euclidean Geometry and Its Physical Basis* (transl. A. Shenitzer). New York: Springer-Verlag, 1979.

Suggested Media

1. "The Gauss-Bonnet Theorem," 25-minute film, distributed by Ward's Modern Learning Aids Division, Rochester, N.Y., 1974.
2. "The Geometry Euclid Didn't Know," 16-minute film, distributed by David Nulsen Enterprises, Santa Monica, Calif., 1979.
3. "Hungarian Rhapsody: Janos Bolyai and his Revolutionary Discovery of Non-Euclidean Geometry," Vol. 3 of *Rebels Who Count*, 30-minute video, Media Center, University of Toronto, 1991.

4. "The Majestic Clockwork," 52-minute video, from *The Ascent of Man* series, distributed by Time–Life Multimedia, New York, 1974.
5. "Non-Euclidean Geometry," 24-minute video, Films for the Humanities and Sciences, Princeton, N.J., 1996.
6. "A Non-Euclidean universe," 25-minute film, distributed by University Media, Solana Beach, Calif., 1978.
7. "Not Knot," 20-minute video and supplement, University of Minnesota Geometry Center, distributed by Jones and Bartlett, Boston, Mass., 1991.
8. "Prince of Mathematicians: Karl Friedrich Gauss' Work in Astronomy, Number Theory, Physics and Geometry," Vol. 8 of *Rebels Who Count*, 30-minute video, Media Center, University of Toronto, 1991.

4
Transformational Geometry

Computers store and reproduce intricate shapes such as the fractal pictured, relying on only a few numbers that somehow encode the shape. To do so they depend on geometric transformations to represent such shapes. In addition, computer-aided design (CAD) depends on transformations to present different views of an object. Transformational geometry underlies these applications and many other aspects of mathematics.

> Geometry is the study of those properties of a set which
> are preserved under a group of transformations on that set.
> —Felix Klein

> A mathematician, like a painter or a poet, is a maker of patterns.
> If his patterns are more permanent than theirs, it is because they
> are made with ideas. —G. H. Hardy

4.1 Overview and History

Moving geometric figures around is an ancient and natural approach to geometry. However, the Greek emphasis on synthetic geometry and constructions and much later the development of analytic geometry overshadowed transformational thinking. The study of polynomials and their roots in the early nineteenth century led to algebraic transformations and abstract groups. At the same time, Augustus Möbius began studying geometric transformations. In the last third of the nineteenth century, Felix Klein and Sophus Lie showed the central importance of both groups and transformations for geometry. This approach enabled Klein and others to unify geometry at a time when new and different geometries seemed to be splitting this ancient discipline into competing theories. Transformations remained the dominant approach to geometry for 50 years. Transformations underlie the modern understanding of symmetry, which is essential in physics and chemistry, as well as mathematics. (See Chapter 5.) Early in the twentieth century physicists realized the power of transformations, starting with Einstein's theory of relativity and then with quantum mechanics. Although many geometric topics now transcend transformational geometry, this aspect of mathematics remains vital for understanding geometry.

We first investigate isometries, or the transformations that preserve distance. The proofs in Sections 4.1 and 4.2 are based on a synthetic approach, although we freely use coordinates in examples. Then the inclusion of linear algebra enables us to extend the types of transformations and to work in higher dimensions. Finally we discuss inversions, which are rich in geometric ideas and have important connections to complex analysis.

Example 1 The top and bottom of the shape depicted in Fig. 4.1 are mirror images. Matching points with their mirror images is one type of transformation, a mirror reflection. If the mirror is the x-axis, we can describe the transformation, say, μ, algebraically by $\mu(x, y) = (x, -y)$. The points on the x-axis remain fixed by μ: $\mu(x, 0) = (x, 0)$. Note that, if we perform the transformation twice, a point's image is mapped back to the original point. That is, $\mu(\mu(x, y)) = \mu(x, -y) = (x, y)$. We say that the entire shape is stable under μ because μ maps the shape to itself. ●

No amount of turning and twisting can turn a left hand into a right hand, even though they mirror each other. This condition is caused by the different *orientation* of an object and its mirror image. An (asymmetrical) three-dimensional object and its mirror image can't be superimposed on each other, even though the object and its

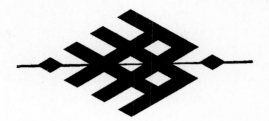

Figure 4.1

image are congruent. We particularly notice this change of orientation if we try to read the mirror image of a book. Rotations do not change orientation. Note that a two-dimensional mirror reflection, as described in Example 1, can be accomplished by a three-dimensional rotation of 180° of the entire space around the axis of fixed points. A two-dimensional mirror reflection switches orientation of points in the plane.

Definition 4.1.1 A *transformation* τ on a set S is a function from S to itself that is one-to-one and onto. That is, (i) [τ is a function] for every point P of S there is a unique point Q that is the image of P under τ: $\tau(P) = Q$, and (ii) [τ is one-to-one and onto] for every point Q of S there is a unique point P for which Q is the image of P under τ. A point P is a *fixed point* of the transformation τ iff $\tau(P) = P$. A subset T of points in S is *stable* under the transformation τ iff the image of the subset T is again T, even if individual points of T move to other points in T. (Stable sets are often called *invariant* sets.)

The fixed points and stable sets of a transformation tell us important information about the transformation. For example, in Section 4.2 you will be able to recognize the type of an isometry by its fixed points and stable lines. Symmetry, the topic of Chapter 5, involves the study of transformations and their stable sets more deeply. In dynamical systems, a new field of mathematics, fixed points and stable sets help explain a much broader family of functions than we present here. (See Abraham and Shaw [1].)

Remark We don't need to separate the properties of one-to-one and onto here, although the difference is often important in mathematics. (See Gallian [5, 16] for a discussion of these concepts.)

Exercise 1 In Example 1 verify that all vertical lines are stable under μ even though individual points move. Verify that the x-axis also is stable under μ. Verify that any other horizontal line is not stable.

Example 2 On $\mathbf{R}^2$ define $\rho(x, y) = (y + 2, 2 - x)$ (Fig. 4.2). Show that ρ is a transformation. (Later we see ρ as a rotation.)

Solution. Each point (x, y) has a unique image $(y + 2, 2 - x)$, so ρ is a function. To show one-to-one and onto, we must start with any point (u, v) and show that there is a unique point (x, y) that ρ sends to (u, v). When we solve $u = y + 2$ and $v = 2 - x$, we find the solution: $(x, y) = (2 - v, u - 2)$. Because there is only one solution, ρ is by definition one-to-one and onto. ●

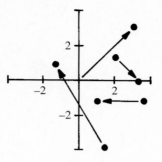

Figure 4.2

Exercise 2 Verify that ρ from Example 2 fixes the point $(2, 0)$. Why does ρ have no other fixed point? Assume that ρ is a rotation of $270°$ around $(2, 0)$. Why are circles with center $(2, 0)$ stable under ρ? Does ρ switch orientation?

Example 3 On $\mathbf{R}^2$ define $\psi(x, y) = (e^x, \sin y)$. Although ψ is a function, it is neither one-to-one nor onto. No matter what value of x we choose, e^x is positive, so ψ cannot be onto all of $\mathbf{R}^2$. Furthermore, the sine function is periodic, so two different points can map to the same point, demonstrating that ψ isn't one-to-one. ●

Example 4 The biologist D'Arcy Thompson used the idea of transformations in his study of comparative anatomy. Figure 4.3 reproduces one of his illustrations, depicting how he compared features of related species. The actual transformations he used go beyond the level of our study. (See Thompson [10].) ●

Example 5 On $\mathbf{R}^2$ let $\rho(x, y) = (y + 2, 2 - x)$ and $\mu(x, y) = (x, -y)$. The composition $\rho \circ \mu$ is given by $\rho \circ \mu(x, y) = \rho(\mu(x, y)) = \rho(x, -y) = (-y + 2, 2 - x) = (2 - y, 2 - x)$. We can show $\rho \circ \mu$ to be a transformation, as in Example 2. ●

Composing functions enables us to build and study a wide variety of functions in calculus, geometry, and other areas of mathematics. Theorem 4.1.1 shows that the composition of two transformations on a set is again a transformation on the set. This

Figure 4.3 Comparing skulls by using transformations that distort coordinates.

property is the first of several general properties about transformations that nineteenth century mathematicians realized were crucial algebraic properties. In later sections we describe how these algebraic properties enable us to prove geometric properties about transformations.

Definition 4.1.2 For two functions f and g from a set S to itself, the *composition* of g followed by f, $f \circ g$ is defined by $f \circ g(P) = f(g(P))$.

Theorem 4.1.1 If α and β are any two transformations on S then $\alpha \circ \beta$ is a transformation on S.

Proof. Let α and β be two transformations on S and $P \in S$. For $\alpha \circ \beta$ to be a function, there must be a unique $R \in S$ such that $\alpha \circ \beta(P) = R$. Now β is a transformation, so there is a unique $Q \in S$ such that $\beta(P) = Q$. Similarly, there is a unique $R \in S$ such that $\alpha(Q) = R$. Hence R is the unique image of P under $\alpha \circ \beta$.

Let $V \in S$. For $\alpha \circ \beta$ to be one-to-one and onto, there must be a unique $T \in S$ such that $\alpha \circ \beta(T) = V$. Because α is a transformation, there is a unique $U \in S$ such that $\alpha(U) = V$. Similarly, there is a unique $T \in S$ such that $\beta(T) = U$. This result implies that T is the only element of S which $\alpha \circ \beta$ takes to V. Thus $\alpha \circ \beta$ is a transformation. ■

Example 6 For any set S, define the *identity* transformation by $\iota(P) = P$. For any transformation α on S, $\alpha \circ \iota = \alpha = \iota \circ \alpha$. The identity may seem of little importance by itself, but its presence simplifies investigations about transformations, just as the number 0 simplifies addition of numbers. ●

Example 7 On $\mathbf{R}^2$, if we compose $\rho(x, y) = (y + 2, 2 - x)$ and $\psi(x, y) = (2 - y, x - 2)$, then $\rho \circ \psi(x, y) = \rho(2 - y, x - 2) = ((x - 2) + 2, 2 - (2 - y)) = (x, y)$. Thus $\rho \circ \psi$ equals the identity, ι: ρ undoes what ψ did. ●

Definition 4.1.3 A transformation β is the *inverse* of a transformation α iff $\alpha \circ \beta = \iota$ and $\beta \circ \alpha = \iota$. We write α^{-1} for the inverse of α.

Exercise 3 In Example 7, verify that $\psi \circ \rho$ is also the identity. In Example 5, verify that $\rho \circ \mu$ is its own inverse.

Theorem 4.1.2 Every transformation α on a set S has a unique inverse, α^{-1}, which is a transformation satisfying $\alpha^{-1}(Q) = P$ iff $\alpha(P) = Q$.

Proof. Note that parts (i) and (ii) of the definition of a transformation (Definition 4.1.1) are closely related. This relationship implies that α^{-1}, as given in Theorem 4.1.2, is a function because α is one-to-one and onto. Similarly, α^{-1} is one-to-one and onto because α is a function. Hence α^{-1} is a transformation. Verify that $\alpha \circ \alpha^{-1} = \iota = \alpha^{-1} \circ \alpha$. To show uniqueness, we suppose that β also is an inverse of α and show that $\beta = \alpha^{-1}$. Let Q be any element of S. Because α is a transformation, there is a unique P such that $\alpha(P) = Q$. Then $\beta(Q) = P = \alpha^{-1}(Q)$. As α^{-1} and β agree everywhere, they are equal. ■

Various sets of transformations correspond to important geometric properties and also form groups, which are structures of great importance in mathematics. We define transformation groups because these are the only groups that we consider.

Remark Associativity plays a key role in group theory, but it always holds for composition of functions. Hence we can omit it from the definition. (See Gallian [5,17].)

Definition 4.1.4 A set **T** of transformations on a set S is a *transformation group* iff the following properties obtain.

   i) (*closure*) The composition of two transformations in **T** is in **T**.
   ii) (*identity*) The identity transformation is in **T**.
   iii) (*inverses*) If a transformation τ is in **T**, then the inverse τ^{-1} is in **T**.

Theorem 4.1.3 The set of all transformations on a set is a transformation group.

Proof. See Theorems 4.1.1 and 4.1.2 and Example 6. ∎

Exercise 4 Verify that $\mu \circ \rho$ does not equal $\rho \circ \mu$ in Example 5.

PROBLEMS FOR SECTION 4.1

1. On **R**, the real numbers, define $\alpha(x) = x^3$ and $\beta(x) = 2x - 1$.
   a) Show that α and β are transformations.
   b) Find $\alpha \circ \beta$ and $\beta \circ \alpha$. Show that $\alpha \circ \beta \neq \beta \circ \alpha$.
   c) Find α^{-1} and β^{-1}. Graph α and α^{-1} together. Repeat for β and β^{-1}. Describe the relationship between the graph of a transformation on **R** and the graph of its inverse.
   d) Find $\alpha^{-1} \circ \beta^{-1}$ and $\beta^{-1} \circ \alpha^{-1}$. Which is the inverse of $\alpha \circ \beta$? Verify your answer.

2. For each of the following functions on $\mathbf{R}^2$, show on a graph what it does to various points, show that it is a transformation, and find its fixed point(s) and stable line(s).
   a) $\omega(x, y) = (2 - x, 4 - y)$.
   b) $\mu(x, y) = (y - 1, x + 1)$.
   c) $\sigma(x, y) = (2x, 2y)$.
   d) $\psi(x, y) = (\frac{1}{2}x + 1, \frac{1}{2}y - 1)$.

3. If α and β are transformations on a set S, prove that both $\alpha^{-1} \circ \beta^{-1}$ and $\beta^{-1} \circ \alpha^{-1}$ are transformations. Which of these two transformations is the inverse of $\alpha \circ \beta$? Prove your answer.

4. a) On $\mathbf{R}^2$ define σ by $\sigma(x, y) = (\frac{x}{2} + 2, \frac{y}{2} - 1)$. Find the fixed point of σ and call it F.
   b) Let P_0 be any point in $\mathbf{R}^2$ and define the sequence $\{P_0, P_1, P_2, \ldots\}$ by $P_{n+1} = \sigma(P_n)$. Graph the sequence $\{P_0, P_1, P_2, \ldots\}$ for several initial choices of P_0. What happens in each case?
   c) Repeat parts (a) and (b) for $\phi(x, y) = (2x + 2, 2y - 4)$.
   d) Repeat parts (a) and (b) for $\rho(x, y) = (3 - y, x - 1)$.

   The study of dynamical systems involves finding the long-term result of repeated application of a function. The fixed point of σ is called an *attracting* (or *stable*) *fixed point* because σ takes all points closer to the fixed point. The fixed point of ϕ is called a *repelling* (or *unstable*) *fixed point* because ϕ sends other points farther from the fixed point. The transformation ρ is said to be *periodic of period 4*, as four applications of ρ give the identity. These terms describe the dynamics of many familiar transformations. (See Abraham and Shaw [1].)

   e) The transformation $\gamma(x, y) = (x + 2, -y)$ doesn't fit any of the preceding dynamics. Describe its long-term dynamics.

4.2 ISOMETRIES

The most important family of transformations, isometries, do not change the distance between points as the transformations move these points. Isometries are the dynamic counterpart to the Euclidean notion of congruence.

Definition 4.2.1 A transformation σ is an *isometry* on a set S with a distance function d iff for all points P and Q in S, $d(P, Q) = d(\sigma(P), \sigma(Q))$. If d is the usual distance on the Euclidean plane, then we call σ a *Euclidean plane isometry*.

Example 1 A rotation is an isometry (Fig. 4.4). It has one fixed point, in this case the origin. Most rotations have no stable lines. Describe the stable lines of a rotation of 180° around a point O. •

Example 2 As shown in Fig. 4.5, σ doubles x-coordinates and halves y-coordinates. It is a transformation but not an isometry. •

Exercise 1 Explain why an isometry takes a circle to a circle.

Example 3 A mirror reflection over a line k is an isometry (Fig. 4.6). Points on the line k are fixed. Any other point P is mapped to the point $\mu(P)$, where k is the perpendicular bisector of $\overline{P\mu(P)}$. The stable lines are the line of fixed points and the lines perpendicular to that line. •

Exercise 2 In Example 3 explain why m is stable provided that $m \perp k$ or $m = k$.

Example 4 The isometry depicted in Fig. 4.7, called a translation, adds 3 to the x-coordinate and 2 to the y-coordinate of any point. A translation has no fixed points. The stable lines of a translation are parallel to the direction of the translation. •

Definition 4.2.2 A Euclidean plane isometry τ is a *translation* iff, for all points P and Q, the points P, Q, $\tau(Q)$, and $\tau(P)$ form a parallelogram. (See Fig. 4.7.) The translation τ is said to be

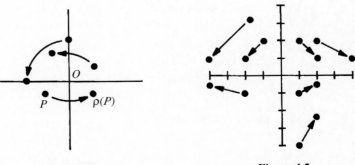

Figure 4.4   **Figure 4.5**

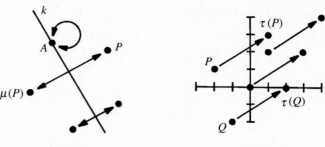

Figure 4.6 Figure 4.7

in the direction of $\overrightarrow{P\tau(P)}$. A Euclidean plane isometry ρ is a *rotation* of $r°$ iff there is a point O such that $\rho(O) = O$ and, for all other P, $m\angle PO\rho(P) = r°$. (See Fig. 4.4.) A Euclidean plane isometry μ is a *mirror reflection* over the line k iff, for every point P, k is the perpendicular bisector of the segment $\overline{P\mu(P)}$. (See Fig. 4.6.)

Exercise 3 Why is the identity both a translation and a rotation?

We now want to characterize all Euclidean plane isometries. Theorems 4.2.3, 4.2.5, and 4.2.7 give important geometric descriptions of isometries. Their proofs depend on the algebraic ideas of transformation groups, as well as on geometric properties. You might benefit from first exploring these ideas visually in Projects 1–4.

Theorem 4.2.1 The isometries of a set form a transformation group.

Proof. For closure, let α and β be isometries on a set S and show that $\alpha \circ \beta$ is an isometry. Let P and Q be any points in S. Then $d(P, Q) = d(\beta(P), \beta(Q)) = d(\alpha(\beta(P)), \alpha(\beta(Q)))$, showing $\alpha \circ \beta$ to be an isometry. Next, the identity, which fixes every point, preserves distance and so is an isometry. Finally, for an isometry α we show that its inverse α^{-1} is also an isometry. For P and Q in S, let $\alpha^{-1}(P) = U$ and $\alpha^{-1}(Q) = V$. We must show that $d(P, Q) = d(U, V)$. Because α is an isometry and the inverse of α^{-1}, $d(U, V) = d(\alpha(U), \alpha(V)) = d(P, Q)$. Thus the isometries form a transformation group. ∎

Theorem 4.2.2 A Euclidean plane isometry that fixes three noncollinear points is the identity.

Proof. Let α be an isometry, and A, B, and C be three noncollinear points fixed by α, and D be any other point. We must show that $\alpha(D) = D$. Wherever $\alpha(D)$ is, it must satisfy three distance equations: $d(A, D) = d(A, \alpha(D))$, $d(B, D) = d(B, \alpha(D))$, and $d(C, D) = d(C, \alpha(D))$. Thus $\alpha(D)$ must be on three circles: one centered at A with radius AD, the second centered at B with radius BD, and the third centered at C with radius CD (Fig. 4.8). Because A and B are distinct, the first two circles intersect in at most two points, one of which is D. If D is the only intersection, we are done. But suppose that there is another point, say, E. Then $\overleftrightarrow{AB}$ is the perpendicular bisector of $\overline{DE}$. However, C is not on $\overleftrightarrow{AB}$. Thus C cannot be the same distance from D and E. Hence $\alpha(D)$ cannot be E, forcing $\alpha(D) = D$. ∎

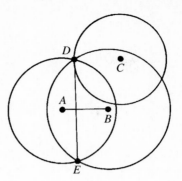

Figure 4.8

Theorem 4.2.3   A Euclidean plane isometry is determined by what it does to any three noncollinear points.

Proof.   Let A, B, and C be any three noncollinear points, α be any isometry, and β be any isometry such that $\alpha(A) = \beta(A)$, $\alpha(B) = \beta(B)$ and $\alpha(C) = \beta(C)$. By Theorem 4.2.1, $\beta^{-1} \circ \alpha$ is an isometry and $\beta^{-1} \circ \alpha(A) = A$, $\beta^{-1} \circ \alpha(B) = B$ and $\beta^{-1} \circ \alpha(C) = C$. By Theorem 4.2.2, $\beta^{-1} \circ \alpha$ is the identity: $\beta^{-1} \circ \alpha = \iota$. When we compose both sides on the left with β, we get $\alpha = \beta$.   ∎

Theorem 4.2.4   For any two distinct points P and Q in the Euclidean plane, there is exactly one mirror reflection that takes P to Q.

Proof.   With two distinct points we can use Euclid I-10 and I-11 to construct a perpendicular bisector. This perpendicular is unique by Hilbert's axiom III-4. Then the definition of a mirror reflection gives a unique mirror reflection, taking one point to the other.   ∎

Theorem 4.2.5   Every Euclidean plane isometry can be written as the composition of at most three mirror reflections.

Proof.   Let α be any Euclidean plane isometry and A, B, and C be any three noncollinear points in the plane. By Theorem 4.2.3 we only have to find a composition of mirror reflections that together take A, B, and C to the same images as α does, say, P, Q, and R (Fig. 4.9). If $A \neq P$, then by Theorem 4.2.4, there is a mirror reflection μ_1 such that $\mu_1(A) = P$. Let $\mu_1(B) = B'$ and $\mu_1(C) = C'$. If $B' \neq Q$, we repeat this process, finding μ_2, which maps B' to Q. However, we need to prove that μ_2 leaves P fixed. Note that $d(P, Q) = d(A, B) = d(P, B')$. Thus P is on the perpendicular bisector of $\overline{QB'}$, which means that $\mu_2(P) = P$. Hence $\mu_1 \circ \mu_2$ moves A to P, B to Q, and C to some point C''. Finally, we need to move C'' to R. Again, we assume that $C'' \neq R$, use a mirror reflection μ_3, and verify that μ_3 leaves P and Q fixed. Thus, in the general case, $\mu_1 \circ \mu_2 \circ \mu_3$ maps A to P, B to Q, and C to R. If $A = P$, $B' = Q$, or $C'' = R$, then we omit μ_1, μ_2, or μ_3, respectively. The only case not covered by the argument presented is the identity transformation, ι. However, $\iota = \mu \circ \mu$, for any mirror reflection

140 **Chapter 4** **Transformational Geometry**

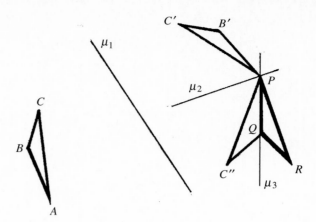

Figure 4.9

μ. Hence every Euclidean plane isometry can be written as the composition of three or fewer mirror reflections. ∎

We can use Theorem 4.2.5 to classify the four possible types of Euclidean plane isometries, which are summarized in Theorem 4.2.7. We have already discussed three kinds of isometries: mirror reflections, rotations, and translations. Later we discuss the other kind of isometry, glide reflections. First, let's relate the three kinds of isometries we know to the composition of one, two, or three mirror reflections in Theorem 4.2.5. The "composition" of just one mirror reflection must be that mirror reflection.

For the composition of two mirror reflections there are three cases for the lines of reflection: The two lines are the same, they are distinct but parallel, and they intersect in a unique point. The first case gives the identity $\mu \circ \mu = \iota$. The second case gives a translation twice as long as the distance between the lines and in a direction perpendicular to them. Figure 4.10 illustrates several subcases. Problem 5 involves the use of congruent triangles to show the arrows in Fig. 4.10 are all parallel and the same length. The third case gives a rotation around the intersection of the two lines, where the angle is twice the angle between the lines. Figure 4.11 illustrates two subcases. Problem 6 involves the use of congruent triangles to show that the rotations in Fig. 4.11 are all the same angle and around the same point. A mirror reflection switches orientation. Because the identity, translations, and rotations are composed from two mirror reflections, they do not switch orientation and we say that they are *direct* isometries.

The remaining option, the composition of three mirror reflections, switches orientation three times and so is an *indirect* isometry, like a mirror reflection. Theorem 4.2.6 shows that three mirror reflections result either in a mirror reflection or a glide reflection, defined as follows.

Definition 4.2.3   A Euclidean plane isometry γ is a *glide reflection* iff there is a line k such that γ is the composition of the mirror reflection over k and a translation parallel to k.

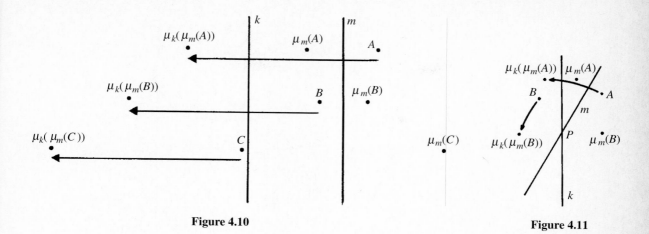

Figure 4.10

Figure 4.11

Example 5 Figure 4.12 indicates that a glide reflection can be written as the composition of a mirror reflection and a translation. ●

Exercise 4 Explain how a mirror reflection is a special case of a glide reflection. Explain why no other glide reflections have fixed points. Explain why only the line k is stable in other glide reflections.

Theorem 4.2.6 The composition of three mirror reflections is either a mirror reflection or a glide reflection.

Proof. By Theorem 4.2.3, the images of three noncollinear points determine an isometry. Let β be the composition of three mirror reflections and take A to A', B to B', and C to C'. Construct the midpoints M_1 and M_2 of the two line segments $\overline{AA'}$ and $\overline{BB'}$.

Case 1 Assume that $M_1 \neq M_2$. Let μ_k be the mirror reflection over the line $k = \overleftrightarrow{M_1 M_2}$ (Fig. 4.13). Let X be the intersection of k with the line through A and $\mu_k(A)$. Then $\triangle AM_1X$ is similar to $\triangle AA'\mu_k(A)$ by Theorem 1.5.4. Theorem 1.5.1 then implies that $\overleftrightarrow{\mu_k(A)A'}$ is parallel to k. Similarly, $\triangle BM_2Y \sim \triangle BB'\mu_k(B)$ and $\overleftrightarrow{\mu_k(B)B'}$ is parallel to k, where Y is the intersection of k with $\overline{B\mu_k(B)}$. Thus $\overleftrightarrow{\mu_k(A)A'}$ is parallel to $\overleftrightarrow{\mu_k(B)B'}$. Note that $\triangle A'B'C' \cong \triangle \mu_k(A)\mu_k(B)\mu_k(C)$ because both are congruent with $\triangle ABC$.

Let τ be the translation taking $\mu_k(A)$ to A'.

Claim. τ takes $\mu_k(B)$ to B' and $\mu_k(C)$ to C', so $\beta = \tau\mu_k$, a glide reflection.

Figure 4.12

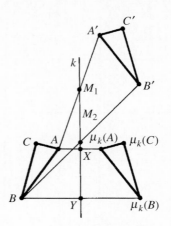

Figure 4.13

For the moment, let τ take $\mu_k(B)$ to B'' and $\mu_k(C)$ to C''. Then $\mu_k(B)$, B', and B'' are on the same line parallel to $\overleftrightarrow{\mu_k(A)P}$. Only two points on $\overleftrightarrow{\mu_k(B)B'}$ have a distance from A' of $d(A', B'')$: B'' or the point Q, for which M_1 is the midpoint of $\overline{BQ}$. However, we assumed for case 1 that $M_1 \neq M_2$, so $B' \neq Q$. Because $d(A', B') = d(A', B'')$, $B' = B''$. By Problem 4 there are only two places for C'' to be, one of which is C'. However, the triangle $\triangle A'B'C''$ has the same orientation as $\triangle A'B'C'$, so $C'' = C'$. Thus β is the composition of μ_k with the translation τ. If $\mu_k(A) = A'$, then the translation is the identity and $\beta = \mu_k$. Otherwise β is a glide reflection.

Case 2 Assume that $M_1 = M_2$. If the midpoint of $\overline{CC'}$ is this same point, then the isometry is a rotation of 180° around the point, but that is a direct isometry and so is not β. Hence we can assume that the midpoint of A and A' differs from the midpoint of C and C'. Now we can apply the reasoning of case 1 here. ∎

Theorem 4.2.7 There are four types of Euclidean plane isometries: mirror reflections, translations, rotations, and glide reflections.

Proof. From Theorem 4.2.5, every isometry can be written as the composition of one, two, or three mirror reflections. Verify that Problems 5 and 6, Theorem 4.2.6, and the preceding discussion cover all the possibilities for isometries. ∎

4.2.1 Klein's definition of geometry

Felix Klein in his famous *Erlanger Programm*, given in 1872, used groups of transformations to give a definition of geometry: Geometry is the study of those properties of a set that are preserved under a group of transformations on that set. Klein realized that we can, for example, investigate the properties of Euclidean geometry by studying isometries. Thus he would say that the area of a triangle is a Euclidean property because area is preserved by isometries. That is, for any $\triangle A$ and any isometry σ, $\triangle A$ and $\sigma(\triangle A)$ have the same area. Under Klein's definition, congruence and measures of lengths and

FELIX KLEIN

At the age of 23 Felix Klein (1849–1925) gave his inaugural address as a professor at the University of Erlangen, the talk for which he is best remembered today. This presentation, the Erlanger Programm, raised transformation groups in geometry from an important concept to a unifying theme. Using transformation groups, Klein showed that the various non-Euclidean geometries, projective geometry, and Euclidean geometry were closely related, not competing subjects.

When Klein was 17 he became the assistant to Julius Plücker, a physicist and geometer. Inspired by Plücker's approach, Klein always emphasized the physical and intuitive aspects of mathematics over rigor and abstraction. After Plücker's death in 1868, Klein went to Berlin to finish his graduate work. There he met Sophus Lie, who became a lifelong friend. They went to Paris in 1870 for further studies. Both men were deeply influenced there by the possibility that group theory could unify mathematics. Indeed, Klein's Erlanger Programm in 1872 is a direct outgrowth of this inspiration.

Klein developed numerous theoretical models in geometry, including the Klein bottle, a curious two-dimensional surface with no inside and requiring four dimensions to realize it. The following figure illustrates that a three-dimensional representation of a Klein bottle intersects itself, unlike the theoretical shape.

A three-dimensional representation of a Klein bottle.

He distinguished single elliptic geometry from spherical geometry and investigated its models and transformations. To connect projective, Euclidean, and non-Euclidean geometries by using transformations, Klein developed the model of hyperbolic geometry named for him. He started to develop what we now call the Poincaré model, but he failed to see its connection to inversions that Henri Poincaré (1854–1912) found.

Klein was impressed with Poincaré's work and corresponded with him. However, their common interest soon became a fierce rivalry. Both produced important mathematics, but Klein suffered a nervous breakdown from the intense strain and felt that he had lost the contest. After recovering, Klein produced some mathematics and wrote several books. However, he focused on new tasks that called on his superior administrative abilities, building up mathematics research and education at his university, throughout Germany, and even in the United States.

angles are Euclidean properties, as is the shape of a figure. However, the orientation of figures isn't a Euclidean property because mirror reflections and glide reflections switch orientation. Also, verticality isn't a Euclidean property because some isometries, such as a rotation of 45°, tilt vertical lines. If we wanted to study orientation or verticality, we would need to use different groups of transformations, and, according to Klein, we would be studying a different geometry.

PROBLEMS FOR SECTION 4.2

1. Suppose that an isometry β takes $(1, 0)$ to $(-1, 0)$, $(2, 0)$ to $(-1, -1)$ and $(0, 2)$ to $(1, 1)$, respectively. Find the images of $(0, 0)$ and $(2, 2)$ and of a general point (x, y). Draw a figure showing these points and their images.

2. Outline the original placement of a small rectangular piece of paper on a larger piece of paper. Label the corners of both the small rectangle and the outline A, B, C, and D so that you can determine the rectangle's movements. Note that the centers of rotation are on the outline and do not move.

   a) Rotate the small rectangle 180° around A and then 180° around C on the outline. Describe the resulting transformation.

   b) Return the small piece of paper to its starting position and repeat part (a) but switch the order of the rotations. Describe how this new transformation differs from the one in part (a).

   c) Repeat parts (a) and (b) but use rotations of 90° at A and C.

   d) Repeat part (c) but rotate the rectangle 90° around A followed by a rotation of $-90°$ around C.

   e) Repeat part (c) with various angles and centers of rotations. Make a conjecture about the resulting transformations.

3. a) If μ_k is a mirror reflection over the line k and τ is a translation in the direction of k, investigate whether $\mu_k \circ \tau = \tau \circ \mu_k$ and justify your answer. [*Hint:* It may help to do this first physically with a triangle placed on a sheet of paper. Draw the line k on the paper. Geometer's Sketchpad or CABRI also will help.]

   b) Find three mirror reflections whose composition is a glide reflection.

   c) What is the composition of a glide reflection with itself? Justify your answer.

4. a) If α is an isometry which fixes two points, prove that α is the identity or the mirror reflection over the line through the fixed points.

   b) If α and β are isometries such that $\alpha(A) = \beta(A)$ and $\alpha(B) = \beta(B)$, prove that $\alpha = \beta$ or $\alpha = \beta \circ \mu$, where μ is the mirror reflection over the line $\overleftrightarrow{AB}$.

5. Let k and m be parallel with a perpendicular distance of d between them and μ_k and μ_m be the mirror reflections over these lines. Prove that $\mu_k \circ \mu_m$ is a translation of length $2d$ in the direction perpendicular to k and m. [*Hint:* In Fig. 4.10 select the midpoint of A and $\mu_m(A)$, as well as another point on m. These points form congruent triangles with A and $\mu_m(A)$. Repeat with the line k. Analyze other cases similarly.] Also prove that $\mu_m \circ \mu_k$ and $\mu_k \circ \mu_m$ are inverses.

6. Let k and m intersect at point P and form an angle of $r°$ and μ_k and μ_m be the mirror reflections over these lines. Prove that $\mu_k \circ \mu_m$ is a rotation of $2r°$ around P. [*Hint:* In Fig. 4.11 let Q be the midpoint of A and $\mu_m(A)$. Use triangles $\triangle PAQ$ and $\triangle P\mu_m(A)Q$. Continue as in Problem 5. Decide what other cases, besides those in Fig. 4.11, can occur.] Also prove that $\mu_m \circ \mu_k$ and $\mu_k \circ \mu_m$ are inverses.

7. Let Q be between P and R on a Euclidean line. Explain why, for any isometry α, $\alpha(Q)$ is between $\alpha(P)$ and $\alpha(R)$ and all three are on a line.

8. Let ρ_1 and ρ_2 be any two rotations. Prove that their composition $\rho_1 \circ \rho_2$ is a translation, a rotation, or the identity. Find the conditions that are necessary and sufficient for the composition $\rho_1 \circ \rho_2$ to be a translation.

9. Let τ_1 and τ_2 be two translations and P and Q be two points. How are $\tau_2 \circ \tau_1$ and $\tau_1 \circ \tau_2$ related? Draw a figure showing P, Q, $\tau_1(P)$, $\tau_1(Q)$, $\tau_2(\tau_1(P))$, and $\tau_2(\tau_1(Q))$. Prove that the composition $\tau_1 \circ \tau_2$ is a translation. [*Hint:* Use SAS.]

10. Prove that **D**, the set of all direct isometries of the Euclidean plane, is a transformation group. Note that **D** preserves orientation in addition to all Euclidean properties.

11. Let **V** be the set of all Euclidean plane isometries that take vertical lines to vertical lines. Describe **V** and prove that it is a transformation group.

12. Define two sets $A = \{A_i : i \in I\}$ and $B = \{B_i : i \in I\}$ to be *congruent*, written $A \cong B$, iff for all $i, j \in I$, $d(A_i, A_j) = d(B_i, B_j)$.

   a) Why are two triangles congruent under this definition also congruent under the usual definition? [*Hint:* Consider the vertices of the triangles.]

   b) Why are any two lines congruent under this definition?

   c) Why are circles with equal radii congruent under this definition?

13. Define two sets A and B to be *isometric* iff there is an isometry α such that $\alpha(A) = B$. The definition of congruent sets in Problem 12 guarantees that isometric sets are congruent. Show the converse in Euclidean geometry: For any two congruent Euclidean plane sets $A = \{A_i : i \in I\}$ and $B = \{B_i : i \in I\}$, there is an isometry taking A to B. [*Hint:* Use Theorem 4.2.3, its proof, and Theorem 4.2.5.]

4.3 ALGEBRAIC REPRESENTATION OF TRANSFORMATIONS

How can a computer display various viewpoints, zooming in and rotating as the user desires? Computer graphics software uses matrices and linear algebra extensively to compute such geometric transformations. (See Section 6.6 for more information, including the use of perspective.) This algebraic representation helped mathematicians, physicists, and others in many fields long before the advent of computers. Matrices and linear algebra also give a deeper insight into transformations. For convenience we often identify a transformation by its matrix.

If M is a matrix and P is a point, we write $Q = M \cdot P$ or $Q = MP$ to indicate that Q is the image of P by the transformation represented by M. This notation fits well with that of functions, $y = f(x)$, and follows standard linear algebra notation. However, it means that points are column vectors—for example $\begin{bmatrix} 2 \\ 3 \end{bmatrix}$, rather than the familiar ordered pair $(2, 3)$. Column vectors are awkward to print in the body of the text, so we will write the column vector $\begin{bmatrix} x_1 \\ x_2 \\ \vdots \\ x_n \end{bmatrix}$ as $(x_1, x_2, \ldots, x_n)$. We use row vectors, such as $[x_1, x_2, \ldots, x_n]$, to represent lines.

Example 1 Recall that $\begin{bmatrix} a & b \\ d & e \end{bmatrix} \begin{bmatrix} x \\ y \end{bmatrix} = \begin{bmatrix} ax + by \\ dx + ey \end{bmatrix}$. Thus the matrix $\begin{bmatrix} a & b \\ d & e \end{bmatrix}$ defines a transformation taking (x, y) to $(ax + by, dx + ey)$. The matrix $R = \begin{bmatrix} 0 & -1 \\ 1 & 0 \end{bmatrix}$ represents the rotation of 90° around the origin (Fig. 4.14). The matrix $D = \begin{bmatrix} 2 & 0 \\ 0 & 2 \end{bmatrix}$ doubles each point's distance from the origin (Fig. 4.15). The matrix $M = \begin{bmatrix} 0.6 & 0.8 \\ 0.8 & -0.6 \end{bmatrix}$ represents the mirror reflection over the line $y = \frac{1}{2}x$ (Fig. 4.16). Select various points, such as $(-1, 1)$ and $(1, 2)$, and find their images under these matrices to verify the preceding statements. ●

146   Chapter 4   Transformational Geometry

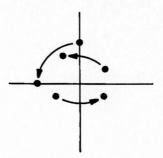

Figure 4.14

Exercise 1   In Example 1 find the product $R \cdot R$ and verify that it represents a rotation of 180° around the origin. Similarly, find $M \cdot M$ and verify that it represents the identity transformation. Find the inverse of D and verify that it halves each point's distance from the origin.

The preceding examples indicate that two-by-two matrices can represent a variety of plane transformations. However, they have a fatal drawback for our purposes: They all fix the origin $(0, 0)$. Thus these matrices cannot represent nontrivial translations and many other transformations. For example, the translation $\tau(x, y) = (x + 3, y + 2)$ takes $(0, 0)$ to $(3, 2)$. The key difference of τ from the preceding examples is the addition of constants. Mathematicians have devised a simple way around this problem by using as their model the plane $z = 1$ in $\mathbf{R}^3$. Clearly, it has the same geometric properties as $\mathbf{R}^2$, which, in effect, is the plane $z = 0$. However, $z = 1$ has the key algebraic advantage

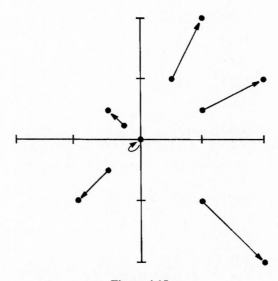

Figure 4.15

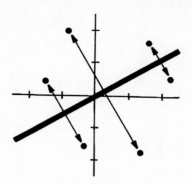

Figure 4.16

that all its points $(x, y, 1)$, including the new "origin" $(0, 0, 1)$, can be moved by 3×3 matrices. The third coordinate of $(x, y, 1)$ does not really "do" anything. For example, the distance between two points still depends only on their first two coordinates.

Interpretation By a *point* of the Euclidean plane we mean any triple $(x, y, 1)$, where x and y are real numbers. The *distance* between $(x, y, 1)$ and $(u, v, 1)$ is $\sqrt{(x-u)^2 + (y-v)^2}$.

Exercise 2 Verify that the matrix $\begin{bmatrix} a & b & 0 \\ d & e & 0 \\ 0 & 0 & 1 \end{bmatrix}$ maps $(x, y, 1)$ to $(ax+by, dx+ey, 1)$ and so corresponds to the same transformation given in Example 1 as $\begin{bmatrix} a & b \\ d & e \end{bmatrix}$. Verify that the matrix $\begin{bmatrix} 1 & 0 & 3 \\ 0 & 1 & 2 \\ 0 & 0 & 1 \end{bmatrix}$ represents the translation $\tau(x, y, 1) = (x+3, y+2, 1)$.

It is no accident that the bottom row of both matrices in Exercise 2 is [0 0 1]. This restriction ensures that a general 3×3 matrix $M = \begin{bmatrix} a & b & c \\ d & e & f \\ g & h & i \end{bmatrix}$ maps the plane $z = 1$ to itself. That is, $M \cdot (x, y, 1)$ must equal $(\_, \_, 1)$. This forces M to have $g = h = 0$ and $i = 1$. A theorem of linear algebra states that a linear transformation is one-to-one and onto the whole space iff its matrix is invertible or, equivalently, the determinant is not zero. Hence an invertible 3×3 matrix represents a plane transformation provided that its bottom row is [0 0 1]. We call the transformations for the plane $z = 1$ *affine transformations* to distinguish them from linear transformations, which leave the origin of the space fixed.

Interpretation By *(plane) affine matrix* we mean any invertible 3×3 matrix whose bottom row is [0 0 1].

Exercise 3 Verify that an affine matrix leaves $(0, 0, 1)$ fixed provided that the last column of the matrix is $\begin{bmatrix} 0 \\ 0 \\ 1 \end{bmatrix}$. Rewrite the matrices of Example 1 as affine matrices. Verify that $ae - bd$ is the determinant of $\begin{bmatrix} a & b & c \\ d & e & f \\ 0 & 0 & 1 \end{bmatrix}$.

Linear algebra provides an elegant interpretation of lines. Usually, a line is the set of points (x, y) satisfying some equation $ax + by + c = 0$. However, with triples $(x, y, 1)$ for points, the 1 has a natural place in that equation: $ax + by + c \cdot 1 = 0$. If we replace the left-hand side by the product of a row vector and a column vector, the equation $[a, b, c] \begin{bmatrix} x \\ y \\ 1 \end{bmatrix} = 0$ suggests that we can represent a line by three coordinates $[a, b, c]$.

In $ax + by + c = 0$ if $b \neq 0$, we can solve the equation for y to get the familiar form (Exercise 4). If $b = 0$ and $a \neq 0$, we have a vertical line. (If $b = 0 = a$, the equation reduces to $c = 0$, which is not a line.)

Exercise 4 Verify that the familiar equation $y = mx + b$ corresponds to $[m, -1, b]$. Verify that vertical lines $x = c$ correspond to $[1, 0, -c]$.

Note that for any nonzero k, $[ka, kb, kc]$ represents the same line as $[a, b, c]$ because $[ka, kb, kc](x, y, 1) = k[a, b, c](x, y, 1)$. Technically, then, a line is "the equivalence class of all row vectors differing by a nonzero scalar." However, for convenience, we use a row vector $[a, b, c]$ as the name of a line. This interpretation of lines ensures that affine transformations map lines to lines and, in addition, that a general transformation of the plane taking lines to lines must be one of these affine transformations. We know that an affine matrix M moves a point P to MP, but where M takes a given line isn't obvious. Experiment with examples before reading the answer in Theorem 4.3.1.

Interpretation A *line* is a row matrix $[a, b, c]$ such that not both a and b are 0. The point $(x, y, 1)$ is *on* the line $[a, b, c]$ iff their product is 0: $ax + by + c \cdot 1 = 0$. Two row vectors represent the same line iff one is the product of the other by a nonzero scalar.

Exercise 5 Verify that the line $[-2, -1, 1]$ is on the points $(-1, 3, 1)$ and $(2, -3, 1)$. Note that the point $(-2, -1, 1)$ is on the lines $[-1, 3, 1]$ and $[2, -3, 1]$. There is a close relationship between points and lines in this model, which we explore more deeply in Chapter 6.

Theorem 4.3.1 The affine matrix M takes $[a, b, c]$ to the line $[a, b, c]M^{-1}$.

Proof. We need to show that, for any point $(x, y, 1)$ on $[a, b, c]$, the new point $M(x, y, 1)$ is on the proposed image of $[a, b, c]$, namely, $[a, b, c]M^{-1}$. The product of the new line and point is $[a, b, c]M^{-1} \cdot M(x, y, 1) = [a, b, c]I(x, y, 1) = [a, b, c](x, y, 1) = 0$. ∎

4.3 Algebraic Representation of Transformations

4.3.1 Isometries

An affine matrix is readily determined by where it takes the three reference points $O = (0, 0, 1)$, $X = (1, 0, 1)$, and $Y = (0, 1, 1)$. Theorem 4.3.2 uses these points to describe which affine matrices are isometries.

Exercise 6 Verify that the images of the points O, X, and Y under the matrix $M = \begin{bmatrix} a & b & c \\ d & e & f \\ 0 & 0 & 1 \end{bmatrix}$ are $O' = (c, f, 1)$, $X' = (a + c, d + f, 1)$, and $Y' = (b + c, e + f, 1)$.

Theorem 4.3.2 An affine matrix $M = \begin{bmatrix} a & b & c \\ d & e & f \\ 0 & 0 & 1 \end{bmatrix}$ is an isometry iff $M = \begin{bmatrix} \cos\theta & -\sin\theta & c \\ \sin\theta & \cos\theta & f \\ 0 & 0 & 1 \end{bmatrix}$ or $M = \begin{bmatrix} \cos\theta & \sin\theta & c \\ \sin\theta & -\cos\theta & f \\ 0 & 0 & 1 \end{bmatrix}$, for some angle θ.

Proof. ($\Rightarrow$) If M is an isometry, then, for the three points O', X', and Y' of Exercise 6, $\triangle O'X'Y' \cong \triangle OXY$ (Fig. 4.17). The distance $d(O', X')$ is $\sqrt{a^2 + d^2}$. As M is an isometry, $a^2 + d^2 = 1$. Then for some angle θ, $a = \cos\theta$ and $d = \sin\theta$. Similarly, the distance $d(O', Y') = \sqrt{b^2 + e^2}$ forces $b = \cos\phi$ and $e = \sin\phi$, for some angle ϕ. Furthermore, $m\angle X'O'Y' = 90°$. As Fig. 4.17 illustrates, $\phi = \theta \pm 90°$. When $\phi = \theta + 90°$, the isometry is direct. When $\phi = \theta - 90°$, the isometry is indirect. Trigonometry gives $\sin(\theta \pm 90°) = \pm \cos\theta$ and $\cos(\theta \pm 90°) = \mp \sin\theta$.

($\Leftarrow$) See Problem 4. ∎

Exercise 7 Verify that the determinant of $\begin{bmatrix} \cos\theta & -\sin\theta & c \\ \sin\theta & \cos\theta & f \\ 0 & 0 & 1 \end{bmatrix}$ is $+1$ and that the determinant of $\begin{bmatrix} \cos\theta & \sin\theta & c \\ \sin\theta & -\cos\theta & f \\ 0 & 0 & 1 \end{bmatrix}$ is -1.

Isometries split naturally into two classes, as given in Theorem 4.3.2. Exercise 7 shows that determinants identify these classes. The first class, with determinant $+1$, contains the direct isometries, for which θ is the angle of rotation. If $\theta = 0$, the direct

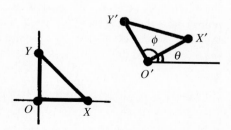

Figure 4.17

Example 2 Find the center of rotation of $\begin{bmatrix} 0 & -1 & 2 \\ 1 & 0 & 3 \\ 0 & 0 & 1 \end{bmatrix}$ and the line of reflection of $M = \begin{bmatrix} 0.6 & 0.8 & 2 \\ 0.8 & -0.6 & -4 \\ 0 & 0 & 1 \end{bmatrix}$. Which lines are stable under M?

Solution. The center of this rotation is a fixed point, say, $(u, v, 1)$, satisfying $\begin{bmatrix} 0 & -1 & 2 \\ 1 & 0 & 3 \\ 0 & 0 & 1 \end{bmatrix} \begin{bmatrix} u \\ v \\ 1 \end{bmatrix} = \begin{bmatrix} u \\ v \\ 1 \end{bmatrix}$. This equation reduces to $-v + 2 = u$ and $u + 3 = v$, or $u = -\frac{1}{2}$ and $v = 2\frac{1}{2}$. We can find the fixed points of M similarly, obtaining the pair of equations $-0.4u + 0.8v + 2 = 0$ and $0.8u - 1.6v - 4 = 0$. The second is a multiple of the first, so we get an infinite family of fixed points, $(u, \frac{1}{2}u - 2\frac{1}{2}, 1)$. That is, all the fixed points are on the line of reflection $[\frac{1}{2}, -1, -2\frac{1}{2}]$ or, more familiarly, $y = \frac{1}{2}x - 2\frac{1}{2}$. The other stable lines of M need a more general approach because for any nonzero multiple λ, $\lambda[a, b, c]$ is the same line as $[a, b, c]$. We need to solve $[a, b, c]M^{-1} = \lambda[a, b, c]$ but first have to find the possible values of λ. We need $[a, b, c]M^{-1} = \lambda[a, b, c] = \lambda[a, b, c]I$ or, equivalently, $[a, b, c](M^{-1} - \lambda I) = 0$. That is, the matrix $(M^{-1} - \lambda I)$ must have a determinant of zero. For $M = \begin{bmatrix} 0.6 & 0.8 & 2 \\ 0.8 & -0.6 & -4 \\ 0 & 0 & 1 \end{bmatrix} = M^{-1}$, $M^{-1} - \lambda I = \begin{bmatrix} 0.6 - \lambda & 0.8 & 2 \\ 0.8 & -0.6 - \lambda & -4 \\ 0 & 0 & 1 - \lambda \end{bmatrix}$ and the determinant is $(1 - \lambda)[(0.6 - \lambda)(-0.6 - \lambda) - (0.8)(0.8)] = (1 - \lambda)(\lambda^2 - 1)$. Thus the possibilities are $\lambda = 1$ (as a double root) and $\lambda = -1$. The root $\lambda = -1$ gives the line of reflection, $[\frac{1}{2}, -1, -2\frac{1}{2}]$. For $\lambda = 1$, we get $[a, b, c] \begin{bmatrix} 0.6 & 0.8 & 2 \\ 0.8 & -0.6 & -4 \\ 0 & 0 & 1 \end{bmatrix} = [a, b, c]$, which reduces to $a = 2b$. This outcome gives a family of parallel lines $[2, 1, c]$ or, equivalently, $y = -2x - c$, all perpendicular to the line of reflection. Note that the double root gives a family of stable lines, whereas the single root gives just one line. Actually, the fact that $\lambda = 1$ is a double root for M ensures that there is a family of fixed points, as we found previously. Problem 7 shows that the only values of λ for any isometry are 1 and -1. In linear algebra the λ are called *eigenvalues*, and the solutions $[a, b, c]$ are called *eigenvectors*. ●

Example 3 Find the matrix M representing a rotation of θ around the point $(u, v, 1)$.

Solution. We build the rotation around $(u, v, 1)$ from the translation $T = \begin{bmatrix} 1 & 0 & u \\ 0 & 1 & v \\ 0 & 0 & 1 \end{bmatrix}$

4.3 Algebraic Representation of Transformations

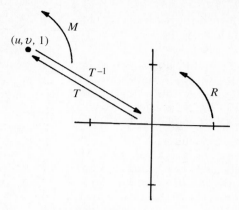

Figure 4.18

that moves $(0, 0, 1)$ to $(u, v, 1)$ and the rotation $R = \begin{bmatrix} \cos\theta & -\sin\theta & 0 \\ \sin\theta & \cos\theta & 0 \\ 0 & 0 & 1 \end{bmatrix}$ of θ around $(0, 0, 1)$. In effect we first move the point $(u, v, 1)$ to $(0, 0, 1)$, rotate there, and move back (Fig. 4.18). That is, we claim that $M = TRT^{-1}$. We verify that TRT^{-1} fixes $(u, v, 1)$ and is a direct isometry because its determinant is 1. Hence it is a rotation. Then we verify that the image of the point $(u + 1, v, 1)$ is $(u + \cos\theta, v + \sin\theta, 1)$. Explain why this solution shows the angle of rotation of TRT^{-1} to be θ. ●

PROBLEMS FOR SECTION 4.3

1. For each matrix, decide whether it is a translation, a rotation, a mirror reflection, or a glide reflection and find its fixed points and stable lines.

$$A = \begin{bmatrix} 0.8 & -0.6 & 2 \\ 0.6 & 0.8 & 0 \\ 0 & 0 & 1 \end{bmatrix} \quad B = \begin{bmatrix} 1 & 0 & -2 \\ 0 & 1 & 7 \\ 0 & 0 & 1 \end{bmatrix}$$

$$C = \begin{bmatrix} 0.8 & 0.6 & -1/3 \\ 0.6 & -0.8 & 1 \\ 0 & 0 & 1 \end{bmatrix}$$

$$D = \begin{bmatrix} 0.8 & 0.6 & 2 \\ 0.6 & -0.8 & 1 \\ 0 & 0 & 1 \end{bmatrix}$$

$$E = \begin{bmatrix} \sqrt{2}/2 & -\sqrt{2}/2 & 2 \\ \sqrt{2}/2 & \sqrt{2}/2 & 0 \\ 0 & 0 & 1 \end{bmatrix}$$

2. Repeat Problem 1 for the matrices $A \cdot B$, $B \cdot A$, $B \cdot C$, and $C \cdot D$.

3. **a)** Find the matrix for the rotation of $30°$ with a center of rotation of $(2, 3, 1)$.

   b) Find the matrix of the mirror reflection over the line $y = 2x$.

   c) Repeat part (b) for the line $y = 2x + 1$.

   d) Find the matrices of all glide reflections over the line $y = 2x$.

4. Let $M_1 = \begin{bmatrix} \cos\theta & -\sin\theta & c \\ \sin\theta & \cos\theta & f \\ 0 & 0 & 1 \end{bmatrix}$,

   $M_2 = \begin{bmatrix} \cos\theta & \sin\theta & c \\ \sin\theta & -\cos\theta & f \\ 0 & 0 & 1 \end{bmatrix}$, $P = (u, v, 1)$, and

   $Q = (s, t, 1)$. Show that $d(P, Q) = d(M_1 P, M_1 Q) = d(M_2 P, M_2 Q)$. [*Hint:* $\cos^2(\theta) + \sin^2(\theta) = 1$.]

5. **a)** Show that $\begin{bmatrix} \cos\theta & \sin\theta & 0 \\ \sin\theta & -\cos\theta & 0 \\ 0 & 0 & 1 \end{bmatrix}$ switches $(1, 0, 1)$ and $(\cos\theta, \sin\theta, 1)$. Explain why this isometry must be the mirror reflection over the line through the origin that makes an angle of $\theta/2$ with the x-axis.

b) Let M and N be mirror reflections over lines through the origin. Show that their product in either order is a rotation around the origin.

c) In Theorem 4.3.2 let M and N be two general indirect isometries with the same angle θ. (Thus the last column of each matrix should have general variables.) What can you say about their product? Interpret this conclusion geometrically.

6. Write general matrices M for a rotation of θ and N for a rotation of ϕ. (Thus the last column of each matrix should have general variables.) Use trigonometry to show that MN and NM both represent rotations of $(\theta + \phi)$, unless $(\theta + \phi)$ is a multiple of $360°$. What happens in this situation?

7. a) Prove that $\lambda = 1$ is always an eigenvalue of an affine matrix.

b) Prove that an indirect isometry has $\lambda = 1$ as a double eigenvalue and $\lambda = -1$ as an eigenvalue.

c) Prove that a matrix representing a direct isometry satisfies one of the following three situations, depending on the value of θ: (i) $\lambda = 1$ is a triple eigenvalue (θ is a multiple of $360°$); (ii) $\lambda = 1$ is an eigenvalue, and $\lambda = -1$ is a double eigenvalue (θ is an odd multiple of $180°$); or (iii) $\lambda = 1$ is an eigenvalue, and the other eigenvalues are complex (all other values of θ). Thus the only possible real eigenvalues of an isometry are $\lambda = 1$ and $\lambda = -1$.

8. The *central symmetry with respect to A* takes a point P to the point P', where A is the midpoint of $\overline{PP'}$ (Fig. 4.19).

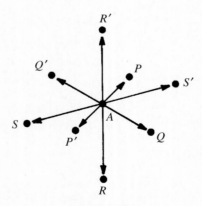

Figure 4.19

9. a) Find the matrix for the central symmetry with respect to $(0, 0, 1)$.

b) Repeat part (a) for $(u, v, 1)$. [*Hint:* Find the image of $(0, 0, 1)$.]

c) Verify that the matrix in part (b) has a determinant of $+1$ and its square is the identity. Explain what these algebraic properties mean geometrically.

d) Find the stable lines of the general central symmetry from part (b).

e) Find the composition of two central symmetries, one with respect to $(u, v, 1)$ and the other with respect to $(s, t, 1)$. Identify the type of isometry for this composition and explain what is happening geometrically. What happens when you switch the order?

f) Prove that a central symmetry sends every line to a line parallel to itself.

g) Prove that an isometry sends every line to a line parallel to itself iff that isometry is either a central symmetry or a translation.

h) Show that the set of translations and central symmetries form a transformation group. [Part (g) shows these isometries preserve the direction of a line.]

9. Let T be the matrix for a translation and M for an isometry. Explore the idea behind Example 3: TMT^{-1} represents a transformation essentially the same as M. (TMT^{-1} is called the *conjugate* of M by T.)

a) If M represents a rotation of $m°$ with any center, prove that TMT^{-1} is a rotation with the same angle.

b) If M represents the mirror reflection over the line k, prove that TMT^{-1} represents a mirror reflection by showing that its determinant is -1 and when multiplied by itself it gives the identity. Explain why these algebraic properties force TMT^{-1} to be a mirror reflection.

c) What can you say about TMT^{-1} if M is a translation?

10. a) Show the set of translations to be a transformation group.

b) Show the set of rotations fixing $(0, 0, 1)$ to be a transformation group.

c) Show the set of rotations fixing a point A to be a transformation group. [*Hint:* Use Example 3.]

d) Show the set of isometries fixing a given point A to be a transformation group.

e) Show the set of isometries leaving a line stable to be a transformation group.

f) Show that the indirect isometries are not a transformation group. Which properties for a transformation group fail?

11. a) Prove that an invertible matrix M never has $\lambda = 0$ as an eigenvalue.

b) If $\lambda \neq 0$ is an eigenvalue of an affine matrix M, prove that $1/\lambda$ is an eigenvalue of M^{-1}.

4.4 SIMILARITIES AND AFFINE TRANSFORMATIONS

4.4.1 Similarities

Intermediate between isometries and affine transformations are *similarities* (also called *similitudes*), the transformations corresponding to similar figures. (See Section 1.5.)

Definition 4.4.1 A transformation σ of the Euclidean plane is a *similarity* iff there is a positive real number r such that for all points P and Q in the plane, $d(\sigma(P), \sigma(Q)) = r \cdot d(P, Q)$. The number r is the *scaling ratio* of σ.

Example 1 In Fig. 4.20, we can transform the smaller triangle into the larger similar triangle by rotating it 90°, translating it, and then using a scaling ratio of $r = 1.5$. ●

Theorem 4.4.1 The set of similarities forms a transformation group.

Proof. See Problem 2. ∎

Example 2 The matrix $\begin{bmatrix} r & 0 & 0 \\ 0 & r & 0 \\ 0 & 0 & 1 \end{bmatrix}$ takes $(x, y, 1)$ to $(rx, ry, 1)$. Thus the origin $(0, 0, 1)$ is fixed, and all points expand or contract with respect to the origin by a scaling ratio of r. ●

Theorem 4.4.2 An affine matrix M represents a similarity iff
$$M = \begin{bmatrix} r \cos\theta & \mp r \sin\theta & a \\ r \sin\theta & \pm r \cos\theta & b \\ 0 & 0 & 1 \end{bmatrix}, \text{ for some } r > 0.$$

Proof. See Problem 4. ∎

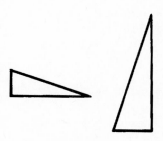

Figure 4.20

Exercise 1 Show that the matrix form of a similarity can be written as the product of an isometry and the matrix given in Example 2.

Theorem 4.4.3 Similarities preserve angle measures and the proportions of all distances. If a similarity has a scaling ratio of r, then the area of the image of a convex polygon is r^2 times the area of the original polygon.

Proof. See Problem 5. ∎

Theorem 4.4.4 A similarity with a scaling ratio of $r \neq 1$ has a unique fixed point.

Proof. Let M be the similarity matrix and $(u, v, 1)$ be a candidate for a fixed point. Theorem 4.4.2 gives one of the following systems of two equations in the two unknowns u and v:

$$\begin{cases} r\cos(A)u - r\sin(A)v + a = u \\ r\sin(A)u + r\cos(A)v + b = v \end{cases} \text{ and } \begin{cases} r\cos(A)u + r\sin(A)v + a = u \\ r\sin(A)u - r\cos(A)v + b = v \end{cases}.$$

The first system becomes $\begin{cases} (r\cos A - 1)u - r\sin(A)v = -a \\ r\sin(A)u + (r\cos A - 1)v = -b \end{cases}$, which has a unique solution when the determinant isn't zero. The determinant is $(r\cos A - 1)^2 + r^2 \sin^2 A = r^2 - 2r\cos A + 1$. By the quadratic formula, $r^2 - 2r\cos A + 1 = 0$ only if $r = (2\cos A \pm \sqrt{4\cos^2 A - 4})/2$. The value under the $\sqrt{}$ is negative except when $\cos^2 A = 1$, which forces $r = \pm 1$. As $r > 0$, the determinant can be 0 only when $r = 1$. Hence, if $r \neq 1$, there must be a unique fixed point.

Similarly, the second system reduces to a system whose determinant is $1 - r^2$. Again, for $r > 0$ and $r \neq 1$, we have a nonzero determinant and so a unique fixed point. ∎

4.4.2 Affine transformations

Theorem 4.4.5 The set of affine transformations is a transformation group. Affine transformations preserve lines, parallelism, betweenness, and proportions on a line.

Proof. We prove betweenness and proportions on a line. (See Problem 9 for the rest.) From Section 2.5, a point R (considered as a vector) is on line $l = \overleftrightarrow{PQ}$ iff $R = P + r(Q - P)$, where r is a real number. (For example, $r = 0$ gives P and $r = 1$ gives Q.) In effect, the values of r give the coordinates for the points on line l, and $|r|$ is the ratio of the length of $\overline{PR}$ to $\overline{PQ}$. The point R is between P and Q iff $0 \leq r \leq 1$. Note that any affine transformation α is a linear transformation, so $\alpha(R) = \alpha(P) + r(\alpha(Q) - \alpha(P))$, which is sufficient to show that α preserves betweenness and proportions on a line. ∎

Example 3 Verify that the affine matrix $\begin{bmatrix} 3 & 0 & 0 \\ 0 & 2 & 0 \\ 0 & 0 & 1 \end{bmatrix}$ triples every x-coordinate, doubles every y-coordinate, has a determinant of 6, and increases the area of every triangle by a factor of 6.

SOPHUS LIE

The Norwegian Sophus Lie (1842–1899) achieved international status in mathematics for his profound work in continuous transformation groups. Lie met Felix Klein in Berlin, and they quickly became lifelong friends. Together they went to Paris in 1870 and studied groups, but the outbreak of the Franco–Prussian War in 1871 ended their studies. Lie decided to spend his enforced vacation hiking in the Alps. As a tall, blond stranger with poor French he was soon arrested as a spy. He spent a month in prison, working on mathematics problems. When he was freed, in part owing to a French mathematician's efforts, he continued his hiking tour.

Both Lie and Klein were deeply influenced by the possibility that group theory could unify mathematical thinking. Lie first applied transformation groups in the field of differential equations, classifying solutions. He searched more broadly for an understanding of all continuous transformation groups, a goal that hasn't yet been achieved. He pioneered the study of these groups, now called Lie groups. He revealed the geometric structure of these groups and went on to develop algebras, also named for him, that matched these groups. Lie groups and Lie algebras are essential ideas in quantum mechanics, a part of physics, as well as in mathematics.

Lie applied his profound understanding of transformations to solve in 1893 a geometry problem posed by Hermann von Helmholtz. Helmholtz sought to use curvature to characterize all continuous homogeneous geometries, that is, geometries in which rigid bodies could be freely moved. Lie used transformations to prove that the only such spaces were Euclidean, hyperbolic, spherical, and single elliptic geometries, in any number of dimensions.

Solution. We leave all but the last part to you. For the last part, note that every triangle without a horizontal side can be split into two triangles, each with a horizontal side (Fig. 4.21). Then each triangle has its base tripled and its height doubled. Hence its area is multiplied by 6. As this example suggests, the absolute value of the determinant of an affine matrix is the scaling factor for areas, as is the case for similarities. ●

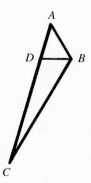

Figure 4.21

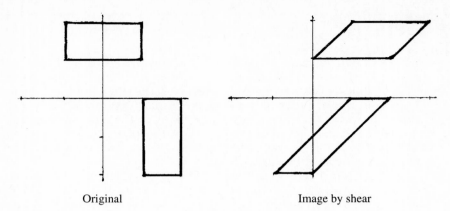

Original   Image by shear

Figure 4.22

Example 4   The affine matrix $\begin{bmatrix} 1 & 1 & 0 \\ 0 & 1 & 0 \\ 0 & 0 & 1 \end{bmatrix}$ is an example of a *shear* transformation, where all points move in parallel, but have different displacements (Fig. 4.22). Note that the determinant of this matrix is 1, preserving area. Geologic shears display the same effect on parts of the earth's crust as geometric shears do on the plane.

Shears and other affine transformations can distort shapes—for example, altering a circle to become an ellipse. Theorem 4.4.6 gives one limit to the amount of distortion of affine transformations: A convex set is always mapped to a convex set. (The converse holds as well: Transformations of the Euclidean plane that preserve convexity are affine transformations.)   ●

Theorem 4.4.6   An affine transformation preserves convexity.

Proof.   Suppose that α is any affine transformation and that A is any convex set. We must show that $\alpha(A)$ also is convex. That is, for X' and Y' any two points in $\alpha(A)$ and Z', any point between them, we must show that Z' is in $\alpha(A)$. By definition of $\alpha(A)$, there are X and Y in A such that $\alpha(X) = X'$ and $\alpha(Y) = Y'$. Furthermore, there is a unique Z such that $\alpha(Z) = Z'$. We need only show that Z is between X and Y because then the convexity of A will guarantee that Z is in A and so Z' is in $\alpha(A)$. Consider α^{-1}, the inverse of α, which is also an affine transformation. By Theorem 4.4.5, α^{-1} preserves betweenness. As Z' is between X' and Y', Z is between X and Y, implying that $\alpha(A)$ is convex.   ∎

4.4.3 Iterated function systems

Whereas the affine image of a convex set is always convex, iterated function systems (abbreviated IFSs) combine affine transformations in an unusual way to yield highly nonconvex sets called *fractals*. The points in a fractal are all the limits of infinitely many applications of the various affine transformations in all possible orders. The figure

4.4 Similarities and Affine Transformations

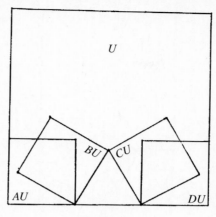

Figure 4.23

introducing this chapter shows such a fractal. A simpler example reveals the idea behind this approach to fractals introduced by Michael Barnsley.

Example 5 Matrices A, B, C, and D shrink the unit square U shown in Fig. 4.23 to the four smaller squares shown. $A = \begin{bmatrix} 1/3 & 0 & 0 \\ 0 & 1/3 & 0 \\ 0 & 0 & 1 \end{bmatrix}$, $B = \begin{bmatrix} \cos 60°/3 & -\sin 60°/3 & 1/3 \\ \sin 60°/3 & \cos 60°/3 & 0 \\ 0 & 0 & 1 \end{bmatrix}$, $C = \begin{bmatrix} \cos 300°/3 & -\sin 300°/3 & 1/2 \\ \sin 300°/3 & \cos 300°/3 & \sqrt{3}/6 \\ 0 & 0 & 1 \end{bmatrix}$, and $D = \begin{bmatrix} 1/3 & 0 & 2/3 \\ 0 & 1/3 & 0 \\ 0 & 0 & 1 \end{bmatrix}$. Thus any starting point P in U must map to some point P' in one of the small squares in Fig. 3.23. In turn, P' is mapped to some point P'' in the 16 even smaller squares in Fig. 4.24. If we continue this process infinitely, any point will end up somewhere along the extremely convoluted Koch curve shown in Fig. 4.25, which fits into all the nested squares. ●

For the practical goal of drawing interesting pictures, we don't need to iterate the process of Example 5 infinitely many times. Even six iterations will shrink the original square to a multitude of squares so small that they look like points. Each iteration quadruples the number of squares and shrinks the sides of each to one-third its previous size. Thus, after six iterations, there will be $4^6 = 4096$ squares, each with a side of length $1/3^6 = 1/729$. Barnsley realized that it is faster still to apply a random sequence of the four defining transformations to one point and plot the resulting images. These images quickly approximate the theoretical curve as closely as the eye can see.

Not every affine matrix can be part of an IFS, for the matrices must contract all distances to ensure limit points to form the fractal. Similarities with a scaling ratio $r < 1$ contract distances, as do some other affine matrices.

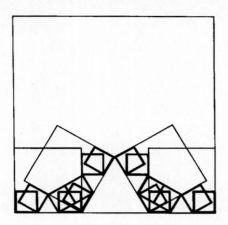

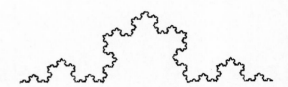

Figure 4.24 **Figure 4.25** The Koch curve.

Example 6 Figure 4.26 shows a distorted Koch curve made from the four matrices
$$A' = \begin{bmatrix} 1/3 & 1/6 & 0 \\ 0 & 1/3 & 0 \\ 0 & 0 & 1 \end{bmatrix}, \quad B' = \begin{bmatrix} \cos 60°/3 & \cos 60°/6 - \sin 60°/3 & 1/3 \\ \sin 60°/3 & \sin 60°/6 + \cos 60°/3 & 0 \\ 0 & 0 & 1 \end{bmatrix},$$
$$C' = \begin{bmatrix} \cos 300°/3 & -\cos 300°/6 - \sin 300°/3 & 1/2 \\ \sin 300°/3 & -\sin 300°/6 + \cos 300°/3 & \sqrt{3}/6 \\ 0 & 0 & 1 \end{bmatrix}, \quad \text{and}$$
$$D' = \begin{bmatrix} 1/3 & -1/6 & 2/3 \\ 0 & 1/3 & 0 \\ 0 & 0 & 1 \end{bmatrix}.$$

Figure 4.27 shows the images of the unit square for each of these matrices, which are products of the shears $\begin{bmatrix} 1 & \pm 1/2 & 0 \\ 0 & 1 & 0 \\ 0 & 0 & 1 \end{bmatrix}$ with the matrices of Example 5. Shears shift rectangles to parallelograms, as shown in Fig. 4.22. ●

Definition 4.4.2 An affine transformation κ is a *contraction mapping* iff there is a real number r with $0 < r < 1$ such that for all points P and Q, $d(\kappa(P), \kappa(Q)) \leq r \cdot d(P, Q)$. An *iterated function system* (IFS) is a finite set of contraction mappings. The set of limit points resulting from infinitely many applications of the contraction mappings in all possible orders is an *IFS fractal*.

Theorem 4.4.7 Every contraction mapping has a unique fixed point. An IFS fractal is a closed and bounded set.

Proof. See Barnsley [2] for a proof. The reasoning of Problem 6(a) applies to any contraction mapping to give a unique fixed point. Note that in Example 4 the fractal is bounded by the sequence of squares. ■

Exercise 2 Explain why the contraction mappings do not form a transformation group.

4.4 Similarities and Affine Transformations

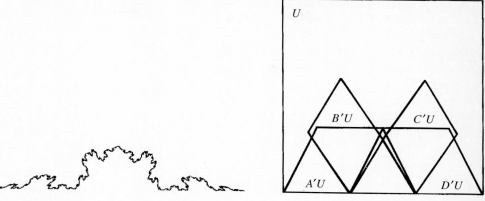

Figure 4.26 A distorted Koch curve.

Figure 4.27

Figures 4.25 and 4.26 and the fractal introducing this chapter were made by using two-place decimals for each matrix entry. In other words only 12 digits per matrix times the number of matrices are needed to encode any such curve. Five matrices, and so 60 digits, are sufficient to produce the intricacies of the fractal introducing this chapter. Barnsley has patented a way to replace an entire picture of a real scene with a collection of IFSs. This method allows him to compress pictorial information into a small set of numbers. A computer program can quickly recover the picture from the set of numbers. Currently, one CD-ROM encyclopedia encodes its 10,000 pictures as IFS. (See Barnsley [2] for a detailed development of IFSs.)

PROBLEMS FOR SECTION 4.4

1. **a)** Let $M = \begin{bmatrix} 0 & -2 & 0 \\ 2 & 0 & 0 \\ 0 & 0 & 1 \end{bmatrix}$ and $X = (1, 0, 1)$. Find and graph the points X, MX, M^2X, and M^3X. Describe the shape of the curve that appears to go through these points.

   b) You can fill in some of the points on the curve of part (a) by finding a matrix S that is the "square root" of M. Verify that $S = \begin{bmatrix} 1 & -1 & 0 \\ 1 & 1 & 0 \\ 0 & 0 & 1 \end{bmatrix}$ satisfies $S^2 = M$. Find and graph on the same axis as part (a) the points SX, S^3X, and S^5X. Do they fill in the curve you described in part (a)?

   c) For $M = \begin{bmatrix} r\cos\theta & -r\sin\theta & 0 \\ r\sin\theta & r\cos\theta & 0 \\ 0 & 0 & 1 \end{bmatrix}$ and $S = \begin{bmatrix} \sqrt{r}\cos(\theta/2) & -\sqrt{r}\sin(\theta/2) & 0 \\ \sqrt{r}\sin(\theta/2) & \sqrt{r}\cos(\theta/2) & 0 \\ 0 & 0 & 1 \end{bmatrix}$, use trigonometry to prove that $S^2 = M$. Explain how S and M move points.

   d) For M in part (c), find a "cube root" C and an "nth root" N and explain why they qualify as a cube root and an nth root of M.

2. Prove Theorem 4.4.1.

3. An affine transformation δ is a *dilation* (or *dilatation*) provided that, for every line k, $k \parallel \delta(k)$.

   a) Prove that the set of dilations form a transformation group.

   b) Show that lines parallel to $[s, t, u]$ have the form $[s, t, w]$ or $[rs, rt, rw]$, for $r \neq 0$.

   c) Show that, if $M = \begin{bmatrix} x & 0 & c \\ 0 & x & f \\ 0 & 0 & 1 \end{bmatrix}$, where $x \neq 0$, then M is a dilation. Describe how M moves points for various x.

d) Show the converse of part (c). Show that dilations are similarities. [*Hint:* Consider the images of the lines [1, 0, 0], [0, 1, 0], and [1, 1, 0].]

4. Prove Theorem 4.4.2. [*Hint:* See Theorem 4.3.2.]

5. Prove Theorem 4.4.3. [*Hint:* Use Exercise 1 and Section 1.5.]

6. Figure 4.28 suggests an alternative proof for Theorem 4.4.4. Let α be any similarity with a scaling ratio $r \neq 1$ and $P = P_0$ be any point. Then WLOG assume that $\alpha(P) \neq P$. Find the fixed point of α as follows.

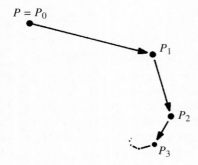

Figure 4.28

a) (Calculus) Case 1: $r < 1$. Let $P_1 = \alpha(P) = \alpha(P_0)$ and, in general, $P_{n+1} = \alpha(P_n)$. What happens to $d(P_{n+1}, P_n)$ as $n \to \infty$? Explain why there must be a point $Q = \lim_{n \to \infty} P_n$. Explain why Q must be a fixed point of α. (A proof of this fact uses analysis and so is beyond the level of this book.)

b) Case 2: $r > 1$. Show that the scaling ratio for α^{-1} is $1/r < 1$. By part (a), α^{-1} has a fixed point. Prove that α has the same fixed points as α^{-1}.

c) Prove that, if $r \neq 1$, there cannot be two distinct fixed points of α.

7. Show that any triangle $\triangle ABC$ can be transformed to any other triangle $\triangle PQR$ by some affine transformation.

a) Give general coordinates for P, Q, and R. Find the matrix M that takes $O = (0, 0, 1)$ to P, $X = (1, 0, 1)$ to Q, and $Y = (0, 1, 1)$ to R. Explain why M is a transformation provided that P, Q, and R aren't collinear.

b) Let A, B, and C be any three noncollinear points. Show that there is an affine matrix N that takes A to O, B to X, and C to Y.

c) Prove that there is an affine matrix taking A to P, B to Q, and C to R.

8. Show that $M = \begin{bmatrix} -1 & 2 & 0 \\ -1 & 1 & 0 \\ 0 & 0 & 1 \end{bmatrix}$ maps the circle $x^2 + y^2 = 1$ to an ellipse as follows. For $(x, y, 1)$ on $x^2 + y^2 = 1$, show that $(u, v, 1) = M(x, y, 1)$ satisfies the equation $2u^2 - 6uv + 5v^2 = 1$. Use Section 2.2 to verify that $2u^2 - 6uv + 5v^2 = 1$ is the equation of an ellipse. Graph this ellipse.

9. a) Prove the rest of Theorem 4.4.5.

b) Prove that affine transformations map rays to rays.

10. Prove that an affine matrix with determinant d changes the area of convex polygons by a factor of $|d|$, as follows.

a) Show that a matrix $A = \begin{bmatrix} 1 & p & 0 \\ 0 & q & 0 \\ 0 & 0 & 1 \end{bmatrix}$ changes the area of any triangle by a factor of $|q|$. [*Hint:* Show that A takes horizontal lines to horizontal lines and use Exercise 2.]

b) Show a claim similar to part (a) for a matrix $B = \begin{bmatrix} r & 0 & 0 \\ s & 1 & 0 \\ 0 & 0 & 1 \end{bmatrix}$.

c) Show that a general affine matrix M can be written as the product TAB (or TBA if the center entry of M is 0), where T is a translation, A is a matrix of the form in part (a), and B is a matrix of the form in part (b).

d) Use part (c) to extend parts (a) and (b) to any affine matrix. [*Hint:* The determinant of a product is the product of the determinants.]

e) Extend part (d) from triangles to convex polygons.

11. Consider the IFS with just two matrices,
$A = \begin{bmatrix} 1/3 & 0 & 0 \\ 0 & 1/3 & 0 \\ 0 & 0 & 1 \end{bmatrix}$ and $D = \begin{bmatrix} 1/3 & 0 & 2/3 \\ 0 & 1/3 & 0 \\ 0 & 0 & 1 \end{bmatrix}$.

a) On a graph, show the unit square U and its images AU, DU, AAU, ADU, DAU, and DDU.

b) Describe and, as best as you can, draw the fractal of this IFS. (It is a very disconnected set called the Cantor set.)

c) The matrices A and D of this problem are two of the four matrices in Example 5. Explain how the limit set of this IFS relates to the one shown in Fig. 4.25.

12. For ease of programming, IFSs are often restricted to maps on the unit square; that is, the points $(x, y, 1)$ satisfying $0 \leq x \leq 1$ and $0 \leq y \leq 1$. Find restrictions on the coefficients of an affine matrix so that it will be a contraction mapping that sends the unit square into itself.

4.5 TRANSFORMATIONS IN HIGHER DIMENSIONS; COMPUTER-AIDED DESIGN

Transformations in three and more dimensions illustrate the power of linear algebra. We utilize the same method we used in two dimensions to move the origin: We add an extra coordinate to the vectors and represent transformations by the corresponding matrices. To generalize the definition of isometries to three and more dimensions we use the isometries of the sphere.

Interpretation By a *point* in three-dimensional affine space, we mean a column vector $(x, y, z, 1)$. By *three-dimensional affine matrix* we mean an invertible 4×4 matrix whose bottom row is $[\,0\ \ 0\ \ 0\ \ 1\,]$.

Example 1 The translation $\tau = \begin{bmatrix} 1 & 0 & 0 & p \\ 0 & 1 & 0 & q \\ 0 & 0 & 1 & r \\ 0 & 0 & 0 & 1 \end{bmatrix}$ moves all points by p in the x-direction, q in the y-direction, and r in the z-direction. Note the final column gives the image of the origin $(0, 0, 0, 1)$. The upper left 3×3 submatrix describes the transformation type. ●

Exercise 1 Explain why the bottom row of an affine transformation must be $[\,0\ \ 0\ \ 0\ \ 1\,]$.

4.5.1 Isometries of the sphere

Any affine transformation that maps the unit sphere to itself necessarily maps the origin to itself. Hence spherical isometries can be represented as 3×3 matrices, in effect the upper left corner of the 4×4 affine transformations. These isometries give insights about isometries in all dimensions and the symmetries of polyhedra.

Example 2 The transformation $\rho = \begin{bmatrix} 0 & 1 & 0 \\ 0 & 0 & 1 \\ 1 & 0 & 0 \end{bmatrix}$ is a rotation of the sphere (and all of $\mathbf{R}^3$). Points of the form $A = (a, a, a)$ are fixed by ρ and so form the axis of rotation. Furthermore, $\rho(a, b, c) = (b, c, a)$, so the composition of ρ three times will take every point back to itself, showing the angle of rotation to be $120°$. In particular, the x-, y- and z-axes map to one another (Fig. 4.29). The determinant of the matrix ρ is 1, just like two-dimensional rotations. ●

Exercise 2 Verify that the transformation $\mu = \begin{bmatrix} 1 & 0 & 0 \\ 0 & 0 & -1 \\ 0 & -1 & 0 \end{bmatrix}$ has determinant -1 and that μ^2 is

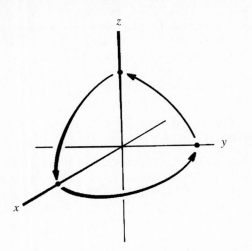

Figure 4.29 A rotation of 120° around $x = y = z$.

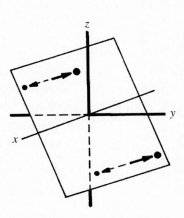

Figure 4.30 A mirror reflection in the plane $y + z = 0$.

the identity, indicating that μ is a mirror reflection. It leaves the plane $y + z = 0$ fixed (Fig. 4.30).

We can compose rotations and mirror reflections to form the only other isometries of the sphere, rotatory reflections, illustrated in Example 3. (Translations and glide reflections have no fixed points, so they aren't isometries of the sphere.)

Example 3 The rotatory reflection $\begin{bmatrix} 0 & -1 & 0 \\ 1 & 0 & 0 \\ 0 & 0 & -1 \end{bmatrix} = \begin{bmatrix} 1 & 0 & 0 \\ 0 & 1 & 0 \\ 0 & 0 & -1 \end{bmatrix} \cdot \begin{bmatrix} 0 & -1 & 0 \\ 1 & 0 & 0 \\ 0 & 0 & 1 \end{bmatrix}$ is the composition of a rotation of 90° around the z-axis followed by a mirror reflection over the plane $z = 0$ (the equator) (Fig. 4.31). The eigenvalues of this matrix are -1, i, and $-i$, which show that there is no fixed point. The opposite points $(0, 0, 1)$ and $(0, 0, -1)$ are mapped to each other. ●

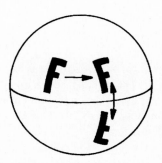

Figure 4.31 A rotary reflection composed of a 90° rotation and a mirror reflection.

4.5 Transformations in Higher Dimensions; Computer-Aided Design

Definition 4.5.1 A *rotation* in three-dimensional Euclidean geometry fixes the points on one line, called the axis of rotation, and rotates all other points through a set angle around that axis. A *mirror reflection* over a plane S in three-dimensional Euclidean geometry maps every point P to the point Q such that S is the perpendicular bisector of $\overline{PQ}$. A *rotatory reflection* in three-dimensional Euclidean geometry is the composition of a rotation with a mirror reflection in a plane perpendicular to the axis of rotation.

A spherical isometry must map the unit basis vectors $(1, 0, 0)$, $(0, 1, 0)$, and $(0, 0, 1)$ to mutually perpendicular unit vectors, leading to the algebraic description of these isometries in Theorem 4.5.1. Theorem 4.5.2 gives a more geometric description of spherical isometries, although its proof depends on linear algebra.

Theorem 4.5.1 The following statements are equivalent.

  i) A 3×3 matrix M is an isometry of the unit sphere.
  ii) The columns of M form an orthonormal basis of $\mathbf{R}^3$.
  iii) $M^{-1} = M^T$, the transpose of M.

Proof. See Problem 4. ∎

Theorem 4.5.2 Every isometry of the sphere has at least two opposite points on the sphere that are either fixed or are mapped to each other. The circle on the sphere midway between these opposite points is stable.

Proof. When we use eigenvalues to find fixed points (and stable lines), we obtain an equation in λ called the *characteristic equation*. The characteristic equation of a 3×3 matrix involves a third-degree real polynomial. All odd-degree real polynomials have a real root, so every isometry of the sphere has at least one real eigenvalue. For that eigenvalue, any eigenvector of length 1 is a point on the sphere. The matrix being considered is an isometry, so the point will be mapped back onto the sphere. Thus the image also has length 1, which means that the real eigenvalue is either 1 or -1. For an eigenvalue of $+1$, the eigenvector (point) is fixed, as is its negative (opposite point). For -1, the point and its opposite change places. In either case, the circle midway between these two points must be stable because the transformation is an isometry. ∎

4.5.2 Transformations in three and more dimensions

We obtain all isometries of $\mathbf{R}^3$ by combining in our general 4×4 matrix the 3×3 submatrix, representing an isometry of the sphere, and the final column of the matrix, representing a translation. In effect, we can build any isometry by composing these two special cases. In addition to the rotations, translations, mirror reflections, and glide reflections from two dimensions, there are two other types of three-dimensional isometries. Rotatory reflections, as we showed, are isometries of the sphere. Figure 4.32 illustrates the other type: screw motions.

Example 4 The matrix $R = \begin{bmatrix} \cos\theta & -\sin\theta & 0 & 0 \\ \sin\theta & \cos\theta & 0 & 0 \\ 0 & 0 & 1 & 0 \\ 0 & 0 & 0 & 1 \end{bmatrix}$ represents a rotation of θ around the z-axis.

Figure 4.32 A screw motion.

A point $(0, 0, k, 1)$ on the z-axis is fixed by R. Also, each plane $z = k$ is stable under R because $R(x, y, k, 1)$ has k for its third coordinate. We define a *screw motion* as the composition of a rotation and a translation in the direction of the axis of rotation—for example, RZ, where $Z = \begin{bmatrix} 1 & 0 & 0 & 0 \\ 0 & 0 & 0 & 0 \\ 0 & 0 & 1 & z \\ 0 & 0 & 0 & 1 \end{bmatrix}$. Verify that $RZ = ZR$. ●

Exercise 3 Give definitions for translations and glide reflections in three dimensions.

Theorem 4.5.3 Every three-dimensional Euclidean isometry can be written as the composition of at most four mirror reflections. There are three types of three-dimensional direct Euclidean isometries: translations, rotations, and screw motions; and three indirect types: mirror reflections, glide reflections, and rotatory reflections.

Proof. See Project 7 and Coxeter [3]. ■

In Section 4.3 we used row vectors $[a, b, c]$ to represent a line, or the set of points satisfying the equation $ax + by + c = 0$. In three dimensions $[a, b, c, d]$ represent a plane, the points satisfying $ax + by + cz + d = 0$.

By now the pattern may be clear: For points in n-dimensional affine space, use column vectors with $n + 1$ coordinates, the last of which is 1. The corresponding row vectors are, in general, called *hyperplanes* and are $(n - 1)$-dimensional. The affine transformations will be $(n + 1) \times (n + 1)$ invertible matrices whose bottom row is $[0 \ \ldots \ 0 \ 1]$. The upper left $n \times n$ corner tells us, up to a translation, what type of a transformation we have. Theorem 4.5.1 leads to the definition of isometries in higher dimensions.

Definition 4.5.2 An $(n + 1) \times (n + 1)$ invertible matrix is an *affine* matrix iff its bottom row is $[0 \ 0 \ \ldots \ 1]$. An $n \times n$ matrix M is *orthogonal* iff $M^{-1} = M^T$, where M^T is the

4.5 Transformations in Higher Dimensions; Computer-Aided Design

transpose of M. An $(n + 1) \times (n + 1)$ affine matrix is an *isometry* iff its upper left $n \times n$ submatrix is orthogonal.

Exercise 4 Verify that two-dimensional isometries satisfy this definition.

Exercise 5 Define translations in n-dimensional space. What does the matrix of a translation in n dimensions look like?

4.5.3 Computer-aided design and transformations

A CAD program stores the various reference points of a design as the columns in a matrix. Matrices quickly provide the images of these points for other views of figures. The transformations of Chapter 6 enable engineers and others to give perspective views of designs by altering the bottom row of affine matrices.

Exercise 6 Draw the quadrilateral in the plane whose four corners have the columns of
$$A = \begin{bmatrix} 1 & 0 & -2 & -1 \\ 0 & 2 & 1 & -3 \\ 1 & 1 & 1 & 1 \end{bmatrix}$$ for their coordinates. For the rotation $\rho = \begin{bmatrix} 0.6 & -0.8 & 1 \\ 0.8 & 0.6 & 0 \\ 0 & 0 & 1 \end{bmatrix}$

of approximately $53°$ around $(0.5, 1, 1)$, the product $\rho A = \begin{bmatrix} 1.6 & -0.6 & -1 & 2.8 \\ 0.8 & 1.2 & -1 & -2.6 \\ 1 & 1 & 1 & 1 \end{bmatrix}$

gives the matrix whose columns are the images under ρ of the four corners of this quadrilateral. Draw the resulting quadrilateral on the same axes as the original one.

Theorem 4.5.4 Let α be an n-dimensional affine transformation and A be an $(n + 1) \times k$ matrix whose columns $A_1, A_2, \ldots, A_k$ are k points in n-dimensional affine space. Then the columns of αA are $\alpha(A_1), \alpha(A_2), \ldots, \alpha(A_k)$.

Proof. See Problem 7. ∎

As a result of Theorem 4.5.4, once the computer has been given the new reference points, it can redraw the various lines, curves, and surfaces among them in the same manner as originally. The analytic geometry of Chapter 2 combined with the linear algebra of this chapter provide the graphics of CAD. Computers also use matrices to present three-dimensional designs as two-dimensional graphics displays and printouts. These matrices aren't transformations since they aren't one-to-one. (See Mortenson [8].)

Example 5 The matrix $\begin{bmatrix} 1 & 0 & 0 & 0 \\ 0 & 1 & 0 & 0 \\ 0 & 0 & 0 & 0 \\ 0 & 0 & 0 & 1 \end{bmatrix}$ maps every point of three dimensions orthogonally onto the xy-plane. Thus two points with the same z-coordinate will be mapped to the same point. Note that the determinant of this matrix is 0. ●

PROBLEMS FOR SECTION 4.5

1. **a)** Explain why $\begin{bmatrix} -1 & 0 & 0 & 0 \\ 0 & -1 & 0 & 0 \\ 0 & 0 & 1 & 0 \\ 0 & 0 & 0 & 1 \end{bmatrix}$ is a 180° rotation around the z-axis.

   b) Find the matrices for rotations of 180° around the x- and y-axes. What is the product of these two matrices? What is the product of one of these matrices with the matrix in part (a)?

   c) Verify your answers in part (b) physically by rotating a cube 180° around the centers of two opposite faces, followed by a 180° rotation around the centers of two other opposite faces. Mark several points on the cube so that you can recognize their starting and ending positions.

   d) Repeat part (c) with 90° rotations and describe the resulting transformation.

   e) Find the matrices for the 90° rotations in part (d) and multiply them. Describe the product.

   f) Describe the matrix for a rotation of θ around the x-axis.

2. The *central symmetry* with respect to Q takes each point P to the point P', where Q is the midpoint of $\overline{PP'}$. In two dimensions, this isometry is a rotation of 180°, often called a *half-turn*.

   a) Find the matrix form of a central symmetry in three dimensions and decide what type of an isometry it is. Explain your answer.

   b) Describe the composition of two three-dimensional central symmetries.

   c) Repeat parts (a) and (b) for four and more dimensions.

3. Find the matrix for a screw motion made of a rotation of θ around the y-axis followed by a translation in the y-direction by y. Verify that you get the same screw motion if you first translate and then rotate.

4. Prove Theorem 4.5.1. [*Hint:* The ijth entry in the product AB is the inner product of the ith row of A and the jth column of B. Recall that, in an orthonormal basis, the vectors are mutually perpendicular and have length 1.]

5. **a)** Prove that the set of spherical isometries is a transformation group.

   b) Prove that the set of orthogonal $n \times n$ matrices is a transformation group.

   c) Use the definition of an orthogonal matrix to prove that its determinant must be either $+1$ or -1. Why does this proof guarantee that the determinant of an n-dimensional isometry must also be $+1$ or -1? (As in two dimensions, the direct isometries have a determinant of $+1$, whereas the indirect isometries have a determinant of -1.)

6. **a)** Define a three-dimensional similarity with a scaling ratio of r.

   b) Explain why every three-dimensional similarity can be written as the product of an isometry and a dilation centered at the origin, represented by $\begin{bmatrix} r & 0 & 0 & 0 \\ 0 & r & 0 & 0 \\ 0 & 0 & r & 0 \\ 0 & 0 & 0 & 1 \end{bmatrix}$.

   c) What is the determinant of a three-dimensional similarity with a scaling ratio of r? What does this determinant tell you about a three-dimensional object, such as a cube, and its image under a similarity?

7. Prove Theorem 4.5.4.

8. Rotations in four dimensions extrapolate properties of rotations in two and three dimensions. For convenience, use 4×4 orthogonal matrices so that the origin $(0, 0, 0, 0)$ is fixed.

   a) Describe what is fixed by a rotation in two dimensions and by a rotation in three dimensions. What should be fixed by a rotation in four dimensions?

   b) Verify the following matrices are orthogonal with determinants of $+1$.
   $$A = \begin{bmatrix} 0 & -1 & 0 & 0 \\ 1 & 0 & 0 & 0 \\ 0 & 0 & 1 & 0 \\ 0 & 0 & 0 & 1 \end{bmatrix}, B = \begin{bmatrix} 0 & 0 & -1 & 0 \\ 0 & 1 & 0 & 0 \\ 1 & 0 & 0 & 0 \\ 0 & 0 & 0 & 1 \end{bmatrix},$$
   and $C = \begin{bmatrix} 1 & 0 & 0 & 0 \\ 0 & 1 & 0 & 0 \\ 0 & 0 & 0 & -1 \\ 0 & 0 & 1 & 0 \end{bmatrix}$.

   c) The preceding matrices are rotations of the four-dimensional sphere. Find all of their fixed points ("axes") and angles of rotation.

d) Find the products and fixed points of AB and AC. Describe how they differ. One of these products is again a rotation. Decide which product is a rotation and what its angle of rotation is. What can you say about the other product?

9. Recall that the standard basis of $\mathbf{R}^n$ is the set $\{(1, 0, 0, \ldots, 0), (0, 1, 0, \ldots, 0), \ldots, (0, \ldots 0, 1)\}$.

a) Prove that an $n \times n$ matrix is orthogonal iff it maps the standard basis vectors to an orthonormal basis of $\mathbf{R}^n$.

b) Explain why orthogonal $n \times n$ matrices will be isometries of the n-dimensional unit sphere.

c) Explain why the matrices we defined to be n-dimensional isometries actually are isometries.

4.6 Inversions and the Complex Plane

Affine transformations map lines to lines, but some important classes of transformations do not do so. Inversions form one important family of such "nonlinear" transformations. In brief, an inversion switches points on the inside of a circle with the points on the outside of a circle (Fig. 4.33). The center of the circle has no Euclidean point for its image, and no Euclidean point can be mapped to the center. This situation creates a problem in terms of transformations because transformations must be one-to-one onto functions. We solve this problem by adding a point to the plane that can switch places with the center. Intuitively, this extra point must be "at infinity," so we call it ∞. This new point is defined to be on every line.

Definition 4.6.1 The *inversive plane* is the Euclidean plane with one additional point, denoted ∞. Let a circle C with center O and radius r be given. The *inversion* ν_c with respect to C maps a Euclidean point P ($P \neq O$) to $\nu_c(P)$ on the ray $\overrightarrow{OP}$, where $d(O, P) \cdot d(O, \nu_c(P)) = r^2$. We define $\nu_c(O) = \infty$ and $\nu_c(\infty) = O$. The *center of the inversion* is O.

Exercise 1 Illustrate the inversion with respect to the unit circle $x^2 + y^2 = 1$. Verify that every line through the origin is mapped to itself, as is the unit circle. Explain why a circle with center $(0, 0)$ is mapped to another circle with the same center. How are the radii of these circles related?

Exercise 2 Explain why any inversion is its own inverse.

Theorem 4.6.1 Let k be any line not through the center of inversion O of ν_c. Then the image of k is a circle through O. Conversely, the image of a circle through O is a line not through O.

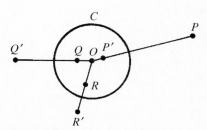

Figure 4.33 The inversion in circle C.

168 Chapter 4 Transformational Geometry

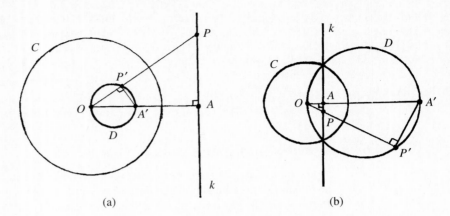

Figure 4.34

Proof. As shown in Figs. 4.34a and 4.34b, let the perpendicular from O to k intersect k at A and let $A' = \nu_c(A)$. Show that the circle D with diameter $\overline{OA'}$ is the inversive image of k. To do so, let P be any point on k and P' the (second) intersection of $\overleftrightarrow{OP}$ with the circle D. (Explain why there must be a second point of intersection.) Because $\overline{OA'}$ is a diameter, $\angle OP'A'$ is a right angle. Hence $\triangle OP'A' \sim \triangle OAP$. By the proportionality of the sides, there is some p such that $d(O, P') = p \cdot d(O, A)$ and $d(O, A') = p \cdot d(O, P)$. Then $d(O, P') \cdot d(O, P) = d(O, A) \cdot d(O, A') = r^2$ and $P' = \nu_c(P)$, as claimed. Thus D is the inversive image of k. For the other direction, note from Exercise 2 that $P' = \nu_c(P)$ iff $\nu_c(P') = P$. Hence this construction shows the inversive image of any circle D through O must be the corresponding line k. ∎

Exercise 1 is a special case of Theorem 4.6.2. Another special case of that theorem, Theorem 4.6.3, is important in the Poincaré model of hyperbolic geometry. Although the proof of Theorem 4.6.2 isn't difficult, it does require lemmas from Euclidean geometry that would sidetrack us.

Theorem 4.6.2 Let D be a circle that does not pass through the center of inversion of ν_c. Then the inversive image of D is another circle that does not pass through the center of inversion.

Proof. See Eves [4, 78]. ∎

Definition 4.6.2 Two circles are *orthogonal* iff the radii of these circles at their points of intersection are perpendicular.

Theorem 4.6.3 If a circle D is orthogonal to the circle of inversion C, then $\nu_c(D) = D$.

Proof. With D orthogonal to C, the radii of C that go to the intersections P and Q of these two circles are tangents to D (Fig. 4.35). Because P and Q are on C, they are fixed by the inversion. Lines $\overleftrightarrow{OP}$ and $\overleftrightarrow{OQ}$ are stable. By Theorem 4.6.2, $\nu_c(D)$ is a circle through P and Q. Furthermore, lines $\overleftrightarrow{OP}$ and $\overleftrightarrow{OQ}$ must still be tangent to $\nu_c(D)$, for each has one point of intersection with this circle. Thus the perpendiculars to these

4.6 Inversions and the Complex Plane 169

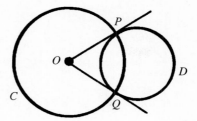

Figure 4.35

lines through P and Q intersect in the center of the circle. However, this center is also the center of D, which shows that $v_c(D) = D$. ∎

In the Poincaré model of hyperbolic geometry, arcs of orthogonal circles are used as lines. (See Section 3.1.) Henri Poincaré had the key insight that certain inversions correspond to mirror reflections in the model now named for him. In Fig. 4.36, the points inside circle H are the points of the hyperbolic plane. Circle C is orthogonal to H, so its arc inside H is a hyperbolic line. By Theorem 4.6.3 the inversion with respect to C maps H to itself. Hence this inversion is a transformation in this model of hyperbolic geometry. For example, it switches the hyperbolic lines $\overleftrightarrow{PQ}$ and $\overleftrightarrow{P'Q'}$. Using the definition of distance given in Section 3.5, Poincaré showed that this transformation is actually a mirror reflection. Euclidean mirror reflections over diameters of circle H also are hyperbolic mirror reflections. Hyperbolic mirror reflections resemble Euclidean mirror reflections in two ways. First, they switch the orientation of figures. Second, all hyperbolic plane isometries can be written as compositions of three or fewer hyperbolic mirror reflections, analogous to Theorem 4.2.5.

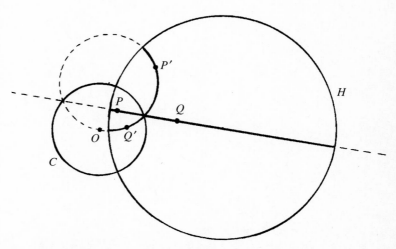

Figure 4.36   A hyperbolic mirror reflection for the Poincaré model.

170 Chapter 4 Transformational Geometry

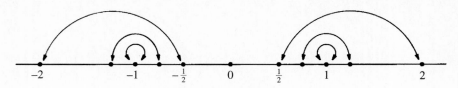

Figure 4.37

To understand hyperbolic isometries in the Poincaré model we need to consider compositions of inversions, which generally are not inversions. Complex numbers provide a convenient way to find formulas for inversions and Möbius transformations, which we discuss shortly.

Inversion is a two-dimensional analog of a function such as $f(x) = 1/x$ (Fig. 4.37), which inverts the real numbers with respect to a "circle" of radius 1 centered at 0. In effect, the real function $f(x) = 1/x$ considers just one line through the center of inversion, and 1 and -1 are the only points on the "circle." Complex numbers, of the form $a + bi$, provide a way to represent inversions on the plane. However, Example 1 shows that we need a more sophisticated function than just $1/x$ to represent inversions in the plane. Just as we needed to add a point, ∞, to the Euclidean plane to make inversions transformations, we need to extend the complex numbers. We write **C#** for the extended complex numbers, or the usual complex numbers and ∞, which is the limit of $a + bi$ as $a \to \infty$ or $b \to \infty$ or both.

Example 1 Consider the function f defined on the complex numbers by $f(z) = 1/z$. Figure 4.38 shows the images of various complex numbers. Verify that $1/(a + bi) = (a - bi)/(a^2 + b^2)$ by multiplying by $a + bi$. Verify that $f(\tfrac{1}{2}i) = -2i$, $f(2 - 2i) = \tfrac{1}{4} + \tfrac{1}{4}i$, and $f(-1 - 2i) = -\tfrac{1}{5} + \tfrac{2}{5}i$. Points inside the unit circle are mapped to points outside

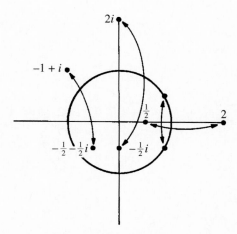

Figure 4.38 $f(z) = \tfrac{1}{z}$.

that circle, and vice versa. However, a point and its image are not on a line with the origin. Instead, there is a mirror reflection over the x-axis (real axis) in addition to the inversion. Fortunately, a fairly simple complex function acts as a mirror reflection over that axis: complex conjugation. Recall that $\overline{a+bi} = a - bi$. A complex number z and its conjugate $\bar{z}$ are mirror images with respect to the x-axis. Hence we can define the inversion with respect to the unit circle as $v(z) = \overline{1/z}$. Verify that $v(a+bi) = (a+bi)/(a^2+b^2)$, a positive scalar multiple of $a+bi$. •

Theorem 4.6.4 The inversion with respect to the circle of radius r and center w is given by $v(z) = \overline{r^2/(z-w)} + w$, for $z \neq w$.

Proof. First, show that this formula works when $w = 0 + 0i = 0$. In this case, $v_0(z)$ reduces to $\overline{r^2/z}$. The product $z \cdot \overline{r^2/z}$ has a length of r^2, showing that $\overline{r^2/z}$ is the correct distance from the origin. To complete this case, show that 0, z, and $\overline{r^2/z}$ are on the same ray. Let $z = a+bi$. Then $1/z = (a-bi)/(a^2+b^2)$, and $\overline{r^2/z} = r^2(a+bi)/(a^2+b^2) = r^2 z/(a^2+b^2)$, a positive real scalar multiple of z.

We use the method in Example 3 of Section 4.3 to develop the general formula. The addition of w to every complex number is a translation τ of the complex plane. The formula $v(z) = \overline{r^2/(z-w)} + w$ is the composition $\tau \circ v_0 \circ \tau^{-1}$, which is an inversion because translations do not alter distance. To verify that this is the desired formula, let z be any point on the circle of radius r and center w. Then $z - w$ is a point on the circle of radius r at the origin. This condition implies that $\overline{r^2/(z-w)} = z - w$ from the first part. The addition of w now takes this point back to z. Thus every point on the desired circle of inversion is fixed, proving the theorem. ∎

Example 2 Find the transformation of the extended complexes **C#** that is the composition of the inversions $\overline{1/z}$ followed by $\overline{4/z}$.

Solution. When we replace the z of $\overline{4/z}$ with $\overline{1/z}$, we get $\overline{4/(\overline{1/z})} = \overline{(\overline{(4z)})} = 4z$. This outcome is a similarity (dilation) with a scaling factor of 4. •

Compositions of inversions can be similarities (as in Example 2), hyperbolic isometries, or other transformations. The transformation group that contains the inversions and their compositions is called the Möbius transformations, after Augustus Möbius, one of the first mathematicians to understand the importance of transformations in geometry. Möbius transformations, which leave a circle H stable, are the isometries for the Poincaré model of hyperbolic geometry. These transformations, as Theorem 4.6.6 shows, preserve angles, an important fact both in complex analysis and in geometry. In complex analysis, transformations preserving angles are called *conformal*.

Definition 4.6.3 A *Möbius transformation* is a function of the extended complex numbers **C#** that is one-to-one, onto, and has one of the two forms $f(z) = (pz+q)/(rz+s)$ or $f(z) = \overline{(pz+q)/(rz+s)}$, where $z \neq -s/r$. We define $f(-s/r) = \infty$ and $f(\infty) = p/r$, or $\overline{p/r}$, in the second form. The constants must satisfy $ps - qr \neq 0$.

AUGUSTUS MÖBIUS

Augustus Möbius (1790–1868) earned his living as an astronomer in Leipzig, Germany, but achieved international recognition as a geometer. His absentminded concentration on mathematics often caused him to forget his keys or other things. He was very shy, which may have led him to avoid a controversy at the time between geometers about which approach, synthetic or analytic, was superior. From a modern vantage point, the quarrel seems pointless because these approaches complement one another. His work built on both approaches. In 1827 he invented barycentric coordinates, which he developed into homogeneous coordinates for projective geometry. (See Sections 2.3 and 6.3.) Julius Plucker (1801–1868) developed these coordinates into the first analytic model for projective geometry, then the cutting edge of synthetic geometry research.

Forty-five years before Klein's Erlanger Programm, Möbius investigated transformations in geometry. Even without the powerful framework of group theory, he was able to initiate the study of isometries, similarities, affine transformations, and the transformations of projective geometry. He also developed the geometry of inversions and investigated the complex transformations that now bear his name. Later in life, he initiated the study of topological transformations. At age 68, he discovered the Möbius strip, a mathematical model with the curious topological property that it has just one side and one edge as shown in the accompanying figure.

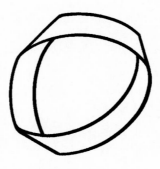

A Möbius strip.

Exercise 3 Explain why $f(-s/r) = \infty$ and why $f(z) = (pz + q)/(rz + s)$ should go to the limit p/r as z gets larger. Explain how to rewrite the inversions of Theorem 4.6.4 as Möbius transformations.

Theorem 4.6.5 The set of Möbius transformations is a transformation group.

Proof. See Problem 8. ∎

Theorem 4.6.6 Möbius transformations preserve angle measure.

4.6 Inversions and the Complex Plane

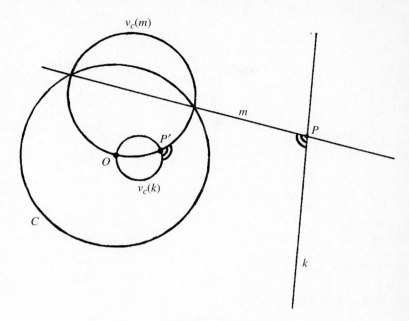

Figure 4.39

Proof. A general Möbius transformation can be written as a composition of similarities and inversions. (See Problem 6.) We already know from Theorem 4.4.3 that similarities preserve angles. Hence we only need to show that inversions preserve angles.

Recall that the angle between two curves is the angle that their tangents make at their intersection. As shown in Fig. 4.39, we know the angle between two tangents at P and are to show that the corresponding angle at P' has the same measure. We simplify this proof by comparing the angles that one of these tangents and its corresponding circle make with $\overleftrightarrow{OP}$ (Fig. 4.40). We show that $\angle PP'Q' \cong \angle OPA$ and then the angles for the other tangent and circle follow. Then we get the desired preservation of angles in Fig. 4.39 by adding the measures of these angles.

In Fig. 4.40, $\angle OP'A'$ is a right angle because $\overline{OA'}$ is a diameter, and $\angle O'P'Q'$ is a right angle because $\overleftrightarrow{P'Q'}$ is tangent to the circle and $\overline{O'P'}$ is a radius. Thus $\angle OP'O' \cong \angle A'P'Q'$. Also $\angle P'OO' \cong \angle OP'O'$ because $\triangle OO'P'$ is isosceles. Thus $\angle A'P'Q' \cong \angle P'OO'$. Angles $\angle OPA$ and $\angle P'OO'$ are complementary because they are in a right triangle. Angles $\angle PP'Q'$ and $\angle A'P'Q'$ together form a right angle, so they are complementary. Hence $\angle PP'Q' \cong \angle OPA$, as required. ∎

The fundamental concept of a transformation has great significance in mathematics, linking geometry and algebra. The transformations presented are important throughout mathematics and its applications. In addition, there are many groups of transformations beyond those discussed here, including far more general ones in topology.

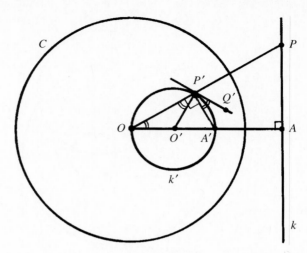

Figure 4.40

PROBLEMS FOR SECTION 4.6

1. **a)** Find distinct circles C and D for which $v_C(4, 0) = v_D(4, 0) = (1, 0)$.

   b) Explain why the center of any circle C or D for part (a) must be on the x-axis. Can every point on the x-axis be the center of such a circle of inversion? Explain.

   c) Find a formula for the radius of the circle from part (b) in terms of the center. [*Hint:* Look first at centers to the left of $(1, 0)$. Consider the distances from the center to $(1, 0)$ and to $(4, 0)$.]

2. **a)** Let P be any point outside the circle of inversion C with center O. Draw the tangents from P to C and let their points of intersection with C be Q and R. Prove that the inversive image of P is the point of intersection of $\overleftrightarrow{OP}$ and $\overleftrightarrow{QR}$. Illustrate your proof.

   b) Give a construction of the inversive image of a point that is inside the circle of inversion. Prove your construction correct.

3. In Problem 2, the line $\overleftrightarrow{QR}$ is perpendicular to $\overleftrightarrow{OP}$ and through the inversive image of P. Such a line is called the *polar* of P, and P is called the *pole* of the line $\overleftrightarrow{QR}$ with respect to the circle C. Note that the definition of a pole and a polar don't depend on the point being outside the circle. Prove that, if a point U is on the polar of a point P, then P is on the polar of U. Draw a diagram for your proof.

4. A circle D passes through a point P and the inversive image P' of P with respect to the circle C, where $P \neq P'$. Prove that D is orthogonal to C. [*Hint:* Assume that three noncollinear points determine a circle and that any path from inside a circle to outside the circle must intersect the circle.]

5. Let P', Q', and R' be the inversive images of P, Q, and R with respect to a circle C. Does Theorem 4.6.6 imply that $\triangle PQR \sim \triangle P'Q'R'$? Explore this question by using a particular circle and points. Explain your answer.

6. This problem investigates similarities as Möbius transformations.

   a) Describe the similarity $f(z) = z + (a + bi)$.

   b) If a is a nonzero real number, what similarity is $f(z) = az$? If $a + bi$ is on the unit circle, what similarity is $f(z) = (a + bi)z$? If $a + bi$ is any nonzero complex number, what similarity is $f(z) = (a + bi)z$?

   c) What similarity is $f(z) = \bar{z}$?

   d) Prove that $f(z) = (pz + q)/s$ and $f(z) = (p\bar{z} + q)/s$ are similarities for any complex numbers p, q, and s, where p and s are not 0.

   e) Prove that any Möbius transformation is the composition of a similarity and at most one inversion.

7. Möbius transformations are used to convert the half-plane model to the Poincaré model and vice versa. As Poincaré knew, this procedure implies that the hyperbolic isometries of the half-plane model are Möbius transformations.

   a) Find the function for the inversion v_C with respect to a circle with radius $\sqrt{2}$ and center at $-2i$.

   b) Find the function for the inversion v_D with respect to a circle with radius 2 and center at $-3i$.

   c) Find a circle E that is the image of the real axis under v_C. Verify that the points above the axis $(a + bi,$ with $b > 0)$ are mapped to the interior of the circle. Draw a picture.

   d) Verify that the unit circle is the image of circle E under v_D. Verify that the points inside circle E are mapped to points inside the unit circle. Draw a picture.

   e) The composition $v_D \circ v_C$ maps the half-plane model to the Poincaré model. For the point $(0, 1) = i$ from the half-plane model find its image in the Poincaré model.

   f) Find a composition of inversions that converts the Poincaré model to the half-plane model.

8. a) Show that the composition of two Möbius transformations is again a Möbius transformation. [*Hint:* First leave out the complex conjugates. Then describe how complex conjugates affect the compositions.]

   b) Show that the inverse of a Möbius transformation is a Möbius transformation. [*Hint:* solve $w = (pz + q)/(rz + s)$ for z.]

   c) Prove that the Möbius transformations form a transformation group.

9. In complex analysis **C#** is often represented on the surface of a sphere. Imagine a sphere of radius 1 with its south pole on the origin of the complex numbers. Then match complex numbers with the points on the sphere as follows. Draw a line from the north pole to a complex number. Where this line intersects the sphere (besides the north pole) is the matching point. The north pole acts as the infinity point ∞. (In reverse, this process is called the *stereographic projection* of the sphere.)

   a) Illustrate the process of matching complex numbers to the points on the sphere.

   b) What on the sphere corresponds to the circle $x^2 + y^2 = 4$? What spherical isometry corresponds to the inversion with respect to this circle?

PROJECTS FOR CHAPTER 4

1. Place two mirrors at an angle facing each other with a shape in their interior. Investigate the multiple images of this shape, with the use of a protractor.

   a) How do the images move as you move the original shape closer to one of the mirrors?

   b) For mirror angles of 90°, 60°, 45° and smaller, count the number of images (plus the original shape) that you can see. Find a formula relating the angle and the number of images.

   c) Use an asymmetric shape so that you can distinguish the orientation of the images. Describe the orientation of successive images of the original shape. For various mirror angles, measure as best as you can with a protractor the angle between the original shape and the first image having the same orientation. How does this angle relate to the mirror angle?

2. Place two mirrors parallel and facing each other with a shape in their interior. Investigate the multiple images of this shape.

   a) How do the images move as you move the original shape closer to one of the mirrors?

   b) Use an asymmetric shape so that you can distinguish the orientation of the images. Describe the orientation of successive images of the original shape. For various distances between the mirrors, measure as best as you can with a ruler the distance between the original shape and the first image having the same orientation. Relate the distance between the mirrors and the distance between the original and this image.

3. Place three mirrors facing each other to make three sides of a square with an asymmetric shape inside the square. Investigate the multiple images of this shape.

a) Make a diagram showing the various images and their orientations.

b) Which images have the same orientation as the original shape and which have the opposite orientation?

c) Indicate on the diagram in part (a) which isometry produces each image.

4. Use the Geometer's Sketchpad or CABRI to experiment with the effects of elementary transformations. Find the image of a triangle under various transformations and their compositions. Investigate Theorems 4.2.3–4.2.6.

5. On a transparency, draw axes and randomly insert numerous small dots (Fig. 4.41). Photocopy this transparency and align the transparency on top of the copy.

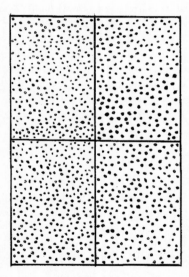

Figure 4.41

a) Rotate by a small angle the transparency relative to the paper around their common origin. Describe the pattern that the two sets of dots form. Translate the transparency relative to the paper and describe the resulting pattern. Try this procedure with other translations and small rotations. What can you say about the composition of a rotation followed by a translation?

b) Switch the order and repeat part (a). What can you say about the composition of a translation followed by a rotation? Do you get the same transformation regardless of the order?

c) Each composition in part (a) has a fixed point. Keep the initial angle of rotation the same and describe what happens to the fixed point as you use increasingly long translations in the same direction.

d) Repeat part (c), using the switched order of part (b). Compare these fixed points with those in part (c).

e) Experiment with two rotations of small angles around different points. Based on the two centers and the angles of rotation, can you predict where the new center of rotation will be?

6. Investigate geometric applications of inversions. (See Eves [4, 84–91] and Greenberg [6, 257–262].)

7. Recall that an *equivalence relation* $\equiv$ on a set S is reflexive, symmetric, and transitive. (See Gallian [5, 13–15].)

a) Suppose that **G** is a group of transformations on a set S and that, for $a, b \in S$, we define $a \equiv b$ whenever there is some $\alpha \in$ **G** such that $\alpha(a) = b$. Prove that $\equiv$ is an equivalence relation.

b) Suppose that $\equiv$ is an equivalence relation on a set S and that we define **G** to be all transformations α on S such that, for all $a \in S$, $a \equiv \alpha(a)$. Prove that **G** is a transformation group.

c) Let S be the set of all lines in the Euclidean plane and interpret $\equiv$ as parallel. What is **G** in part (b)?

d) Let **G** be the transformation group of isometries of the Euclidean plane and S be the set of all line segments. Describe $\equiv$.

e) Repeat part (d) with S being the set of all triples of points.

f) Describe other groups of transformations and equivalence relations.

8. Investigate transformations in CAD programs. (See Mortenson [8].)

9. Investigate iterated function systems (IFSs). (See Barnsley [2] and the software, "The Desktop Fractal Design System," [6] under Suggested Media.)

10. Prove the three-dimensional analogs of Theorems 4.2.2–4.2.5

a) A three-dimensional Euclidean isometry fixing four points not all in the same plane is the identity.

b) Any three-dimensional Euclidean isometry is determined by where it maps four points not all in the same plane.

c) Given any two distinct points in three-dimensional Euclidean space, there is a unique mirror reflection switching these two points.

d) Every three-dimensional Euclidean isometry can be written as the composition of at most four mirror reflections.

11. State the analogs of Theorems 4.2.2–4.2.7 in n-dimensional Euclidean geometry.

12. Define similarities in three dimensions. State the analogs of Theorems 4.4.1–4.4.3 for three-dimensional similarities and prove them.

13. Define similarities in n-dimensions. State the analogs of Theorems 4.4.1–4.4.4 for n-dimensional similarities.

14. Investigate Theorem 4.4.6 on convex sets in n dimensions.

15. Investigate stereographic projections and other mappings of a sphere to a plane. (See Hilbert and Cohn-Vossen [7, 248–263].)

16. (Calculus) Vertical lines play a special role in calculus because functions have just one y-value for any x-value.

   a) Prove that an affine matrix maps vertical lines to vertical lines iff it is of the form $\begin{bmatrix} a & 0 & c \\ d & e & f \\ 0 & 0 & 1 \end{bmatrix}$. Prove such matrices form a transformation group.

   b) Investigate what happens to the points $(x, x^2, 1)$ on the parabola $y = x^2$ under the matrix $\begin{bmatrix} 2 & 0 & 1 \\ 1 & 1 & 0 \\ 0 & 0 & 1 \end{bmatrix}$. Graph the resulting curve. Does the minimum point on the original curve get mapped to the minimum point on the image? Find the equation of the new function h. [*Hint:* If $w = 2x + 1$, write $y = h(w)$ in terms of x and then replace x by $(w-1)/2$. Explain why this hint works.]

   c) Show that the affine matrix in part (a) transforms the function $y = g(x)$ to the function $h(ax + c) = dx + e \cdot g(x) + f$.

   d) If the function g in part (c) has a derivative at every point, what can you say about the derivative of the function h? Can you find the relative maxima and minima of h? Be sure that your answers match with what you found in part (b). Experiment with other polynomials for g. What can you say about the second derivatives of g and h from part (c)?

17. Investigate hyperbolic geometry transformations. (See Greenberg [6].)

18. Investigate dynamical systems. (See Abraham and Shaw [1].)

19. Investigate topological transformations. Describe properties preserved by transformations that are more general than affine. (See Smart [9, Chapter 8].)

20. Investigate the use of transformations in biology. (See Thompson [10].)

21. Investigate the history of transformational geometry. (See Yaglom [11].)

22. Write an essay discussing Klein's definition of geometry in light of the variety in groups of transformations presented in this chapter.

Suggested Readings

[1] Abraham, R., and C. Shaw (eds.). *Dynamics—The Geometry of Behavior* (4 vols.). Santa Cruz, Calif.: Ariel Press, 1982–1988.

[2] Barnsley, M. *Fractals Everywhere*. New York: Academic Press, 1988.

[3] Coxeter, H. *Introduction to Geometry*, 2d Ed. New York: John Wiley & Sons, 1969.

[4] Eves, H. *Fundamentals of Modern Elementary Geometry*. Boston: Jones and Bartlett, 1992.

[5] Gallian, J. *Contemporary Abstract Algebra*. Lexington, Mass.: D. C. Heath, 1994.

[6] Greenberg, M. *Euclidean and Non-Euclidean Geometries*. San Francisco, Calif.: W. H. Freeman, 1980.

[7] Hilbert, D., and S. Cohn-Vossen. *Geometry and the Imagination*. New York: Chelsea, 1952.
[8] Mortenson, M. *Geometric Modeling*. New York: John Wiley & Sons, 1985.
[9] Smart, J. *Modern Geometries*, 3d Ed. Pacific Grove, Calif.: Brooks/Cole, 1988.
[10] Thompson, D. *On Growth and Form*. New York: Cambridge University Press, 1992.
[11] Yaglom, I. *Felix Klein and Sophus Lie: Evolution of the Idea of Symmetry in the Nineteenth Century*. Boston: Birkhäuser, 1988.

Suggested Media

1. "Central Similarities," 10-minute film, International Film Bureau, Chicago, 1966.
2. "Geometric Transformations," 10-minute film, Ward's Modern Learning Aids Division, Rochester, N.Y., 1969.
3. "Inversive Geometry," 24-minute video, Films for the Humanities and Sciences, Princeton, N.J., 1996.
4. "Isometries," 26-minute film, International Film Bureau, Chicago, 1967.
5. "Mr. Klein Looks at Geometry," 25-minute film, University Media, Solana Beach, Calif., 1978.
6. "The Desktop Fractal Design System," software and handbook by M. Barnsley, Academic Press, Boston, 1989.
7. "Transformations and Matrices," 25-minute film, University Media, Solana Beach, Calif., 1978.

5
Symmetry

The rules of symmetry restrict how an artist can fit a repeating motif (such as this Iranian design) together to make a design. Archaeologists and anthropologists have started using symmetry in their study of designs to provide greater insight into cultures. Chemists and physicists use symmetry to organize new discoveries, to analyze empirical evidence, and to suggest fruitful lines for future inquiries. Knowing the fundamental concepts and mathematics of symmetry increases understanding in many subjects.

> Symmetry, as wide or as narrow as you may define its meaning, is one idea by which man through the ages has tried to comprehend and create order, beauty and perfection.   —*Hermann Weyl*
>
> The investigation of the symmetries of a given mathematical structure has always yielded the most powerful results.
> —*Emil Artin*

5.1 Overview and History

Repeated patterns abound in nature, and artists in virtually every culture and time have used repeated patterns in their designs. The repetition of a motif underlies symmetry, whether in a bilateral Mexican design (Fig. 5.1) or the intricate atomic structure of a diamond crystal (Fig. 5.2). Symmetry combines aesthetic and practical values which happily augment one another.

The bodies of most animals illustrate *bilateral symmetry;* that is, a mirror reflection interchanges the two sides of the animal. Hunting lions as well as hunted antelopes need the ability to turn left as readily as right and to hear from each side equally well. However, feet are useful only underneath an animal, so there is no evolutionary advantage to a symmetry between up and down. Similarly, running backward isn't important for either predator or prey, so there is no symmetry between the front and the back of animals. Hence the practical needs of most animals require symmetry between right and left, but no other.

Figure 5.1   A Mexican design.

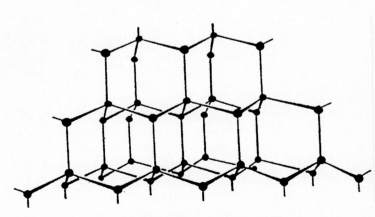

Figure 5.2   A diamond crystal.

5.1 Overview and History

Exercise 1 Some animals, such as jellyfish, have more than just bilateral symmetry. What makes this extra symmetry advantageous for these animals?

Evolution can explain symmetry in animals, but it provides no insight into the widespread aesthetic appreciation people have for symmetry. The unity and balance of symmetric objects seems to appeal to people of all cultures. Hermann Weyl in his classic book on symmetry [22, 3] writes:

> *Symmetric means something like well-proportioned, well-balanced, and symmetry denotes that sort of concordance of several parts by which they integrate into a whole. Beauty is bound up with symmetry.*

Human beings have thought about symmetry in artistic terms for thousands of years. However, a systematic and mathematical study of symmetry required a shift from a static viewpoint to a dynamic one. Group theory and transformational geometry provided the mathematics needed to study symmetry. In particular, the symmetries of a figure are the transformations under which the figure is stable, and these transformations always form a group. Thus the evolution of the concept of a group in the nineteenth century is inseparable from symmetry in algebra and geometry.

Joseph Louis Lagrange (1736–1813) started investigating transformations (symmetries) of the roots of polynomials in 1770. Evariste Galois (1811–1832) pursued these ideas and in the process developed many important theorems and concepts of group theory. Chemistry provided another impetus for the study of symmetry and groups, especially the classification in 1849 of all possible types of chemical crystals. The French physicist Auguste Bravais (1811–1863) developed this classification by using groups long before x-ray crystallography in 1912 confirmed these mathematical conclusions. Indeed, his classification predicted possible crystal types that were only later discovered and one that hasn't yet been found. Camille Jordan (1838–1922) united the algebraic work of Galois and others and the geometric work of Bravais in the first book on group theory in 1870. Klein and Lie extended the use of transformations and groups throughout geometry.

The classification of possible symmetry groups has supplied scientists and others with a clear understanding of the possible patterns that can be found in their areas. Symmetric patterns, especially in physics, are often formal rather than visual. Even so, the same geometric intuition underlies symmetry in that context. In turn, questions from other disciplines have stretched the notion of symmetry and raised new mathematical questions. The beauty of the mathematics of symmetry and the beauty of symmetric objects have inspired the study of symmetry.

Definition 5.1.1 A transformation σ is a *symmetry* of a subset T of a geometric space iff $\sigma(T) = T$. (Individual points in T can move to other points in T, but T is stable.) A *motif* of a design is a basic unit from which the entire design can be obtained as the images of that motif under the symmetries of the design.

The possible symmetries of a figure are usually limited to the isometries of the larger space. Problem 8 and Section 5.6 consider some other possibilities. Theorem 5.1.1 shows that the symmetries of a set form a transformation group, called the *symmetry group* of that set.

Theorem 5.1.1 The symmetries of a subset T form a transformation group.

Proof. Recall from Section 4.1 that a group of transformations needs to have closure, identity, and inverses. If α is a symmetry of T, then $\alpha(T) = T$. For closure, let α and β be symmetries of T. Then $\alpha \circ \beta(T) = \alpha(\beta(T)) = \alpha(T) = T$. Thus $\alpha \circ \beta$ is also a symmetry of T. The identity ι clearly takes any subset T to itself. Finally, let α be any symmetry of T and α^{-1} its inverse, as guaranteed by Theorem 4.1.2. Then $\alpha^{-1}(T) = \alpha^{-1}(\alpha(T)) = \iota(T) = T$, showing α^{-1} also to be a symmetry of T. ∎

Example 1 The symmetry group of the design shown in Fig. 5.1 contains only the identity and a mirror reflection. Either half of the design can be considered as the motif. The symmetry group of the diamond crystal shown in Fig. 5.2 has infinitely many symmetries, if we assume that the design continues forever in all directions. For example, many translations will slide the entire crystal over to coincide with itself. Rotations of 120° around the lines representing the bonds also are symmetries. The motif in this case is one carbon atom and the bonds attaching it to its neighbors. ●

Exercise 2 Describe the symmetries of the Iranian design at the beginning of this chapter.

The variety of artistic motifs is limited only by artists' imaginations. However, these motifs can fit symmetrically in relatively few ways, which we will classify. In this chapter we discuss symmetry in two and three dimensions and present applications of symmetry in the sciences and other fields. In Section 5.6 we investigate fractals, a new area of mathematics and science that stretches the idea of symmetry in an intriguing direction. (Yaglom [23] provides further historical information; Weyl [22] is a classic study of the ideas of symmetry.)

PROBLEMS FOR SECTION 5.1

1. a) Classify the letters of the alphabet in terms of their symmetries. Consider uppercase (capital) and lowercase letters in any font you choose.

    b) Find the longest word you can that has vertical bilateral symmetry (a vertical mirror reflection); repeat for horizontal bilateral symmetry. Find some words with other symmetry.

    c) Make up a sentence that is a palindrome; that is, it has the same order of letters backward as forward. How does the symmetry of a palindrome differ from vertical bilateral symmetry?

2. If a figure has both horizontal and vertical mirror reflections as symmetries, must it have any other symmetry? Illustrate and explain.

3. Describe the symmetries of each design shown in Fig. 5.3.

4. Find the next few shapes in the sequence shown in Fig. 5.4.

5. a) Make two squares of the same size. Attach one corner of one of the squares to the center of the second square in such a way that the first square rotates in the plane freely around the

5.1 Overview and History

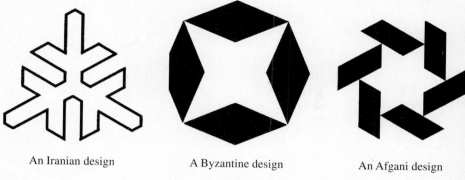

An Iranian design   A Byzantine design   An Afgani design

Figure 5.3

center of the other (Fig. 5.5). Find the maximum and minimum percentages of the area of second square covered by the first square. Explain your answer. [*Hint:* Symmetry is the key.]

b) Repeat part (a) for two regular hexagons.

c) Repeat part (a) for two equilateral triangles. [*Hint:* This answer differs from the others.]

6. Let T and U be subsets of a space S. Prove that the set of transformations of S that are symmetries for both T and U form a group of transformations.

7. a) The designs shown in Fig. 5.6 are intended to continue forever in a line. Describe the types of symmetries of each design.

b) The symmetry groups for the designs shown in Fig. 5.6 differ. Create other designs having translational symmetry that have different symmetry groups. How many different types of these designs can you create?

8. a) The designs shown in Fig. 5.7 have symmetries that are affine transformations but not isometries. Describe the symmetries for these designs.

b) Create designs that have symmetries besides isometries. Identify the symmetries in each design.

Figure 5.4

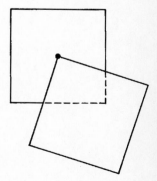

Figure 5.5

184　Chapter 5　Symmetry

A Mexican design

A Chinese design

Figure 5.6

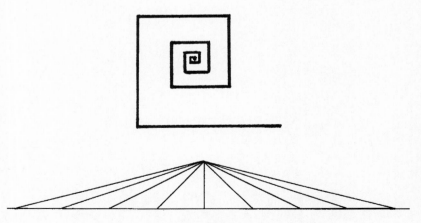

Figure 5.7

5.2 FINITE PLANE SYMMETRY GROUPS

Snowflakes possess the same number of symmetries, even if no two are exactly alike. Figure 5.8 illustrates the six rotations and six mirror reflections that form the group of symmetries of a snowflake. The swastika (Fig. 5.9), a religious symbol in ancient India long before the Nazis appropriated it, has four symmetries, all rotations of multiples of 90°. In this section we classify the finite groups of plane symmetries. Leonardo da Vinci (1452–1519) realized that, in modern terms, all designs in the plane with finitely many symmetries have either rotations and mirror reflections like those in Fig. 5.8 or just rotations like those in Fig. 5.9. The two types of symmetry groups in the classification presented in Theorem 5.2.2 are called dihedral and cyclic. *Dihedral* means "two faces" and refers to the fact that the symmetries in this group can be found by using two mirrors at an angle. (See Project 1 of Chapter 4.) The argument in Theorem 5.2.2 shows how algebraic reasoning can be used to turn geometric intuition into proof.

Definition 5.2.1 The *cyclic group* $\mathbf{C}_n$ contains n rotations, all with the same center. The angles of rotation are the multiples of $360°/n$, where n is any positive integer. The *dihedral group* $\mathbf{D}_n$ contains the n rotations of $\mathbf{C}_n$ and n mirror reflections over lines passing through the center of the rotations. The angles between the lines of the mirror reflections are multiples of $180°/n$. (See Fig. 5.8.)

Exercise 1 What are the symmetry groups for Figs. 5.8 and 5.9?

Exercise 2 Find the symmetry group of a regular n-sided polygon.

Theorem 5.2.1 The isometries of a finite plane symmetry group must fix some point and so are rotations around this fixed point or are mirror reflections over lines through this fixed point.

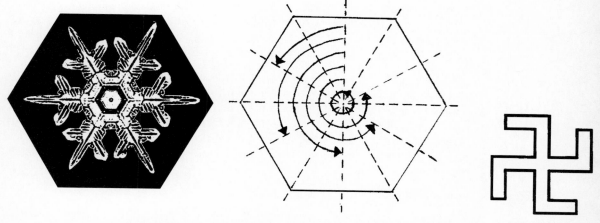

Figure 5.8 The symmetries of a snowflake.

Figure 5.9

Proof. We adapt a proof from Gallian [6, 404]. Let **G** be a finite symmetry group of plane isometries and assume that the plane has coordinates. For a point A, let $S = \{\gamma(A) : \gamma \in \mathbf{G}\}$; that is, S is the set of images of A under the isometries of **G**. Because **G** is finite, so is S, whose elements can be listed as $(x_1, y_1), (x_2, y_2), \ldots, (x_n, y_n)$. The center of gravity of these n points, $(\overline{x}, \overline{y}) = (\frac{x_1+x_2+\cdots+x_n}{n}, \frac{y_1+y_2+\cdots+y_n}{n})$, must be fixed by every γ in **G**. Each (x_i, y_i) is the image of A by at least one γ_i in **G**: $(x_i, y_i) = \gamma_i(A)$. Now, for any γ in **G**, $\gamma \circ \gamma_i$ is another element of **G**, so γ will move the points of S around among themselves. Thus their center of gravity, $(\overline{x}, \overline{y})$, is fixed by each γ. From Section 4.2, we know that the only plane isometries that fix a given point are rotations around that point and mirror reflections over lines through the point. ∎

Theorem 5.2.2 A finite symmetry group containing only Euclidean plane isometries is either a cyclic group or a dihedral group.

Proof. We need to show that the possible rotations and mirror reflections from Theorem 5.2.1 always fit exactly as cyclic and dihedral groups require. First, consider the rotations. If there is only one, it is the identity, a rotation of $0°$. Otherwise, let the smallest positive angle of rotation be $A°$. From Chapter 4 the composition of rotations of $B°$ and $C°$ is a rotation of $B° + C°$. Thus by closure there are rotations by all multiples of $A°$. The number of rotations is finite, so A divides some multiple of 360. Moreover, A divides 360. Let kA be the largest multiple of A less than 360. Then $360 \leq (k+1)A < 360 + A$. If $(k+1)A°$ is greater than $360°$, it is the same angle as $((k+1)A - 360)°$. However, this last angle would be positive and smaller than $A°$, which is a contradiction. Hence, A divides 360, say, $A = 360/n$. Thus we have at least the n rotations whose angles are multiples of $A°$.

Claim. There are no others. Suppose that there were a rotation of $B°$, not a multiple of $A°$. Let jA be the largest multiple of A less than B. Then there would be a rotation of $(B - jA)°$, which would be less than $A°$, which is a contradiction. Hence the rotations form $\mathbf{C}_n$.

Next consider the mirror reflections of this symmetry group. If there are none, we have $\mathbf{C}_n$. If there is at least one, its compositions with the n rotations give n different mirror reflections. Problem 6 of Section 4.2 showed that the composition of two mirror reflections over lines meeting at an angle of $C°$ is a rotation of $2C°$. As these angles of rotation must be multiples of $A°$, the angles between the lines must be multiples of $\frac{1}{2}A° = (180/n)°$. Thus there are just n lines and the symmetry group is $\mathbf{D}_n$. ∎

Theorems 5.2.3 and 5.2.4 apply more generally than just to Euclidean plane geometry. Theorem 5.2.3 shows how to count the number of symmetries of a design without finding them individually. It is an application of LaGrange's theorem in group theory, but we prove it directly. (For those who have studied abstract algebra, the classes in the proof are the cosets of $\mathbf{G}_P$, which is a subgroup of **G**. Furthermore, the symmetries fixing a point are the *stabilizer* of the point, and the points to which that point can be moved are its *orbit*.)

Example 1 Count the symmetries of a pentagonal dipyramid.

Solution. The pentagonal dipyramid shown in Fig. 5.10(a) has seven vertices, but they can't all be mapped to one another. Figure 5.10(b) shows the polyhedron from

5.2 Finite Plane Symmetry Groups

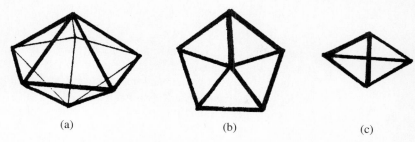

Figure 5.10 Three views of a pentagonal dipyramid.

the top vertex. The symmetries fixing the top vertex form the dihedral group $\mathbf{D}_5$, so 10 symmetries fix this point. The only other vertex symmetric to the top vertex is the bottom. Hence, when we apply Theorem 5.2.3 using the top vertex, we find a total of $10 \times 2 = 20$ symmetries. We can arrive at this same number of symmetries by using one of the other vertices. Figure 5.10(c) shows that the symmetries fixing one of these other vertices form the dihedral group $\mathbf{D}_2$, which has four elements. For five such vertices, Theorem 5.2.3 again yields $4 \times 5 = 20$ symmetries. ●

Theorem 5.2.3 The number of symmetries of a figure, if finite, equals the product nk, where n is the number of symmetries of the entire figure that leave a given point fixed and k is the number of points to which that point can be moved by symmetries.

Proof. Let P be any point of the figure, $\mathbf{G}$ the symmetry group, $\mathbf{G}_P$ be the set of symmetries that fix P, and $\mathbf{G}_p$ have n elements. We collect the symmetries of $\mathbf{G}$ into disjoint classes, show the classes to be the same size, n, and count the number of classes, k. Two symmetries α and β are in the same class iff they map P to the same point Q: $\alpha(P) = \beta(P) = Q$. The number of classes, k, is the number of points to which P can be moved. To complete the proof we need to show all the classes to be the same size, n, as $\mathbf{G}_P$, the class that maps P to P. Suppose that $\alpha(P) = Q$ and consider $[\alpha]$, the class of α. For every $\gamma \in \mathbf{G}_P$, $\alpha \circ \gamma$ is another element of $[\alpha]$ because $\alpha(\gamma(P)) = \alpha(P) = Q$. Hence any other class has at least as many elements as $\mathbf{G}_P$. Conversely, for every β in $[\alpha]$, $\alpha^{-1} \circ \beta$ is in $\mathbf{G}_P$ because $\alpha^{-1}(\beta(P)) = \alpha^{-1}(Q) = P$. (By Problem 4, the symmetries $\alpha \circ \gamma$ and $\alpha^{-1} \circ \beta$ are all distinct.) Hence the classes are all the same size, n. Thus the number of symmetries in $\mathbf{G}$ is nk. ■

Theorem 5.2.4 In a finite symmetry group, either all the isometries are direct or exactly half of them are direct.

Proof. Let $\mathbf{D}$ be the set of the direct isometries and $\mathbf{I}$ the set of the indirect isometries, if any, in the symmetry group. If $\mathbf{D}$ is the entire symmetry group, we are done. If $\gamma \in \mathbf{I}$, then $\gamma \mathbf{D} = \{\gamma \circ \delta : \delta \in \mathbf{D}\}$ is a subset of $\mathbf{I}$ because γ switches orientation but δ does not. Furthermore, distinct δ give distinct products $\gamma \circ \delta$ by Problem 4. Hence $\mathbf{I}$ has at least as many elements as $\mathbf{D}$. A similar argument with $\gamma \mathbf{I} = \{\gamma \circ \beta : \beta \in \mathbf{I}\}$ shows that $\mathbf{D}$ has at least as many elements as $\mathbf{I}$. Hence they have the same number of elements. ■

Example 2 Describe the symmetries of a pentagonal dipyramid.

Solution. By Theorem 5.2.4, we know that half of the 20 symmetries are rotations, including the identity. Four rotations around the axis through the top and bottom vertices have angles of rotation that are multiples of 72°. (See Fig. 5.10(a).) The five remaining rotations, each of 180°, are around axes that go through one of the remaining five vertices on the pentagonal "equator." There are five vertical mirror reflections and one horizontal mirror reflection. Theorem 5.2.4 guarantees four more indirect isometries, which are rotatory reflections. (See Section 4.5.) They can be written as compositions of the horizontal mirror reflection with the rotations around the vertical axis. ●

PROBLEMS FOR SECTION 5.2

1. a) Classify the symmetry group of each of the designs shown in Fig. 5.11.

b) Classify the different types of quadrilaterals (parallelogram, kite, rhombus, and so on) by their symmetry groups.

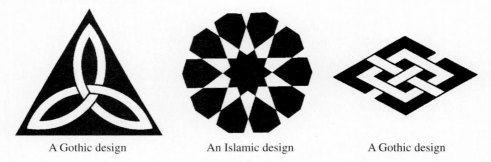

A Gothic design   An Islamic design   A Gothic design

Figure 5.11

2. a) Explain why the symmetries of the rectangle shown in Fig. 5.12(a) and the symmetries of the triangle shown in Fig. 5.12(b) are symmetries of the surrounding hexagons.

b) Draw an analogous design that combines two other dihedral groups.

c) Draw an analogous design that combines two cyclic groups.

d) Draw an analogous design that combines a cyclic group and a dihedral group.

e) For each of the preceding designs give the two symmetry groups. What pattern did you find between each pair of symmetry groups? Make a conjecture about these symmetry groups and try to prove it.

3. This problem introduces the idea of color symmetry. (Because this is a black and white book, we use stippling and cross-hatching to represent colors other than black and white.)

a) For the two-color (black and white) design shown in Fig. 5.13, describe the color-preserving

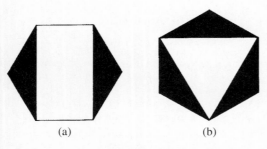

(a)   (b)

Figure 5.12

symmetries—the symmetries that take each region to another region of the same color. Do these symmetries form a transformation group? If so, which one? Explain.

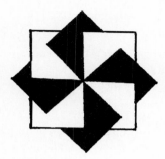

Figure 5.13

b) Describe the symmetries of this design that switch colors. Do these symmetries form a transformation group? If so, which one? Explain.

c) Call the union of the symmetries from part (a) and part (b) the *color symmetries* of the design. Do the color symmetries form a transformation group? If so, which one? Explain.

d) Make other two-color designs having different symmetries. Repeat parts (a), (b), and (c) for each of these designs.

e) Repeat parts (a), (b), and (c) for the solid black, stippled, and cross-hatched design shown in Fig. 5.14.

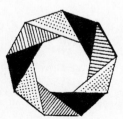

Figure 5.14

f) Make a design having at least three colors. Repeat parts (a), (b), and (c) for this design. Be sure that your color-switching symmetries take all regions of one color to the regions of a second color so that the underlying relationships of the colors are preserved.

g) Make a conjecture based on the results obtained in parts (a)–(f).

4. Show for all symmetries α, β, and γ, if $\beta \neq \gamma$, then $\alpha \circ \beta \neq \alpha \circ \gamma$ and $\beta \circ \alpha \neq \gamma \circ \alpha$. [*Hint:* Use α^{-1}.]

5. If *symmetry* is changed to *rotation* throughout Theorem 5.2.3, is the theorem still correct? If so, prove this revised theorem. If not, explain why it is false.

6. Count the symmetries of each polyhedron named.
   a) The five regular polyhedra: cube, tetrahedron, octahedron, dodecahedron, and icosahedron. (See Fig. 1.44.)
   b) A triangular prism and a square prism.
   c) Generalize part (b) to a prism with regular n-gons for the top and bottom and n rectangles for the sides.
   d) The tetrahedron has half as many symmetries as the cube. Are all the symmetries of the tetrahedron also symmetries of the cube? Explain.

7. Alter and prove Theorem 5.2.1 for isometries in three dimensions.

8. Theorem 5.2.3 can be modified to address faces rather than points. For each of the polyhedra in Problem 6, pick a face, count the symmetries taking that face to itself and the number of faces to which that face can go. Verify that the product of these numbers is the same as the number of symmetries that you found in Problem 6.

9. Modify Theorem 5.2.3 and its proof to count the symmetries of polyhedra, using faces. (See Problem 8.)

10. Relate the group of symmetries of a circle to the dihedral groups $\mathbf{D}_n$.

5.3 SYMMETRY IN THE PLANE

Although real designs cannot contain infinitely many copies of a motif, many designs from various cultures convey that impression. To analyze these patterns we assume that the motif does repeat infinitely often, with translations either in just one direction

Figure 5.15 An Indian frieze pattern.

(Fig. 5.15) or in more than one direction (Fig. 5.16). The classification of the finitely many symmetry groups for frieze and wallpaper patterns provides anthropologists and archaeologists a way to analyze designs of other cultures that transcends cultural boundaries.

Definition 5.3.1 Repetitions of a bounded motif are *discrete* iff there is a minimum positive distance between translations of the motif. A *frieze pattern* is a discrete plane pattern that has translations in just one direction as symmetries. A *wallpaper pattern* is a discrete plane pattern that has translations in more than one direction as symmetries.

To classify frieze patterns we first find which isometries can be symmetries of frieze patterns and then determine how they combine to form groups of symmetries. WLOG we assume that the translations are always horizontal. By Problem 4 frieze patterns

Figure 5.16 A Bornean wallpaper pattern.

5.3 Symmetry in the Plane

always have a horizontal line that must be stable under every symmetry of the frieze. This line, which we call the *midline*, provides one way to limit the possible isometries for frieze patterns. The frieze pattern shown in Fig. 5.15 has all the symmetries described in Theorem 5.3.1, showing that all of them are possible.

Theorem 5.3.1 The only symmetries of a frieze pattern with horizontal translations are horizontal translations, vertical mirror reflections, glide reflections, and mirror reflections over the midline of the frieze pattern, and rotations of 180° with centers on the midline of the frieze pattern.

Exercise 1 Draw figures to illustrate the proof of Theorem 5.3.1.

Proof. Show that all other isometries map the midline to a different line and so cannot be symmetries of the frieze pattern. Translations in a direction other than horizontal lift or lower the midline, so they are eliminated. Rotations other than 180° (or 0°, the identity) tilt the midline. Rotations of 180° whose centers are not on the midline shift it to a different horizontal line. Mirror and glide reflections over lines that are not vertical or horizontal also tilt the midline. Mirror and glide reflections over horizontal lines other than the midline, as well as glide reflections over vertical lines, shift the midline to a different horizontal line. By Theorem 4.2.7, we have eliminated all other isometries.

The analysis of how these possible symmetries fit to form groups requires a deeper understanding of groups than needed in Section 5.2 because now the groups are infinite. (We often simply say *group* instead of symmetry group or transformation group.) We need to find a small number of symmetries for each frieze pattern from which we can generate all the others.

Definition 5.3.2 A *subgroup* **H** of a group **G** is a subset of **G** that is a group in its own right, using the same operation as **G**. The elements $g_1, g_2, \ldots, g_n$ of **G** *generate* a subgroup **H** iff these elements are in **H** and every element h of **H** can be written in terms of these elements and their inverses, in some order and with any finite number of repetitions. We write $\langle g_1, g_2, \ldots, g_n \rangle$ to indicate the subgroup generated by these elements.

Example 1 Show that the dihedral group $\mathbf{D}_n$ is generated by two neighboring mirror reflections, μ_1 and μ_2.

Solution. The composition of μ_1 and μ_2 gives the smallest rotation: $\mu_1 \circ \mu_2 = \rho$. Repetitions of this composition, for example, $\mu_1 \circ \mu_2 \circ \mu_1 \circ \mu_2 = \rho^2$, give other rotations. (See Fig. 5.8.) We generate the mirror reflections as compositions of the form $\mu_1 \circ \rho^i = \mu_1 \circ (\mu_1 \circ \mu_2)^i$, for some power i. ●

Exercise 2 Explain why one translation generates the group of symmetries T of the frieze pattern shown in Fig. 5.17.

Example 2 A translation, a vertical mirror reflection, and the horizontal mirror reflection generate the group of symmetries of the frieze pattern shown in Fig. 5.15.

192 Chapter 5 Symmetry

Figure 5.17 A Mexican frieze pattern.

Solution. We use properties from Chapter 4 to analyze compositions. The smallest translation generates all the others. The compositions of a vertical mirror reflection with the translations give all the vertical mirror reflections. Similarly, the horizontal mirror reflection and the translations generate the glide reflections. The composition of the vertical and horizontal mirror reflections is a rotation of 180°. The other rotations are obtained by composing that rotation with the translations. ●

Theorem 5.3.1 reveals that the symmetry group for the frieze shown in Fig. 5.15 is in some sense the largest such group. The group in Exercise 2 must in the same sense be the smallest, because every frieze pattern must have translations and this frieze pattern has only translations. To be completely rigorous, we need to be more careful. Let T' be the translations that shift each motif an even number of positions in Fig. 5.17. Then T' is a group of symmetries even smaller than group T in Exercise 2. However, T' doesn't differ in any substantial way from T. In algebraic terms, the groups are isomorphic. (See Section 1.4.) The geometric difference is that the distance between repetitions using T' is twice the distance between repetitions using T, which is irrelevant for finding different types of frieze patterns. WLOG we assume that any two friezes have the same smallest translation to the right. Theorem 5.3.2 shows that there are just seven types of frieze patterns.

Example 3 Figure 5.18 shows the seven types of frieze patterns, each having a group of symmetries different from the others. These patterns are traditional patterns from the pottery of the San Ildefonso pueblo in New Mexico. (See Crowe and Washburn [3].) Beside each pattern is the name of the group of symmetries for that pattern. The names of the groups **pxyz** tell us what symmetries they have. If **x** = **m**, there are vertical mirror reflections. If **y** = **m**, there is a horizontal mirror reflection, and, if **y** = **g** there are horizontal glide reflections but not a horizontal mirror reflection. If **z** = **2**, there are rotations of 180°. A **1** in any of these positions indicates that the group doesn't have this type of symmetry. ●

Theorem 5.3.2 There are exactly seven groups of symmetries for frieze patterns, up to isomorphism.

Proof. Example 3 shows that there are at least seven frieze groups. To show there are no others, we consider the possible sets of generators for frieze groups chosen from the isometries described in Theorem 5.3.1. We use τ for the smallest translation to the right,

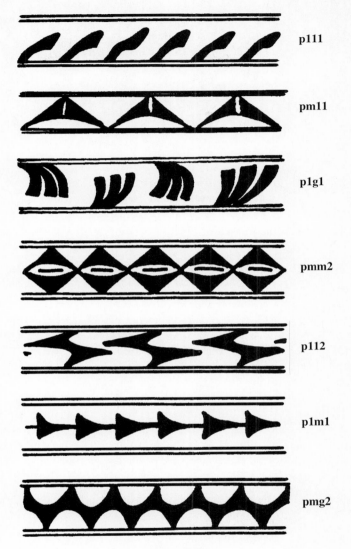

Figure 5.18

ρ for a rotation, η for the horizontal mirror reflection, ν for a vertical mirror reflection, and γ for a glide reflection.

By Problem 6, we do not need to consider every possible such set because of the following observations. All possible rotations are generated by any one rotation and τ. All possible vertical mirror reflections are generated by any one vertical mirror reflection and τ. The horizontal mirror reflection (or a glide reflection) and τ generate all the glide reflections. Finally, the composition of a vertical mirror reflection and a rotation can give two different types of symmetry. If the center of rotation is on the line of reflection, the composition is the horizontal mirror reflection. Otherwise the composition is a glide reflection.

Thus we need to consider generators of the form $\langle \tau, ? \rangle$, where ? is replaced by some of the following general symmetries: ρ, ρ', ν, η, or γ, where we assume the center of ρ to be on the line of reflection of ν but the center of ρ' not to be. (If ν is not one of the generators, it does not matter whether we use ρ or ρ'.) Then $\langle \tau \rangle = \mathbf{p111}$, $\langle \tau, \rho \rangle = \mathbf{p112}$, $\langle \tau, \nu \rangle = \mathbf{pm11}$, $\langle \tau, \eta \rangle = \mathbf{p1m1}$, and $\langle \tau, \gamma \rangle = \mathbf{p1g1}$. We obtain $\mathbf{pmg2}$ from $\langle \tau, \rho', \nu \rangle$, $\langle \tau, \rho, \gamma \rangle$, $\langle \tau, \nu, \gamma \rangle$, or $\langle \tau, \rho', \nu, \gamma \rangle$. We obtain $\mathbf{p1m1}$ from $\langle \tau, \eta, \gamma \rangle$ or $\langle \tau, \eta \rangle$. In Problem 6 you are asked to show that all other sets of generators yield $\mathbf{pmm2}$. ∎

The classification of wallpaper patterns is more complicated than that of the frieze patterns. Figure 5.19 shows that different rotations are possible. However, Theorem 5.3.3 reveals that the angles of rotation illustrated in Fig. 5.19 are the only possible angles. This result is called the crystallographic restriction because it is also crucial in the classification of three-dimensional crystals.

Theorem 5.3.3 **The Crystallographic Restriction** The minimal positive angles of rotations that can be symmetries of a wallpaper pattern are 60°, 90°, 120°, and 180°, and 360°. All other angles of rotations for a given wallpaper pattern are multiples of the minimum angle.

Proof. Let A be a center of rotation for a wallpaper pattern and let B be a point closest to A for which some symmetry takes A to B. Then B must, by symmetry, also be a center of rotation for the wallpaper pattern with the same angles as at A. No two other images of A can be any closer together than are A and B. WLOG assume that A is to the left of B. Let ρ be the smallest positive rotation with center at A and ϕ be the smallest negative rotation with center at B. Now consider $A' = \phi(A)$ and $B' = \rho(B)$ (Fig. 5.20). By symmetry, both A' and B' must be centers of rotation like A. If ρ rotates less than 60°, then $d(A', B')$ will be less than $d(A, B)$, which is impossible. Similar reasoning (see Problem 7) eliminates other minimum angles except 90°, 120°, 180°, or 360°. Hence only the specified angles are compatible with wallpaper patterns. ∎

Theorem 5.3.4 There are exactly seventeen groups of symmetries for wallpaper patterns, up to isomorphism.

Proof. See Crowe [2]. ∎

The Russian chemist and mathematician Vyatseglav Fedorov in 1891 first stated and proved Theorem 5.3.4, but his proof wasn't widely noted. Several other mathematicians, including Felix Klein, independently found and proved this classification. The proof of Theorem 5.3.4 is based on group theory, but it isn't difficult to suspect geometrically that mirror and glide reflections can fit particular angles of rotation in only finitely many ways. Thus the number of wallpaper patterns is finite, even if the number 17 remains somewhat mysterious. The flowchart presented in Fig. 5.21 compresses the mathematics of the proof into a methodical way of classifying wallpaper patterns. The geometer Don Crowe developed such flow charts to aid archaeologists and anthropologists.

5.3 Symmetry in the Plane

Figure 5.19 Wallpaper patterns.

The names of the wallpaper groups in Fig. 5.21 aren't as simple as the names of the frieze groups. The numbers **2**, **3**, **4**, and **6** refer to the maximum number of rotations around a center of rotation, as do the groups $\mathbf{C}_n$ and $\mathbf{D}_n$. The letters **m** and **g** refer to mirror and glide reflections. The letter **c** stands for a rhombic lattice, instead of a rectangular lattice. The difference is explained in Example 4. Then Example 5 considers two groups that are often as difficult to distinguish as are their names, **p31m** and **p3m1**.

196   Chapter 5   Symmetry

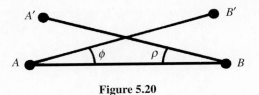

Figure 5.20

Example 4   Classify the patterns shown in Fig. 5.22.

Solution.   Neither pattern has any rotations, so we follow the *none* branch in Fig. 5.21. Both have reflections, so we need to look at the glide reflections. In the Zairean design, the motifs stack like boxes, so the glide reflections line up with the mirror reflections. In the Chinese design the motifs alternate, like bricks in a wall, enabling new glide reflection axes. The Zairean has the group **pm**, while the Chinese has **cm**. ●

Example 5   Classify the patterns shown in Fig. 5.23.

Solution.   Both designs have rotations of 120° but not 60°. Each has some mirror reflections that pass through centers of rotation. Indeed, Fig. 5.23(a) has mirror reflections

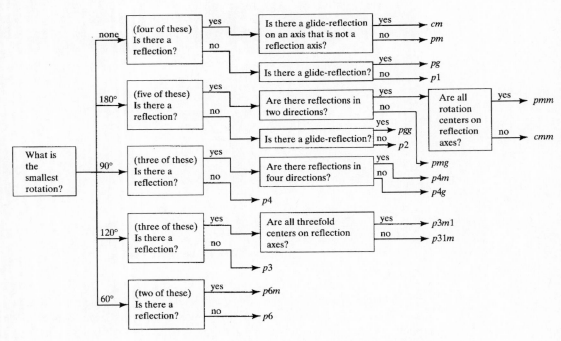

Figure 5.21   Flow chart for the 17 wallpaper groups.
Source: Used with permission from Gallian, J. *Contemporary Abstract Algebra.* Lexington, Mass.: D. C. Heath, 1994.

5.3 Symmetry in the Plane

A Zairean design   A Chinese design

Figure 5.22

through every center of rotation. Hence its group is **p3m1**. However, the centers of the triangles in Fig. 5.23(b) don't have mirror reflections, so it has group **p31m**. ●

The anthropologist Dorothy Washburn teamed with the geometer Don Crowe to pioneer the use of symmetry groups in cross-cultural studies. Before their work, researchers had tried to analyze the varying motifs of different cultures. Although some characteristics seem apparent, the great variety of motifs made an analysis of motifs culturally subjective. However, the symmetry groups are independent of culture. For many cultures, the types of frieze and wallpaper designs used by their artists remain the same over long periods of time. This stability provides people studying cultures another marker in the study of societies and their interactions. A new design can indicate the

(a)   (b)

Figure 5.23   Two Japanese designs.

198   Chapter 5   Symmetry

influence of trade with another region. Cultures that emphasize weaving tend to utilize designs from other media, such as ceramics, which have the symmetry patterns that can be obtained with weaves. For an ancient culture for which no traces of weaving may remain, patterns on ceramics, which can endure millennia, can provide indirect evidence about weaving. (See the Spanish and Mongolian patterns in Fig. 5.19.)

Many cultures create patterns with two-color symmetry or multiple-color symmetry. Color symmetry involves two groups: the color-preserving group and the color group, a larger group that includes the color-switching symmetries. We can analyze each just as we do any other symmetry group. (Again, stippling and cross-hatching are used, as necessary, to represent additional colors.)

Example 6   Classify the colored patterns shown in Fig. 5.24.

Solution.   First consider the color-preserving symmetries shown in Fig. 5.24(a), a Peruvian design. All the black "staircases" are upright, implying no color-preserving rotations. The rows of black staircases alternate facing left and right, indicating no mirror reflections but indicating glide reflections preserve colors. From Fig. 5.21, the color-preserving group is **pg**. All the white staircases are upside down, so 180° rotations can switch colors. Horizontal mirror reflections between rows switch colors, but no vertical mirrors work. Figure 5.21 then tells us the color group is **pmg**. We write the pair of groups with the color group on top: **pmg/pg**.

The pattern shown in Fig. 5.24(b) is a three-color frieze pattern. The central vertical mirror reflection preserves the black parts but switches the white and stippled parts. The only symmetries preserving all three colors are translations and glide reflections, so the color-preserving group is **p1g1**. The color-switching group includes vertical mirror reflections and rotations, but not horizontal mirror reflections. Hence the color group is **pmg2**, and the classification is **pmg2/p1g1**. Note that some translations in **pmg2** are shorter than the translations that preserve colors.   ●

Definition 5.3.3   A symmetry of a design is *color-preserving* iff every repetition A of the motif in the design is mapped to a copy that is the same color as A. A symmetry κ of a design is

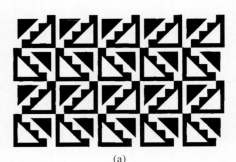
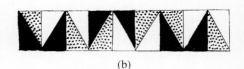

(a)   (b)

Figure 5.24

5.3 Symmetry in the Plane

a *color-switching symmetry* iff, whenever the repetitions A and B of the motif are the same color, then $\kappa(A)$ and $\kappa(B)$ are the same color. A *color symmetry* of a design is either a color-preserving or a color-switching symmetry.

Theorem 5.3.5 The color-preserving symmetries of a design form a subgroup of the color symmetries of the pattern.

Proof.   See Problem 8.   ∎

The widespread use of frieze and wallpaper patterns throughout the world shows the appeal of symmetry across barriers of time, language, and race. It also shows the geometric understanding needed to join motifs to make these patterns. However, there is no evidence that any individual or society explicitly considered finding all types of these patterns until the late nineteenth century. It is no accident that Fedorov and the others who thought about classification were trained in formal mathematics, especially group theory. Mathematics provides new ways to see the world, enriching understanding.

PROBLEMS FOR SECTION 5.3

1. Classify the frieze patterns shown in Fig. 5.25.

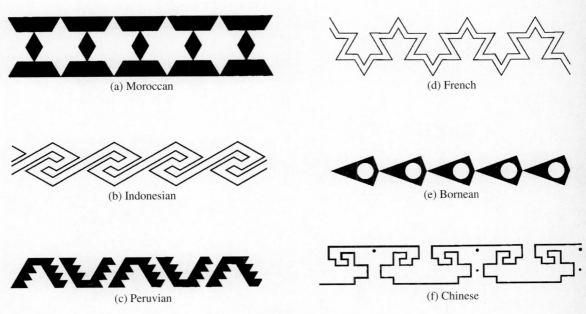

(a) Moroccan

(d) French

(b) Indonesian

(e) Bornean

(c) Peruvian

(f) Chinese

Figure 5.25

200 Chapter 5 Symmetry

2. Classify the two-color frieze patterns shown in Fig. 5.26. These designs represent all 17 types of two-color frieze patterns. The first 14 designs (parts a–n) are from the pottery of the San Ildefonso pueblo. Crowe and Washburn [3] designed the last three (parts o–q) to complete the set.

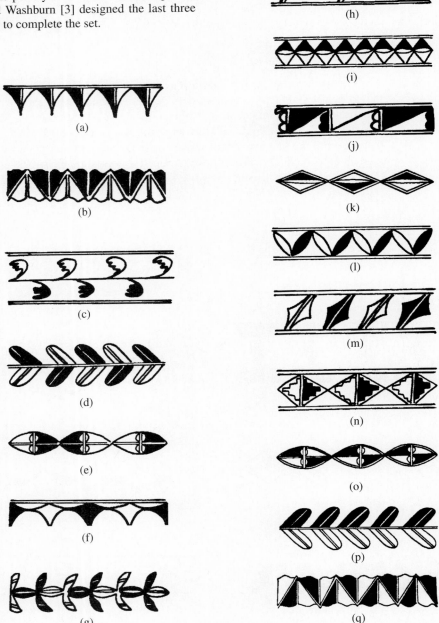

Figure 5.26

3. Classify the wallpaper patterns shown in Fig. 5.27.

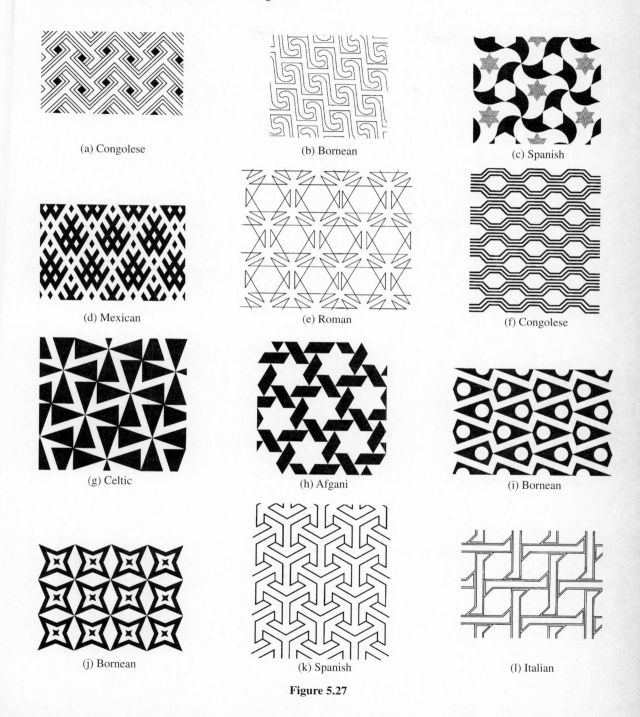

Figure 5.27

4. Prove that in every frieze pattern with horizontal translations there must be at least one stable horizontal line. [*Hint:* Suppose that no such stable line existed; show that a translation would also be possible in some other direction.]

5. a) Make a flow chart to classify frieze patterns.
   b) Describe which symmetry groups for frieze patterns are subgroups of the others.

6. Complete the proof of Theorem 5.3.2.

7. Complete the proof of Theorem 5.3.3.

8. a) To prove Theorem 5.3.5, prove that both sets of symmetries form groups.
   b) (Group theory) Prove that the color-preserving group is a normal subgroup of the color group.

9. Find as many types of wallpaper patterns as you can with a motif of a rectangle twice as long as it is wide. The rectangles need not all line up the same way, although they shouldn't overlap or have gaps. Which symmetry groups can't be realized with this motif? Explain.

10. a) Find all regular wallpaper patterns. To be regular, the motif must be a regular polygon, there can be no gaps between or overlaps of the polygons, and two polygons with more than a point in common must share an entire edge.
    b) Find all eight semiregular wallpaper patterns. A semiregular pattern differs from a regular pattern in that the motif must be two or more regular polygons and all vertices must have the same pattern of polygons around them. [*Hint:* The sum of the angles at each vertex must add to 360°. There are at least three polygons at each vertex.] (Johannes Kepler (1571–1630) was the first to find these patterns.)

11. Draw the design for a plain weave, where each horizontal thread alternately goes over and under the vertical threads. Classify the symmetry group of this design. Classify the Spanish design shown in Fig. 5.19 and verify that all its symmetries are symmetries of a plain weave.

12. Classify the two-color wallpaper patterns of Fig. 5.28.

(a) Moroccan　　　(b) Brazilian　　　(c) Egyptian

(d) Iranian　　　(e) Japanese　　　(f) Zairean

Figure 5.28

5.4 Symmetries in Higher Dimensions

5.4.1 Finite three-dimensional symmetry groups

Mathematicians and others since the time of ancient Greece have enjoyed the elegance of polyhedra, especially the five regular polyhedra. However, not until the nineteenth century was polyhedral symmetry analyzed. Auguste Bravais first published the classification of the finite three-dimensional symmetry groups (Theorem 5.4.2) in 1848, although Johann Hessel had proved it in 1830.

Some of the three-dimensional symmetry groups extend the families $\mathbf{C}_n$ and $\mathbf{D}_n$ discussed in Section 5.2. A prism (Fig. 5.29) has twice as many symmetries as the corresponding regular polygon that forms its top face. The group of symmetries of a prism is called $\overline{\mathbf{D}'_n}$, where n is the number of vertices of the top face. Theorem 5.2.4 guarantees that half the symmetries in $\overline{\mathbf{D}'_n}$ are rotations. They form a group $\mathbf{D}'_n$, which is structurally the same as $\mathbf{D}_n$. In place of the n mirror reflections of $\mathbf{D}_n$, $\mathbf{D}'_n$ has n rotations of 180° around horizontal axes. The overbar in $\overline{\mathbf{D}'_n}$ and the other finite groups indicates a doubling of the size of the group to include mirror reflections and rotatory reflections. The groups $\mathbf{C}_n$ and $\mathbf{D}_n$ may be regarded as subgroups of $\overline{\mathbf{D}'_n}$, which also has other subgroups. (See Project 9.)

In addition to the subgroups of $\overline{\mathbf{D}'_n}$, there are a few other finite three-dimensional symmetry groups. These other groups all come from the symmetries of the regular polyhedra. Although there are five regular polyhedra, they determine only three different groups of symmetries. (See Fig. 1.44.) The tetrahedron has 24 symmetries, forming the tetrahedral group, $\overline{\mathbf{T}}$. These symmetries include rotations of 120° and 180°. The cube and octahedron have the same group of 48 symmetries, the octahedral group, $\overline{\mathbf{W}}$. The angles of rotation in this group are multiples of 90° and 120°. The icosahedron and dodecahedron have the same group of 120 symmetries, the icosahedral group, $\overline{\mathbf{P}}$. The angles of rotation in this group include multiples of 72°, 120° and 180°. These three groups have various subgroups, most notably their subgroups of rotations $\mathbf{T}$, $\mathbf{W}$, and $\mathbf{P}$. Many interesting polyhedra, including 11 of the 13 Archimedean solids have for their symmetry groups $\overline{\mathbf{T}}$, $\overline{\mathbf{W}}$, or $\overline{\mathbf{P}}$. The two remaining Archimedean solids have only rotations, and their groups of symmetries are $\mathbf{W}$ and $\mathbf{P}$. (See Wenninger [21] for pictures of these polyhedra.)

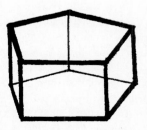

Figure 5.29

Figure 5.30

Example 1 Find the group of symmetries of the great stellated dodecahedron (Fig. 5.30).

Solution. This polyhedron has axes of rotation other than those of a prism. The axis facing out allows rotations of 72°, so the group can't be $\overline{\mathbf{T}}$ or $\overline{\mathbf{W}}$ or their subgroups. Because this polyhedron has mirror reflections, by Theorem 5.4.2 below, it must have $\overline{\mathbf{P}}$ for its group of symmetries. ●

Theorem 5.4.1 The three-dimensional isometries in a finite symmetry group must fix a point and so form a subgroup of the symmetries of a sphere.

Proof. See Problem 3. ■

Theorem 5.4.2 A finite group of three-dimensional isometries must be one of the following or one of their subgroups: $\overline{\mathbf{D}'_n}$, $\overline{\mathbf{T}}$, $\overline{\mathbf{W}}$, and $\overline{\mathbf{P}}$.

Proof. See Weyl [22, 149ff]. ■

5.4.2 The crystallographic groups

The beauty of crystals, especially gems, have fascinated people for centuries. However, prior to 1849 no chemical explanation existed of the visible regularities and other properties of crystals. The mathematical classification by Bravais in 1849 of the 32 types of crystals spurred the study of geometric arrangements of atoms in crystals. (Chemists sometimes say that there are 33 types; two are mirror reflections of each other.) These crystals are the three-dimensional analogs to frieze patterns and wallpaper patterns. A mathematical crystal is a discrete pattern having translations in at least three directions, not all in the same plane. The subgroup of translations takes a point to a three-dimensional lattice of points (Fig. 5.31). To classify crystals, Bravais used the crystallographic restriction (Theorem 5.3.3) which gives the rotations possible for two-dimensional wallpaper patterns. Theorem 5.3.3 applies to crystals as follows. Consider a three-dimensional rotation about an axis k and a point P in the lattice not on k. The rotation moves P to another point in the plane perpendicular to k and through P. The

5.4 Symmetries in Higher Dimensions

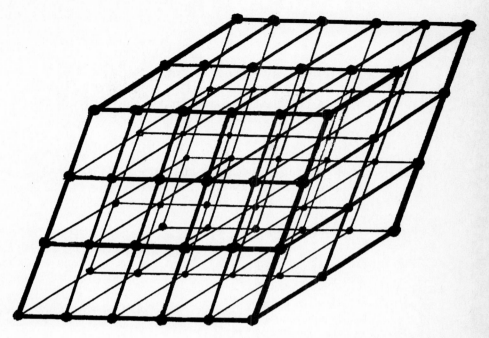

Figure 5.31

points of the lattice lying in this plane form a wallpaper pattern. Thus the rotation must be a rotation of the wallpaper pattern. Hence Theorem 5.3.3 applies to mathematical (and chemical) crystals. However, crystals can combine these possible rotations in more ways than wallpaper patterns can. (See Senechal [17].)

5.4.3 General finite symmetry groups

Finite groups appear in many contexts, many of which don't correspond in any obvious way to isometries in Euclidean geometry in any number of dimensions. Nevertheless, any finite group may be thought of as a symmetry group of some figure. In effect, we may think of the group as a transformation group moving points around. In particular Theorem 5.4.4 allows us to think of these points as embedded in Euclidean space of an appropriate dimension and the transformations of the group as isometries.

Definition 5.4.1 The group of all transformations on a set of n distinct points is the *symmetric group* $\mathbf{S}_n$. If n is a positive integer, $n!$ (read "n factorial") is the product of the numbers from 1 to n.

Theorem 5.4.3 For any positive integer n, the group $\mathbf{S}_n$ has $n!$ elements.

Proof.   See Problem 7.   ∎

Theorem 5.4.4 The group of isometries of the $(n-1)$-dimensional regular simplex is $\mathbf{S}_n$.

Proof. See Problem 8. (A regular simplex is defined in Section 2.5.) ■

Example 2 The symmetry group of an equilateral triangle is $\mathbf{D}_n$ by Theorem 5.2.2 and is $\mathbf{S}_3$ by Theorem 5.4.4. These groups are isomorphic, that is, structurally identical. Similarly, the regular tetrahedron shows the two groups $\overline{\mathbf{T}}$ and $\mathbf{S}_4$ to be isomorphic. ●

PROBLEMS FOR SECTION 5.4

1. Explain why the symmetries of a rectangular box form a subgroup of the symmetries of a square prism, which in turn form a subgroup of the symmetries of a cube. Draw figures. Use Theorem 5.2.3 to find the number of symmetries of each of these shapes. Describe all the symmetries of a rectangular box, which form the group $\overline{\mathbf{D}'_2}$.

2. An *antiprism* has two regular polygons for its bases, but, unlike a prism, the vertices of the top polygon alternate with the vertices on the bottom so that the sides are isosceles triangles (Fig. 5.32). Count the number of symmetries of an antiprism with regular n-gons for its bases. Compare the symmetries of an antiprism with the symmetries of the prism with regular n-gons for bases.

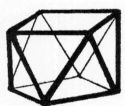

Figure 5.32

3. Prove Theorem 5.4.1. [*Hint:* See the proof of Theorem 5.2.1.]

4. Show that the tetrahedral group is a subgroup of the octahedral group. [*Hint:* Fit a tetrahedron in a cube.]

5. Let ζ be the three-dimensional central symmetry with respect to the origin. (See Problem 2 of Section 4.5.) Write ζ as a 3×3 matrix and show that it commutes with every spherical isometry. That is, if σ is a spherical isometry, $\sigma \circ \zeta = \zeta \circ \sigma$.

6. a) Describe the rotations and mirror reflections of a cube. Explain why the regular octahedron has the same symmetries. How many rotatory reflections does a cube have?

   b) Describe the rotations and mirror reflections of a regular tetrahedron. How many rotatory reflections does a tetrahedron have?

   c) Repeat part (a) for a regular icosahedron and dodecahedron.

7. Use induction to prove Theorem 5.4.3. [*Hint:* For the set $\{1, 2, \ldots, n\}$, why are there n possible places where 1 can go? Use the reasoning followed in Theorem 5.2.3.

8. Prove Theorem 5.4.4.

9. Classify the symmetry groups of the 13 semiregular polyhedra (the Archimedean solids). (See Wenninger [21] for pictures of these polyhedra.)

10. a) For which n is the group $\overline{\mathbf{D}'_n}$ a subgroup of $\overline{\mathbf{T}}$?

    b) Repeat part (a) but replace $\overline{\mathbf{T}}$ with $\overline{\mathbf{W}}$.

    c) Repeat part (a) but replace $\overline{\mathbf{T}}$ with $\overline{\mathbf{P}}$.

    d) Repeat part (a) but replace $\overline{\mathbf{T}}$ with $\overline{\mathbf{D}'_k}$, where $n \leq k$.

5.5 SYMMETRY IN SCIENCE

5.5.1 Chemical structure

Chemists benefit greatly from a geometric understanding of the arrangement of the atoms (and ions) in chemical compounds. (For simplicity we refer to the parts of compounds as *atoms*, ignoring the distinction between atoms and ions. Similarly, we avoid

5.5 Symmetry in Science

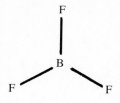

Figure 5.33 Boron trifluoride.

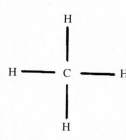

Figure 5.34 An incorrect arrangement of methane bonds.

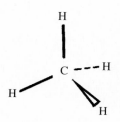

Figure 5.35
Methane (Δ denotes a bond extending from the page toward the reader, and --- denotes a bond receding into the page away from the reader.)

discussion of electrons, orbitals, and quantum mechanics, even though symmetry plays a vital role at that level.) Chemists represent the atoms in a compound by vertices and the bonds between atoms by edges. For example, boron trifluoride (BF_3) has three fluorine atoms bonded to a boron atom (Fig. 5.33). Atoms bonded to a particular atom tend to arrange themselves as far from one another as possible, while staying the appropriate distance from that atom. The arrangement shown in Fig. 5.33 is as symmetric as possible: The six (two-dimensional) symmetries form the group $\mathbf{D}_3$. Thus symmetry frequently allows analysis of chemical structure.

Carbon usually bonds with four atoms. The simplest carbon compound, the gas methane (CH_4), has just one carbon atom and four hydrogen atoms. You might first imagine that the hydrogens would arrange themselves as some chemistry textbooks display this compound for simplicity (as in Fig. 5.34, which has the symmetry group $\mathbf{D}_4$). Although this arrangement has considerable symmetry, chemistry involves three dimensions. Figure 5.35 represents the actual placement of the hydrogens as the vertices of a regular tetrahedron with the carbon at the center, which was deduced in 1874. Hence the hydrogen atoms are farther apart than depicted in Fig. 5.34. Indeed, the angle of two bonds, as shown in Fig. 5.35, is approximately $109.5°$, rather than $90°$, shown in Fig. 5.34. Furthermore, the tetrahedral group $\overline{\mathbf{T}}$ has 24 symmetries, or more than the 8 two-dimensional symmetries of Fig. 5.34. (As a three-dimensional shape, Fig. 5.34 has 16 symmetries, or still fewer than shown in Fig. 5.35.) An increase in symmetry corresponds to a lower, more stable energy state.

One form of pure carbon, a diamond crystal, extends the symmetry of Fig. 5.35. In a perfect diamond, each carbon atom is bonded to four other carbon atoms that form a regular tetrahedron (Fig. 5.36). To analyze a crystal, we assume that it continues in all three dimensions forever, which is a reasonable simplification. For example, a one-carat diamond has approximately 10^{22} carbon atoms, which means millions of repetitions of the pattern shown in Fig. 5.36 along any axis. The symmetries of the crystal include translations in three dimensions and the local isometries of the tetrahedral group. The variety of the directions of the bonds and their uniformity makes diamond the hardest naturally occurring substance. The angles of the bonds also determine the possible angles at which gem diamonds can be cut. For example, you will never see a diamond cut as a cube.

Carbon forms another crystal, graphite, whose very different physical and chemical properties reflect the different geometry of the crystal. The atoms of graphite form layers one atom thick with only weak bonds between layers. In each layer the carbon atoms form a pattern of hexagons (Fig. 5.37). These layers slide easily over one another, making graphite an excellent lubricant. The "lead" in pencils also contains graphite.

Exercise 1 Classify the wallpaper pattern of a layer of graphite.

Salt (NaCl) exhibits a third crystalline structure (Fig. 5.38). The cubic arrangement at the atomic level ensures that salt grains always have rectangular faces. Potassium chloride (KCl) has the same crystalline form as salt and is extremely close chemically to salt. Indeed, KCl is a salt substitute for those restricting their intake of sodium (Na).

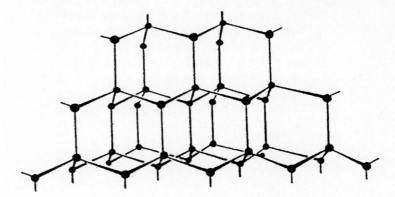

Figure 5.36 The crystal structure of diamond.

The presence of two different kinds of atoms in either salt or potassium chloride affects the group of symmetries. For example, a translation of one bond's length along the x-, y-, or z-axis switches the sodium and chlorine atoms. This switch corresponds to the color symmetries discussed in Section 5.3. The crystal sphalerite, made of zinc and sulfur, is a two-color version of the diamond crystal, with each zinc atom bonded to four sulfur atoms and conversely. (See Senechal [17] for more on crystals.)

Exercise 2 Describe a rotation of the salt crystal that switches sodium and chlorine atoms. [*Hint:* the axis isn't one of the lines for the bonds.]

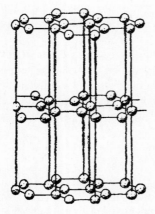

Figure 5.37 The crystal structure of graphite. Each layer consists of a pattern of regular hexagons.

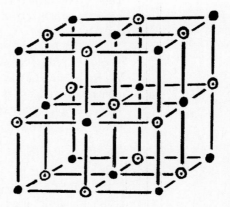

Figure 5.38 The crystal structure of salt.

MARJORIE SENECHAL

Marjorie Senechal (1939–) is a leading researcher in mathematical crystallography and a professor at Smith College, a distinguished college for women. Although she was fascinated with patterns from a young age, not until much later did she realize that "mathematics is the science of patterns." She believes visual thinking to be vital for all mathematics and particularly crystallography.

After completing her Ph.D. in number theory in 1965, Senechal shifted her focus on patterns to mathematical crystallography and, more generally, to discrete geometry. She has taught and conducted research at Smith College since 1966, interspersed with extended research stays in seven countries. Senechal has published six books (with several in preparation) and dozens of articles, and has contributed chapters in more than a dozen books. In addition, she has organized numerous conferences, given many featured addresses, and served several professional organizations as a committee or board member, and has made many other less visible contributions to modern mathematical research.

Senechal eagerly responded to the discovery of quasicrystals in 1984 and the ensuing heightened activity in mathematics, chemistry, and physics. In addition to traditional research in this area, Senechal has worked with the advanced computing facilities at the Geometry Center of the University of Minnesota. Powerful computers are beginning to provide needed visual and analytic insight into the mathematical structure of quasicrystals, which lack the well-understood repetitions of crystals.

5.5.2 Quasicrystals

In Section 5.3 we proved Theorem 5.3.3, the crystallographic restriction, describing the angles of rotation compatible with translations in two or more dimensions. This mathematical result matched chemists' experimental data on x-ray diffraction patterns perfectly until 1984. That year a team of chemists found a compound that gave sharp diffraction patterns with the angle of 72°, which was inconsistent with Theorem 5.3.3. Sharp patterns had previously only been seen with crystals, so these new compounds were called *quasicrystals*. Mathematicians, chemists, and others have explored the rapidly growing field of quasicrystals. Chemists found that, as the crystallographic restriction assures us, the arrangement of atoms in these quasicrystals is not periodic.

Ten years before quasicrystals were discovered, Roger Penrose devised patterns, such as that depicted in Fig. 5.39, which he showed could cover the plane but had no translational symmetry. (See Project 15.) By the early 1980s, Penrose and others had generalized these tilings to three dimensions. Some of these three-dimensional tilings corresponded mathematically to the diffraction patterns of quasicrystals, providing one approach to the mathematical analysis of quasicrystals.

Another mathematical insight of the 1970s brought symmetry into the study of quasicrystals. Mathematicians showed that a six-dimensional "hypercrystal" could be sliced into a three-dimensional cross section that would look like a quasicrystal. The symmetries (including translations) of the six-dimensional model and the angle of the cut

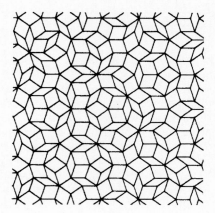

Figure 5.39 A Penrose tiling.

determine the properties of the three-dimensional cross section. We still don't know the chemical relevance of six mathematical dimensions, but researchers Bak and Goldman "emphasize that thinking of [a quasicrystal] as a periodic structure in six dimensions is not merely an amusing mathematical abstraction." (Jaric [10, 146]) (See Peterson [13, 200–212] and Senechal [17] for more information on quasicrystals.)

5.5.3 Symmetry and relativity

Albert Einstein's theory of relativity transformed physics. In geometric language, the special theory corresponds to a four-dimensional group of symmetries that preserve physical properties. Although physicists had already used time as a fourth dimension, Einstein (1879–1955) realized that space and time interacted. The simpler concept of Galilean relativity is helpful both to illustrate the role of symmetry and to clarify Einstein's contribution.

Galileo Galilei (1564–1642) explained why we don't feel the motion of the earth as it travels through space. The motion of everything on the earth includes the speed of the earth, so our measurements (and senses) detect only the relative differences in the speeds of objects, not their absolute speeds. That is, the laws of physics remain the same at different velocities. In other words, this principle of relativity holds that constant velocities in any direction are symmetries for the laws of physics.

Sir Isaac Newton (1642–1727) and physicists following him added another principle: The measurement of time and distance is absolute. That is, clocks record the same amount of time passing and rulers measure the same lengths, regardless of the speed and direction they are traveling relative to one another. This absolute measurement of space and time leads to the additivity of velocities. For example, suppose that an observer on the ground measures a train moving 20 m/sec and an observer in the train observes someone else in the train walking in the direction of the train at 1.5 m/sec. For the observer on the ground, the walker would be moving at 21.5 m/sec. At modest speeds, the additivity of velocities matches our experience.

The Mickelson–Morley experiments conducted at the end of the nineteenth century revealed a problem with Newton's assumption. These experiments sought to measure the influence of the motion of the earth on the velocity of light. Relative to the sun the earth is moving approximately 18 mi/sec, a tiny part of light's velocity of approximately 186,000 mi/sec. However, Mickelson and Morley devised an experiment accurate enough to detect the small difference between 186,000 and 186,018 mi/sec. Regardless of the direction the light was sent, they found its velocity always to be the same, contradicting the additivity of velocities. Later experiments have confirmed that the speed of light in a vacuum is constant. In 1906, Einstein showed how to combine the constancy of the speed of light and Galileo's principle of relativity by dropping the absolute measurement of space and time and so the additivity of velocities. Surprisingly, Eq. (5.1), the formula for combining velocities, corresponds to composing hyperbolic translations, which we discuss in Section 6.5. For ease, we write velocities as fractions of the speed of light, so 1 is the speed of light. For the simple case of velocities along a line, we replace the addition of velocities, $x + y$, with

$$x \oplus y = \frac{x + y}{1 + xy}. \tag{5.1}$$

Example 1 Suppose that observer A is moving at a velocity of 0.4 with respect to observer B and that an object is moving in the same direction at a velocity of 0.5 with respect to A. Then the object is moving at a velocity of $0.4 \oplus 0.5 = 0.75$ with respect to observer B, or less than $0.4 + 0.5 = 0.9$. The velocities of a spacecraft and the earth are much smaller, or approximately 0.00003 and 0.0001. For these velocities, $0.00003 \oplus 0.0001 = 0.0001299999996$, which for all practical purposes is $0.00003 + 0.0001$, or 0.00013. Thus NASA doesn't need to use relativity theory to plan space missions. ●

Exercise 3 Verify that two observers find the same speed of light ($y = 1$ in Eq. 5.1) regardless of their relative velocities, x.

Example 2 Suppose that two people A and B each find the coordinates of two points C and D by using different axes, as illustrated in Fig. 5.40. The Pythagorean theorem in Euclidean geometry guarantees that they will obtain $\Delta x_A^2 + \Delta y_A^2 = \Delta x_B^2 + \Delta y_B^2$ for the square of the distance between two points. ●

In the theory of relativity the measurements of elapsed time and distance by different observers are related much as distances are in Example 2. Suppose that two observers each record two events taking place using suitable units. Observer A finds the difference in time between the events to be Δt_A and the differences in the x-, y-, and z-directions to be Δx_A, Δy_A and Δz_A. Similarly, Δt_B, Δx_B, Δy_B, and Δz_B are observer B's measurements. The theory of relativity guarantees that

$$\Delta x_A^2 + \Delta y_A^2 + \Delta z_A^2 - \Delta t_A^2 = \Delta x_B^2 + \Delta y_B^2 + \Delta z_B^2 - \Delta t_B^2. \tag{5.2}$$

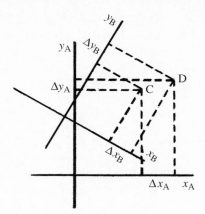

Figure 5.40

Hermann Minkowski developed a four-dimensional geometry by using Eq. 5.2 as a "distance" formula. The Lorentz transformations preserve Eq. 5.2 and so are the symmetries for the special theory of relativity. (In Section 6.6 we discuss these transformations. See Taylor and Wheeler [19] for more on the theory of relativity.)

Example 3 Suppose that person A flashes a light twice, 1 s apart according to A's measurement. If person B is traveling in the x-direction at some velocity with respect to A and observes these light flashes, Δx_B and Δt_B will differ from $\Delta x_A = 0$ and $\Delta t_A = 1$, but $\Delta x_B^2 - \Delta t_B^2 = \Delta x_A^2 - \Delta t_A^2 = 0 - 1 = -1$. The faster B is traveling with respect to A, the larger both Δx_B and Δt_B will be. Note that if B flashes the light twice, the situation is reversed. Thus, although A and B think that their clocks are running at different rates, it doesn't make sense to say whose clock "actually" is slower. Similarly, we can't say that one has a shorter unit of x-distances than the other. ●

The use of symmetry in quantum mechanics, although important, goes beyond the level of this text. See Project 19.

PROBLEMS FOR SECTION 5.5

Two possible arrangements of the atoms for a molecule are chemically equivalent provided that a direct isometry exists that converts one arrangement to the other. If no such direct isometry exists, the possible arrangements are *isomers*.

1. **a)** Describe the two-dimensional symmetries of the ethene molecule (C_2H_4), which has a double bond between the carbon atoms. Find the group of symmetries of C_2H_4 (Fig. 5.41). Note that all six atoms lie in a plane.

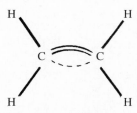

Figure 5.41 Ethene.

b) Dichloroethene ($C_2H_2Cl_2$) replaces two of the hydrogen atoms of ethene with chlorine atoms. Draw the three isomers of this molecule and describe the symmetries of each isomer. Verify that the three groups of symmetries of these isomers are subgroups of the symmetries of ethene.

2. The six carbon atoms of benzene (C_6H_6) form a ring (Fig. 5.42). These atoms exhibit *resonance*, the nonlocalized sharing of electrons, illustrated as a circle. Thus just three atoms are directly attached to each carbon. Assume that the 12 atoms of benzene are in a plane. (Actually, the molecule is three-dimensional.)

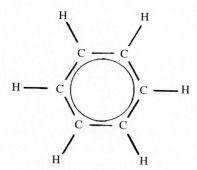

Figure 5.42 Benzene.

a) Describe the two-dimensional symmetries of the model shown in Fig. 5.42. Find the group of symmetries of the model.

b) Repeat Problem 1(b) for dichlorobenzene ($C_6H_4Cl_2$) which replaces two of the hydrogen atoms with chlorine atoms.

c) Repeat Problem 1(b) for trichlorobenzene ($C_6H_3Cl_3$) which replaces three of the hydrogen atoms with chlorine atoms. Consider "color" symmetries that switch the hydrogen and chlorine atoms, as well as those that don't.

3. Draw isomers for a molecule with a central carbon atom and four different atoms (W, X, Y, and Z) attached to it. How are these isomers different geometrically? Chemically, these molecules polarize light differently.

4. Find the symmetry group of buckminsterfullerene, C_{60}, a recently found third form of pure carbon (Fig. 5.43).

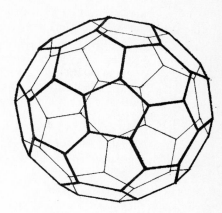

Figure 5.43 Buckminsterfullerene.

5. Each layer of carbonates (CO_3) in a crystal of calcite ($CaCO_3$) looks like the structure depicted in Fig. 5.44. Analyze the wallpaper pattern of this layer. (The calcium atoms lie in different layers not shown. Marble is one form of calcite.)

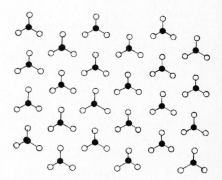

Figure 5.44 The structure of a layer of carbonates in a crystal of calcite.

6. Use the operation given in Eq. 5.1. Explain your answers.

a) Two streams of protons are approaching each other head-on in a cyclotron, each with a velocity of 0.9 of the speed of light relative to an observer. How fast is one stream moving relative to the other stream?

b) Suppose that you observe a particle moving at a velocity of 0.7 relative to you, followed by a second particle moving in the same direction with a velocity of 0.8 relative to you. How fast is the second particle moving relative to the first?

c) Suppose that a first space ship is moving at a velocity of 0.5 relative to you and can launch a second space ship in the same direction moving at a velocity of 0.5 relative to it. In turn, the second space ship can launch a third space ship at a velocity of 0.5 relative to it, the third can do the same, and so on. How many space ships must there be for the fastest one to be moving at a velocity greater than 0.9 relative to you? Repeat for 0.99 and 0.999.

7. Show that the numbers strictly between -1 and 1 form a group with the operation $\oplus$ given in Eq. 5.1 as follows.

   a) What velocity is the identity for $\oplus$? Prove your choice correct.
   b) Find the inverse velocity of x. Prove your choice correct.
   c) Prove that $\oplus$ is associative. That is, $a \oplus (b \oplus c) = (a \oplus b) \oplus c$. (This operation isn't necessarily a geometric transformation, so you need to verify this property of groups, which automatically holds for composition of functions.)
   d) (Calculus) Prove closure. [*Hint:* Let a be any constant between -1 and 1. Verify that the derivative of $f(x) = (x+a)/(1+xa)$ with respect to x is always positive. Find $f(1)$ and $f(-1)$.]

8. Discuss unusual properties of Eq. 5.2 as a distance formula.

9. a) On a graph with axes Δx and Δt, draw and describe the curve of points $(\Delta x, \Delta t)$ such that $\Delta x^2 - \Delta t^2 = -1$. This curve represents the possible measurements that person B could obtain in Example 3. Events with a negative difference are called *timelike* because every observer can determine which event occurred first.

   b) Repeat part (a) for $\Delta x^2 - \Delta t^2 = 1$. This difference could occur if person A flashed two different lights at the same time and at a distance of 1 apart in A's measurement. Events with a positive difference are called *spacelike*. Describe how person B could observe either light flash first.

   c) Repeat part (a) for $\Delta x^2 - \Delta t^2 = 0$. Describe how person A could send two different light flashes to give this difference. Describe how person B could observe these two events as happening at the same time. Events with a difference of zero are called *lightlike*.

Symmetries for relativity must take the curves where $\Delta x^2 - \Delta t^2$ is constant to themselves. The value of b in parts (e) and (f) depends on the velocity of B with respect to A.

   d) Define $\cosh a = (e^a + e^{-a})/2$ and $\sinh a = (e^a - e^{-a})/2$. Verify $\cosh^2 a - \sinh^2 a = 1$.

   e) Explain why you can write the coordinates for the difference between two timelike events as $(r \sinh a, r \cosh a)$, for some a and r in $\mathbf{R}$. Verify that $f(r \sinh a, r \cosh a) = (r \sinh(a+b), r \cosh(a+b))$ preserves $\Delta x^2 - \Delta t^2$ for timelike events. Describe what f does to points on the curve of part (a).

   f) Modify and repeat part (e) for spacelike events.

5.6 Fractals

Historically, geometry has focused on relatively simple, ideal shapes: circles, triangles, polyhedra, and the like. However, even a cursory glance at nature reveals a vast array of shapes unrelated to these traditional objects. Benoit Mandelbrot, the originator of fractals, found geometric structure underlying complicated natural shapes. In 1975 he coined the word fractal to describe the convoluted curves and surfaces that can be used to model natural shapes that had previously seemed beyond mathematical study.

Mathematicians initiated the abstract study of curves related to fractals before 1900. In 1904 Helge von Koch defined the Koch curve (Fig. 5.45) as the limit of an infinite process, illustrated in Fig. 5.46. Starting with the motif at the top of Fig. 5.46, we replace

5.6 Fractals 215

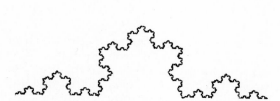

Figure 5.45 The Koch curve.

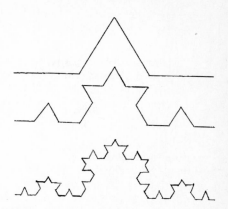

Figure 5.46 Iterations leading to the Koch curve.

each of its line segments with smaller copies of the motif. The middle curve shows the first *iteration*, and the bottom curve shows the second iteration. After infinitely many iterations we arrive at the Koch curve, which has the property of *self-similarity*: The entire curve is similar to a part of the curve.

Example 1 Show the Koch curve is infinitely long.

Solution. Suppose that the original motif in Fig. 5.46 has a length of 1 unit. After one iteration, there are four copies of the motif, each one-third as long, for a length of $4/3 \approx 1.3$ units. The second iteration has length $(4/3)^2 \approx 1.8$ units, and, in general, the nth iteration has length $(4/3)^n$ units, increasing in length as n increases. The length of the Koch curve is $\lim_{n \to \infty} (4/3)^n = \infty$. Note that the Koch curve is enclosed in a finite area despite its infinite length. ●

Exercise 1 Explore why the method of IFS in Section 4.4 produces the same curve as the method Koch used.

Mandelbrot noticed that many real phenomena, such as coastlines, mountains, and lungs, have a roughly self-similar shape: The smaller features of these objects have the same overall bumpiness as the larger features. Of course, no part of the coastline of France will exactly replicate the entire coastline. Furthermore, no real shape can exhibit even approximately self-similar shape at the subatomic level. Thus exact self-similarity is too limited to model nature. Mandelbrot uses the term *statistical self-similarity* to describe approximate similarity over a range of scales. He avoids an exact mathematical definition of this concept because such a definition would apply only to mathematical objects, defeating his purpose. Computers can readily draw statistically self-similar shapes by modifying the iterations at smaller scales with randomly generated fluctuations. The resulting graphics often look strikingly realistic and support the usefulness of statistical self-similarity (Fig. 5.47). However, Mandelbrot realized the need for more

BENOIT MANDELBROT

> Intuition is not something that is given. I've trained my intuition to accept as obvious things which were initially rejected as absurd and I find everyone can do the same. [Fractals] provide a handle to representing nature, and intuition can be changed and refined and modified to include them.   —*Benoit Mandelbrot*

Benoit Mandelbrot (1924–) is the leading proponent of fractal geometry, which he pioneered. The chaos of World War II disrupted his life and his education, but his well-developed geometric intuition helped him to complete a doctorate in mathematics. His interests ranged over a variety of unusual aspects of mathematics, physics, and engineering, including noise in electrical transmissions, which others had thought was simply random. Mandelbrot found that the frequency of noise measured in second-long intervals reassembled the frequency at longer intervals. Slowly he found other phenomena with a uniformity under change of scale, or what he called statistically self-similar or fractal. He collected examples of fractal behavior much the way naturalists collect specimens.

Mandelbrot's interest in the application of fractals is coupled with an intense interest in mathematical ideas, although he is much less interested in mathematical proof. He draws on the results of others, coupled with his remarkable visual intuition and stunning computer graphics, to build new mathematical ideas and conjectures. He helped pioneer the use of computers to draw fractals and to approach mathematics as an experimental field.

structure if the mathematics of self-similarity is to lead to new insights, not just interesting graphics.

Developed in 1919, the Hausdorff dimension provides a measure of how convoluted mathematical shapes are, and Mandelbrot modified it to measure real objects. Felix Hausdorff (1868–1942) noted a relationship between dimensions and the growth in the

Figure 5.47   A computer-generated "mountain."

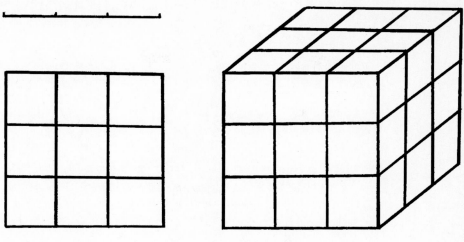

Figure 5.48

number of units as the measuring scale decreases. Figure 5.48 illustrates the intuition behind Hausdorff's approach. If we divide each side into three equal pieces, a line segment has $3 = 3^1$ smaller segments, a square has $9 = 3^2$ smaller squares, and a cube has $27 = 3^3$ smaller cubes. The dimension appears as the exponent. The smaller pieces are similar to the originals by a scaling ratio of $r = \frac{1}{3}$. (Hausdorff's technical definition uses analysis. See Falconer [5].)

Exercise 2 Create drawings analogous to Fig. 5.48, with each side divided into four smaller units, so $r = \frac{1}{4}$. Verify that the line segment has 4^1 smaller segments, the square has 4^2 small squares, and the cube has 4^3 smaller cubes.

We can generalize Fig. 5.48 and Exercise 2: The number of units is proportional to $n = (1/r)^d$, where d is the dimension and r is the scaling ratio. If we take logarithms and solve for d, we obtain the equation for the dimension:

$$d = \log n / \log(1/r). \tag{5.3}$$

Exercise 3 Verify that Eq. 5.3 for the dimension d holds for the line segment, square, and cube of Fig. 5.48 for any choice of r.

Hausdorff applied his notion to more interesting shapes than line segments, squares, and cubes. Thus, in Fig. 5.49, we can approximate the circumference of a circle with increasingly smaller units. As a curve, a circle is essentially a one-dimensional object. Thus halving the length of the units approximately doubles the number of units. None of the values $\log n / \log(1/r)$ exactly equals 1, but as r goes to 0, the limit equals 1. For simple shapes, such as a circle or even the Koch curve, Eq. 5.3 (or its limit as $r \to 0$) is sufficient to determine the Hausdorff dimension. More complicated sets of

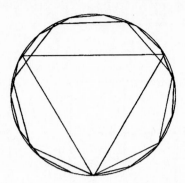

Figure 5.49 Estimating a circumference with different unit lengths. As the unit length decreases ($r \to 0$), the estimate $r \cdot n$ approaches $2\pi R$.

points require Hausdorff's exact definition. Using analysis, Hausdorff proved that every nonempty subset of $\mathbf{R}^k$ has a unique Hausdorff dimension of at most k.

Example 2 Find the Hausdorff dimension of the Koch curve.

Solution. In Example 1 a scaling factor of $r = \frac{1}{3}$ gives $n = 4$ times as many units. Then Eq. 5.3 gives $d = \log 4/\log 3 \approx 1.262$. Verify that a scaling ratio $r = \frac{1}{9} = (\frac{1}{3})^2$ gives the same value of d. The Koch curve is too convoluted to be measured by the one-dimensional unit of length because it is infinitely long. However, the curve has an area of 0, so the curve isn't two-dimensional. The value $\log 4/\log 3$ is the Hausdorff dimension, indicating that the convoluted Koch curve is between a line and a surface. ●

Unfortunately, the Hausdorff dimension doesn't apply to real shapes. In 1961, Lewis Richardson published a study of the estimated length of coastlines according to maps of different unit lengths. The estimated lengths differed widely, but Richardson found a function of the lengths in terms of the unit lengths. Mandelbrot recognized that Richardson's equation fits the intuition behind the Hausdorff dimension. (Richardson apparently was unaware of Hausdorff's work.) In Richardson's equation, we replace the original uninterpreted exponent with $1 - d$ to relate it to d, the Hausdorff dimension, to get

$$L(f) = c(f^{1-d}),$$

where f represents the unit length, $L(f)$ is the estimated length (or area) of the object for that value of f, and c is a constant, depending on the amount of the object measured.

Example 3 In the original motif of the Koch curve (see Fig. 5.46) let each segment have a length of $f_1 = \frac{1}{4}$. Then the four segments give $L(f_1) = 1$ as an estimated length of the Koch curve. In the first iteration when $f_2 = \frac{1}{3} \cdot \frac{1}{4}$, the 16 segments give $L(f_2) = \frac{4}{3}$. In the second iteration when $f_3 = \frac{1}{9} \cdot \frac{1}{4}$, $L(f_3) = \frac{16}{9}$, and so on. We solve for d in Richardson's equation by using two values of f. With f_1 we get $1 = c(\frac{1}{4})^{1-d}$, and with f_2 we get

$\frac{4}{3} = c(\frac{1}{3} \cdot \frac{1}{4})^{1-d} = c(\frac{1}{3})^{1-d}(\frac{1}{4})^{1-d}$. Substituting the first equation into the second yields $\frac{4}{3} = (\frac{1}{3})^{1-d} = \frac{1}{3} \cdot (3)^d$ or $4 = 3^d$. Then $d = \log 4 / \log 3$, as before. ●

The exponent $1 - d$ in Richardson's equation isn't as mysterious as it may seem. For any unit f, the estimated length $L(f)$ of a self-similar curve should be the number of units times their length. Moreover, the number of units is proportional to $n = (\frac{1}{r})^d = (\frac{1}{f})^d$. That is, $L(f) = c \cdot n \cdot f = c \cdot (1/f)^d \cdot f = c \cdot f^{1-d}$. We want to solve Richard's equation for d, based on empirical values for f and $L(f)$. With two unknowns, c and d, we need two values of f and $L(f)$. From $L(f_1) = c(f_1)^{1-d}$ and $L(f_2) = c(f_2)^{1-d}$, we find $L(f_1)/L(f_2) = (f_1/f_2)^{1-d}$ and finally

$$d = 1 - \frac{\log[L(f_1)/L(f_2)]}{\log(f_1/f_2)}. \qquad (5.4)$$

Example 4 Figure 5.50 shows the coastline of England and Wales from Bristol to Liverpool. If we estimate the length of this coastline by using $f_1 = 57$-mi segments, we need 5 segments, so $L(f_1) \approx 285$ mi. For $f_2 = 28.5$ mi, we need $12\frac{1}{2}$ segments, so $L(f_2) \approx 356$ mi; $f_3 = 14.25$ mi gives 32 segments and $L(f_3) \approx 456$ mi. As we shorten the unit length, we can follow the contour better, taking into account more and more of the multitude of peninsulas and bays. When we use f_1 and f_2, Eq. 5.4 gives $d = 1 - \log(285/356) / \log(57/28.5) \approx 1.32$. ●

Exercise 4 Verify that $d \approx 1.36$ when you use f_2 and f_3 and $d \approx 1.34$ when you use f_1 and f_3.

The values of d that Richardson found for coastlines are remarkably stable over a range of scales f. (Such values can't be stable over all possible scales unless the object

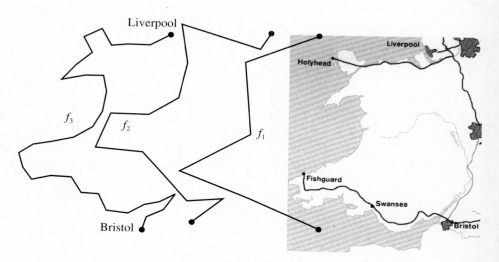

Figure 5.50 The mapped coastline of Wales and part of England and approximate outlines produced by using segments of different lengths.

is perfectly self-similar. The range of scales depends on the object studied.) Mandelbrot calls this empirical value the *fractal dimension* to distinguish it from Hausdorff's abstract definition. Mandelbrot and others have estimated fractal dimensions for a variety of natural curves and surfaces. He avoids defining a fractal, but the following provisional definitions are helpful. Example 2 illustrates how self-similar shapes, such as the Koch curve, can have unexpected dimensions. Self-similarity gives an exact number of copies, n, for an appropriate scaling ratio, r. More convoluted shapes generally have more small copies at a given scale and so have higher dimensions.

Provisional Definition A *fractal curve* has fractal dimension greater than 1. A *fractal surface* has fractal dimension greater than 2.

Fractals provide a good model for the lungs. The trachea splits into the bronchial tubes, which in turn split into shorter and narrower tubes. In addition, the embryonic development of the lung is an iterative process (Fig. 5.51). The convoluted surface of the lung greatly increases its area while keeping its overall volume small. The large surface is biologically essential because the amount of carbon dioxide and oxygen that the lungs

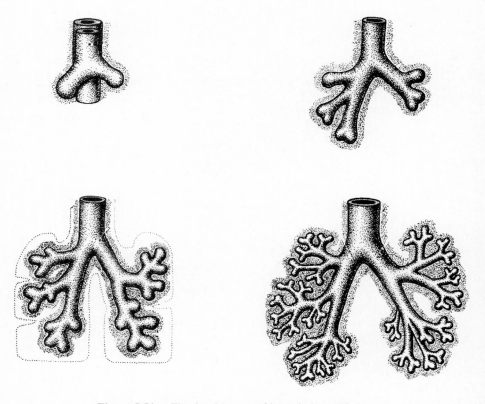

Figure 5.51 The development of lungs in human beings.

can exchange is roughly proportional to their surface area. Using a light microscope, biologists found approximately 80 m² of surface area in a lung (roughly the floor space of a small house). The higher magnification of an electron microscope yielded approximately 140 m². This increase in area at higher magnification corresponds to the measurements given in Example 4. Scientists have estimated the fractal dimension of a lung to be 2.17. Blood vessels, kidneys, the liver, and other organs have good fractal models.

Although fractals provide insightful models, scientists are hoping for more than explanations of already known facts. For example, why is the fractal dimension of a lung 2.17? After all, a higher dimension would give even more surface area. Perhaps higher dimensions impede the free passage of air or blood in the lung. Questions such as these provide ample challenges for research in the applications of fractals. (See Mandelbrot [12].)

PROBLEMS FOR SECTION 5.6

1. For each motif of the fractal curve given, sketch several iterations.
    a) A stylized tree, where each branch splits into two others half as long.
    b) A stylized tree, where each branch splits into three others half as long.
    c) A Cantor set, whereby you divide a line segment into three equal pieces and remove the middle piece and iterate with the remaining pieces.
    d) A modified Koch curve, with a square on the middle third of a line segment, rather than a triangle.
    e) A modified Koch curve, whereby you divide a line segment into fourths and construct squares on the alternate sides of the two middle fourths.
    f) A Sierpinski gasket, whereby you divide a triangle into four smaller triangles by connecting the midpoints of the sides, remove the middle triangle, and iterate with the remaining triangles.

2. Find the Hausdorff dimension of the fractals in Problem 1.

3. Investigate what happens to the figures and the fractal dimension when you increase the number of smaller copies in the fractals in Problem 1(b) and 1(e).

4. For each motif for a fractal surface given, sketch the first and second iterations and find the fractal dimension.
    a) A "Koch pyramid," whereby you divide a triangle into four smaller triangles, as in 1(f), but construct a triangular pyramid on the middle triangle and iterate with each of the smaller triangles and the faces of the pyramid.
    b) A "Koch cube," whereby you divide a unit square into nine smaller squares, construct a cube on the middle square, and iterate as in part (a).
    c) A Menger sponge, whereby you divide a unit cube into 27 smaller cubes, remove the center cube, and iterate with each of the 26 remaining smaller cubes.

5. a) Use a geometric series to find the total area of the infinitely many squares in Problem 1(d).
    b) Repeat part (a) for the total volume of cubes in Problem 4(b).
    c) Repeat part (a) for Problem 4(c) to find the volume remaining in the Menger sponge.

6. You can estimate the fractal dimension of the coastline of Norway using the map shown in Fig. 5.52 by using different length line segments. Start at Oslo and lay out first 1-in., then $\frac{1}{2}$-in., and then $\frac{1}{4}$-in. line segments along the coast to Bergen. Then use Eq. 5.4 to calculate the fractal dimension. On this map, 1 in. ≈ 50 mi.

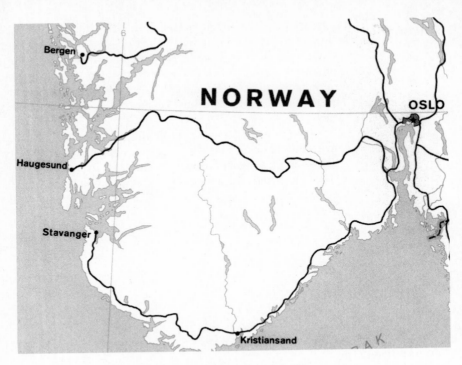

Figure 5.52 The mapped coastline of southern Norway.

PROJECTS FOR CHAPTER 5

1. Design your own wallpaper patterns by hand or use software, such as TesselMania [9] under Suggested Media.

2. Which types of frieze patterns and wallpaper patterns can be formed with mirrors? Arrange the mirrors in various ways facing one another. Place symmetric and asymmetric designs in the region between the mirrors.

3. Investigate symmetry in the art of M. C. Escher. (See Schattschneider [16].)

4. Investigate symmetry in weaving. (See Pizzuto [14].)

5. Find examples of actual wallpaper and classify them. Certain symmetry types of wallpaper patterns predominate. Investigate the reasons, including aesthetic ones, that underlie the choice of symmetry in wallpaper.

6. Investigate symmetry in the art of various cultures. (Wade [20] provides a large variety of examples. For Islamic art, see El-Said and Parman [4].)

7. Investigate tilings of the plane, which generalize wallpaper patterns. (See Grünbaum and Shephard [8].)

8. Investigate the symmetries of continuous (nondiscrete) frieze and wallpaper patterns. Try to classify types of such patterns.

9. In many cultures circular friezes appear as ornaments on cups and other cylindrical objects. Circular friezes replace the translations of friezes with rotations about a vertical axis. The seven frieze groups become seven subgroups of $\overline{\mathbf{D}'_n}$.

   a) Draw a frieze with symmetry group **pmm2** on a sheet of paper. Roll the paper into a cylinder to create a circular frieze. Describe how the symmetries of the frieze become symmetries of the circular frieze. Explain why this circular frieze has $\overline{\mathbf{D}'_n}$ for its group of symmetries. What determines the value of n that a circular frieze has? Why do the symmetries of other circular friezes form subgroups of some $\overline{\mathbf{D}'_n}$?

b) For each of the six other types of friezes, make a corresponding circular frieze pattern. Describe the rotations, mirror reflections, and rotatory reflections of these circular friezes.

c) Which of the circular friezes in (b) has the symmetry group C_n? Which has the symmetry group D_n? (Actually, this group is called $D'_n C_n$ for technical reasons.) Which has the symmetry group D'_n?

d) One of the remaining circular frieze has a horizontal mirror reflection. Its group is $\overline{C_n}$. Find it.

e) The remaining circular friezes have rotatory reflections of angles half as large as the rotations about the vertical axes. Find the one whose rotations are all around a vertical axis; this circular frieze has group $C_{2n}C_n$. Describe the additional rotations of the other, whose group is $D'_{2n}D'_n$.

f) Make a table to match the frieze and circular frieze groups.

g) Classify the group of an antiprism. (See Problem 2 of Section 5.4.)

h) Place a mirror face up on a table. Place two mirrors perpendicular to the first mirror, facing each other at various angles. Which of the circular frieze patterns can you make by placing symmetric and asymmetric designs in the region between the mirrors?

10. Investigate the symmetry of knots and braids.

11. Investigate symmetry in fugues and 12-tone music. (See Senechal and Fleck [18].)

12. Investigate color symmetry. (See Loeb [11].)

13. Borrow a set of hand bells to investigate symmetry and change ringing. (See Senechal and Fleck [18, 47].)

14. Investigate symmetry in crystals. (See Senechal [17].)

15. Build a Penrose tile by using the two types of tiles shown in Fig. 5.53. Match sides with dots on them to ensure that the tiling will be nonperiodic. Investigate Penrose tiles. (See Gardner [7], Grünbaum and Shephard [8], and Peterson [13].)

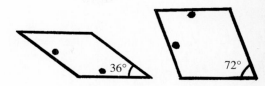

Figure 5.53 Two rhombi that can make a Penrose tiling. Match sides with dots and match sides without dots.

16. Investigate quasicrystals. (See Jaric [10], Peterson [13], and Senechal [17].)

17. Design fractals and investigate the Mandelbrot set and Julia sets. (See "The Desktop Fractal Design System" [4] under Suggested Media.)

18. Estimate the fractal dimension of real shapes. Determine the scale at which the fractal nature of your examples breaks down. The fractal dimension of a surface can be estimated by adding 1 to the fractal dimension of a typical cross section. (See Mandelbrot [12].)

19. Investigate the role of symmetry in quantum mechanics. (Bunch [1] and Rosen [15] provide elementary expositions and bibliographies.)

20. Investigate other ideas in symmetry. (See Bunch [1], Hargittai [9], Rosen [15], and Senechal and Fleck [18].)

21. Write an essay considering the relationship of symmetry and culture. Is a classification of designs by their symmetries culturally objective?

22. Write an essay discussing the application of abstract mathematics to the world around you. For example, why do proofs about infinite, perfect mathematical crystals tell you anything about real, finite crystals?

23. Write an essay discussing the notion of mathematics as an experimental science. Consider the following questions. Are proofs essential to mathematics? In what ways is experimental evidence appropriate in mathematics?

Suggested Readings

[1] Bunch, B. *Reality's Mirrors: Exploring the Mathematics of Symmetry*. New York: John Wiley & Sons, 1989.

[2] Crowe, D. *Symmetry, Rigid Motions, and Patterns*. Arlington, Mass.: COMAP, 1986.

[3] Crowe, D., and D. Washburn. Groups and geometry in the ceramic art of San Ildefonso. *Algebras, Groups and Geometries*, 1985, 2(3):263–277.
[4] El-Said, I., and A. Parman. *Geometric Concepts in Islamic Art*. Palo Alto, Calif.: Dale Seymour, 1976.
[5] Falconer, K. *The Geometry of Fractal Sets*. New York: Cambridge University Press, 1986.
[6] Gallian, J. *Contemporary Abstract Algebra*. Lexington, Mass.: D. C. Heath, 1994.
[7] Gardner, M. *Penrose Tilings to Trapdoor Ciphers*. New York: W. H. Freeman, 1989.
[8] Grünbaum, B., and G. Shephard. *Tilings and Patterns*. New York: W. H. Freeman, 1989.
[9] Hargittai, I. (ed.). *Symmetry: Unifying Human Understanding*. Elmsford, N.Y.: Pergamon Press, 1986.
[10] Jaric, M. (ed.). *Introduction to the Mathematics of Quasicrystals*. Boston: Academic Press, 1988.
[11] Loeb, A. *Color and Symmetry*. New York: John Wiley & Sons, 1971.
[12] Mandelbrot, B. *The Fractal Geometry of Nature*. New York: W. H. Freeman, 1982.
[13] Peterson, I. *The Mathematical Tourist: Snapshots of Modern Mathematics*. New York: W. H. Freeman, 1988.
[14] Pizzuto, J. *101 Weaves in 101 Fabrics*. Pelham, N.Y.: Textile Press, 1961.
[15] Rosen, J. *Symmetry Discovered: Concepts and Applications in Nature and Science*. New York: Cambridge University Press, 1975.
[16] Schattschneider, D. *Visions of Symmetry: Notebooks, Periodic Drawings and Related Work of M. C. Escher*. New York: W. H. Freeman, 1990.
[17] Senechal, M. *Crystalline Symmetries: An Informal Mathematical Introduction*. Philadelphia: Bristol, 1990.
[18] Senechal, M., and G. Fleck (eds.). *Patterns of Symmetry*. Amherst: University of Massachusetts Press, 1977.
[19] Taylor, E., and J. Wheeler. *Spacetime Physics*, San Francisco: W. H. Freeman, 1966.
[20] Wade, D. *Geometric Patterns and Borders*. New York: Van Nostrand Reinhold, 1982.
[21] Wenninger, M. *Polyhedron Models*. New York: Cambridge University Press, 1971.
[22] Weyl, H. *Symmetry*. Princeton, N.J.: Princeton University Press, 1952.
[23] Yaglom, I. *Felix Klein and Sophus Lie: Evolution of the Idea of Symmetry in the Nineteenth Century*. Boston: Birkhäuser, 1988.

Suggested Media

1. "Adventures in Perception," 22-minute film, BFA Educational Media, Santa Monica, Calif., 1973.
2. "Chaos, Fractals and Dynamics: Computer Experiments in Mathematics," 63-minute video, American Mathematical Society, 1989.
3. "The Counting Theorem," 24-minute video, Films for the Humanities and Sciences, Princeton, N.J., 1996.
4. "The Desktop Fractal Design System," software and handbook by M. Barnsley, Academic Press, Boston, 1989.
5. "Dihedral Kaleidoscopes," 13-minute film, International Film Bureau, Chicago, 1966.
6. "Maurits Escher: Painter of Fantasies," $26\frac{1}{2}$-minute film, Coronet Films, Chicago, 1970.
7. "Symmetry Counts," 24-minute video, Films for the Humanities and Sciences, Princeton, N.J., 1996.
8. "Symmetries of the Cube," $13\frac{1}{2}$-minute film, International Film Bureau, Chicago, 1971.
9. "TesselMania," software, Minnesota Educational Computing Corporation, Minneapolis, 1995.

6
Projective Geometry

The artists of the Renaissance uncovered the rules of perspective, as illustrated by Albrecht Dürer's St. Jerome in His Study. Projective geometry provides the foundation for perspective both in art and computer graphics. We explore these aspects of projective geometry and more.

> Metrical geometry is thus a part of [projective] geometry, and [projective] geometry is all geometry. —*Arthur Cayley*

6.1 Overview and History

Projective geometry grew out of the study of perspective in Renaissance art. Albrecht Dürer (1471–1528), Leonardo da Vinci (1452–1519) and other artists worked out the geometric rules of perspective. These rules weren't particularly complicated, but they transformed the appearance of paintings. One key insight clearly contradicting Euclidean geometry was that formerly parallel lines met at "ideal points" on the horizon, an ideal line intuitively infinitely far away. However, these painters were *not* challenging Euclidean geometry. (Indeed Euclid had written about such effects of perception in his work, *The Optics*.) The attempt to paint the way people saw the world emerged from a philosophy that took people as "the measure of all things," rather than viewing the world from a divine point of view.

Girard Desargues (1593–1662) was the first person to prove geometric properties beyond what was needed for painting. Unfortunately, his notes were quite hard to read and had unusual conventions, such as using flowers and trees as the names of geometric objects. Only a few mathematicians, including Blaise Pascal (1623–1662), built on Desargues's efforts and saw the unifying power of what we now call projective methods. Johannes Kepler (1571–1630), who used conics effectively in his astronomical work, realized that circles stretch continuously to ellipses and then to parabolas and hyperbolas. The shadow of a lamp shade cast on a wall illustrates this process, suggesting the notion of projective transformations.

However, the progress made by these pioneers was overshadowed and mostly forgotten because of the marvelous advances of analytic geometry and then calculus, which quickly dominated mathematical thought. Led by Gaspard Monge, the reconsideration and rediscovery of projective ideas started after 1800. One of his students, Jean Victor Poncelet (1788–1867) developed this subject extensively and accelerated others' research with his book, published in 1822, emphasizing a synthetic approach to geometry that featured properties that didn't depend on distance. Poncelet realized the importance of duality, which says that points and lines function the same way. At that time analytic methods seemed useless in this geometry because projective geometry added "ideal points," whereby parallel lines meet on the horizon, and these points had no obvious coordinates. The success of the projective approach launched a rivalry between geometers who advocated an analytic approach and those, including Poncelet, who pursued projective geometry synthetically.

Augustus Möbius and Julius Plücker developed coordinates for projective geometry, and gradually geometers realized that the synthetic and analytic approaches complemented each other. Möbius initiated the study of transformations in projective geometry and other geometries. Karl Van Staudt (1798–1867) showed how to derive projective coordinates independent of Euclidean geometry. Arthur Cayley (1821–1895) and Felix Klein showed how to develop Euclidean, hyperbolic, and single elliptic geometries within projective geometry. Later mathematicians showed that the geometry of the spe-

Figure 6.1

cial theory of relativity is also a part of projective geometry. Klein realized the vital role projective transformations played in unifying geometry. In the twentieth century mathematicians have regularly used analytic projective geometry in various fields, including computer-aided design, algebraic geometry, and statistical design theory.

6.1.1 Projective intuitions

Perspective painting depends on the idea of projecting a scene onto the plane of the easel with respect to the eye of the painter. Figure 6.1 illustrates such a projection, showing parallel lines that appear to meet at the horizon. Axiomatically we characterize projective geometry by the axiom that every two distinct lines in the plane have at least one point on both lines. The definition of a perspectivity captures the idea of perspective drawing.

Definition 6.1.1 A *perspectivity with respect to a point T* is a mapping of the points U_i on one line to the points V_i on another line so that V_i is the image of U_i iff U_i, V_i, and T are on the same line (Fig. 6.2).

In Euclidean geometry, the distance between two points is a fundamental measurement. As the spacing of the telephone poles in Fig. 6.1 and the drawing in Fig. 6.2(a) suggest, perspective distorts distances and proportions. Moreover, the ordering of points on a line can change, as indicated in the drawing in Fig. 6.2(b). That is, projective transformations can move points around so that one no longer needs to lie between the other two. (Artists never need to choose a perspective that switches the order.)

In Problem 2 you are asked to show that a pair of perspectivities can map any three points on one line to any three points on another line. That is, properties involving only two or three points on a line, such as distance or betweenness, are not projective properties. Similarly, angle measure is not a projective property. Geometers have found properties involving four or more points, built on straight-line constructions, and preserved by all projective transformations. The most important of these properties, a

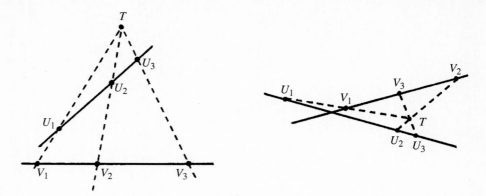

Figure 6.2

harmonic set, is built from a complete quadrangle: Four points, no three on the same line, such as T_1, T_2, T_3, and T_4 in Fig. 6.3, and the six lines they determine form a *complete quadrangle*. The points P, Q, R, and S form a harmonic set.

Definition 6.1.2 Four distinct points P, Q, R, and S on a line k form a *harmonic set*, denoted $H(PQ, RS)$, iff there is a complete quadrangle with six lines different from k such that P is on two of the lines, Q is on two other lines, R is on one other line, and S is on the remaining line.

Example 1 From three points P, Q, and R on a line, the following construction determines the unique fourth point S to form a harmonic set, as shown in Fig. 6.3. Although not obvious, every such construction will give the same fourth point. (See Problem 4.) Indeed, this uniqueness is an axiom in Section 6.2.

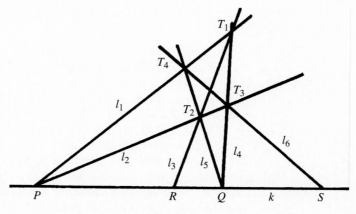

Figure 6.3

6.1 Overview and History

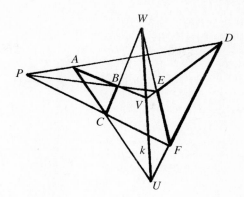

Figure 6.4

Let P, Q, and R be any three points on a line k. Draw two arbitrary lines l_1 and l_2 through P and another arbitrary line l_3 through R. (All the lines l_i are distinct and differ from k.) Let T_1 and T_2 be the intersections of l_3 with l_1 and l_2, respectively. Let l_4 and l_5 connect Q with T_1 and T_2, respectively. Then l_4 intersects l_2, say, at T_3, and l_5 intersects l_1, say, at T_4. Finally, draw the line connecting T_3 and T_4 and find its intersection S with the original line k. The complete quadrangle with points T_i and lines l_j satisfies the conditions of the definition, showing $H(PQ, RS)$. ●

Exercise 1 A *complete quadrilateral* is a set of four lines, no three on the same point, and the six points they determine. Draw a complete quadrilateral.

Example 2 Two triangles $\triangle ABC$ and $\triangle DEF$ are said to be *perspective from the point* P whenever P is on the three lines $\overleftrightarrow{AD}$, $\overleftrightarrow{BE}$, and $\overleftrightarrow{CF}$. Figure 6.4 illustrates this situation, as well as the following closely related concept. Two triangles $\triangle ABC$ and $\triangle DEF$ are said to be *perspective from the line* k whenever the corresponding sides of the triangles intersect on k. That is, $\overleftrightarrow{AB}$ and $\overleftrightarrow{DE}$, $\overleftrightarrow{AC}$ and $\overleftrightarrow{DF}$, and $\overleftrightarrow{BC}$ and $\overleftrightarrow{EF}$ have their three points of intersections on k. Desargues used three-dimensional Euclidean properties to show Desargues's theorem: If two triangles are perspective from a point, then they are perspective from a line. (See Project 2.) Desargues's theorem is often taken as an axiom, and then properties of harmonic sets are proven from it. ●

(See Kline [4, Chapters 14 and 35] for more on the history of projective geometry.)

PROBLEMS FOR SECTION 6.1

1. Draw a picture similar to Fig. 6.1 wherein the telephone poles are spaced geometrically. For example, let the first pole be 4 in. from the ideal point on the horizon where the poles meet, and the succeeding poles be $3 = 4 \cdot \frac{3}{4}$, $2.25 = 4 \cdot (\frac{3}{4})^2$, ... in. from the ideal point. Do the distances between poles look like they are getting smaller or larger as they go toward the ideal point? Try other distance relationships between the poles. What relationship describes the placement of equally spaced items drawn in perspective, as in Fig. 6.1?

2. **a)** For any two distinct points P and Q on a line k and any two distinct points P' and Q' on another line k', show how to find a perspectivity from some point T that takes P to P' and Q to Q'.

b) For any three distinct points P, Q, and R on a line k and any three distinct point P', Q', and R' on another line k', show how to use one or two perspectivities to take P to P', Q to Q', and R to R'. [*Hint:* Use the first perspectivity to take R to R'.]

3. Use a lamp with a light bulb casting sharp shadows on a wall.

  a) Cut out a triangle. Can you make the shadow of any of its angles be acute? right? obtuse? What happens to the shadow angles if you hold one side of the triangle parallel to the wall and rotate the triangle about that side?

  b) Describe the shapes that the shadow of a cut out circle can be.

4. a) Use a dynamic geometry construction program to model the construction of Example 1. Keep P and Q fixed and move R along line k. Note how S moves in response to R.

  b) Keep P, Q, and R fixed but move the lines through them. Does S move?

  c) Have the program measure the distances $d(P, Q)$, $d(P, R)$, and $d(P, S)$. Set $d(P, Q)$ at some convenient distance, such as 1 or 10, and investigate the relationship of $d(P, R)$ and $d(P, S)$.

5. Use graph paper to construct a harmonic set $H(PQ, RS)$, where P, Q, and R are on the x-axis at coordinates 0, 1, and 3. What is the coordinate of S? Choose different initial lines and construct S from the same three points. Do your two constructions yield (nearly) the same point?

6. a) Prove that, if $H(PQ, RS)$, then $H(PQ, SR)$, $H(QP, RS)$, and $H(QP, SR)$.

  b) Suppose that $H(PQ, RS)$. Use constructions to explore which, if any, of $H(PR, QS)$, $H(PS, QR)$, and $H(RS, PQ)$ can be harmonic sets.

7. A harmonic set of lines $H(jk, lm)$ relates to a complete quadrilateral in the same way that a harmonic sets of points relates to a complete quadrangle. The roles of points and lines are simply interchanged.

  a) Define a harmonic set of lines.

  b) Convert Example 1 to a construction of a harmonic set of lines. Carry out this construction twice, once where l is a bisector of j and k and once where the lines seem unrelated. In the first case how is m related to the other lines?

8. Let P, Q, R, and S be points on the x-axis with coordinates 0, 1, a, and $h(a)$, respectively, with $a \neq 2$ and $a \neq \frac{1}{2}$. Assume that $H(PQ, RS)$ and look for a formula for $h(a)$.

  a) Let l_1, l_2, and l_3 be the lines $x = 0$, $y = x$, and $y = -x + a$. Verify that P is on l_1 and l_2 and R is on l_3. Find the points T_1 and T_2, as denoted in Fig. 6.3.

  b) Verify that l_4 and l_5 are $y = -ax + a$ and $y = ax/(a-2) - a/(a-2)$.

  c) Verify that $T_3 = (\frac{a}{1+a}, \frac{a}{1+a})$ and $T_4 = (0, \frac{a}{2-a})$.

  d) Find the equation of l_6. Verify that $S = (h(a), 0)$, where $h(a) = a/(2a - 1)$.

9. a) In Problem 8, alter one of the lines (l_1, l_2, or l_3) to show that, when $a = 2$, $h(2) = \frac{2}{3}$.

  b) In Problem 8, for $a = \frac{1}{2}$ find the equation of l_6. What happens to S in this case? How does this situation fit with the equation in Problem 8 (d)?

10. In Fig. 6.5 the circle has the equation $(x - \frac{1}{2})^2 + y^2 = 0$, with $P = (0, 0)$, $Q = (1, 0)$, $R = (a, 0)$, and $0 < a < 1$, $a \neq \frac{1}{2}$. Find the coordinates of S, where $\overleftrightarrow{SA}$ and $\overleftrightarrow{SB}$ are tangents to the circle and A and B have a for their x-coordinate. Note that Problem 8 shows that $H(PQ, RS)$.

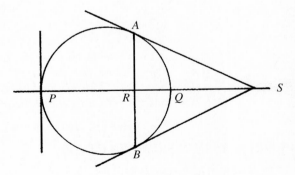

Figure 6.5

11. a) In Fig. 6.4 verify that $\triangle ABP$ and $\triangle UWF$ are perspective from the point C and find the line from which they are perspective.

  b) Find two other triangles, a point, and a line such that the two triangles are perspective from both the point and the line.

6.2 AXIOMATIC PROJECTIVE GEOMETRY

In this section we develop axiomatically some elementary properties of projective planes. (For a more thorough axiomatic development, see Coxeter [2] or Tuller [9].) The first five axioms describe the minimum relations of points, lines and harmonic sets. (Recall that we defined harmonic sets in Section 6.1 in terms of points and lines.) The separation axioms, which appear later in this section, provide a structure of the points on a line analogous to the notion of order. The undefined terms are *point*, *line*, *on*, and *separate*.

Axioms 6.2.1
- i) Two distinct points have exactly one line on them.
- ii) There are at least four points with no three on the same line.
- iii) Every two distinct lines have at least one point on both lines.
- iv) Given three distinct points P, Q, and R on a line k, there is a unique point S on k, distinct from P, Q, and R, such that $H(PQ, RS)$.
- v) If $H(PQ, RS)$, then $H(RS, PQ)$.

Theorem 6.2.1
- i) Two distinct lines have exactly one point on both lines.
- ii) Every line has at least four distinct points on it.
- iii) If $H(PQ, RS)$, then $H(PQ, SR)$.

Proof. See Problem 1 for parts (i) and (iii). For part (ii), let k be any line. By Axiom (ii) there are four points A, A_1, A_2, and A_3 and at least one of them, say, A, is not on k. Consider the lines k_i on A and A_i, for $i = 1, 2, 3$. By Axiom (ii) these lines are distinct. By part (i) these lines each have one point in common with k, which gives three distinct points on k. Axiom (iv) guarantees the fourth point. ∎

Definition 6.2.1 The point on the lines k and l is denoted $k \cdot l$. Points on the same line are *collinear*, and lines on the same point are *concurrent*.

We follow Karl Van Staudt's method of using harmonic sets to construct and coordinatize infinitely many points on a line and the plane without any dependence on distance. The subscripts reflect Problem 9 (b) of Section 6.1, which showed that two given points, their Euclidean midpoint, and the ideal point formed a harmonic set.

Definition 6.2.2 Given three distinct collinear points X_0, X_1, and X, define X_2 to be the point such that $H(XX_1, X_0X_2)$. Given X_n and X_{n+1}, define X_{n+2} to be the point such that $H(XX_{n+1}, X_nX_{n+2})$. Given X_a and X_b, define $X_{(a+b)/2}$ to be the point such that $H(XX_{(a+b)/2}, X_aX_b)$ (Fig. 6.6).

Exercise 1 Define points X_{-n}, for n a positive integer.

Example 1 Figure 6.6 illustrates the placement of various points X_a. If X is the ideal point on the horizon of a perspective painting, this construction shows how to place points so that they look equally spaced. By Problem 4, the sequence of Euclidean distances $d(X_n, X)$ forms a harmonic sequence, such as $1, \frac{1}{2}, \frac{1}{3}, \frac{1}{4}, \ldots$. ●

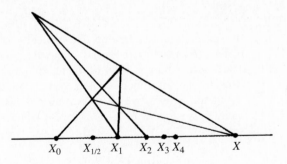

Figure 6.6

The first five axioms don't guarantee that all the points X_a from Definition 6.2.2 are distinct. (Indeed, in Chapter 7 we demonstrate that a finite projective plane with just four points on a line satisfies all five axioms.) Theorem 6.2.3 depends on the following separation axioms to ensure that the X_a are distinct. We write $PQ//RS$ to denote that P and Q separate R and S. Separation in projective geometry takes the role of Euclidean geometry's betweenness.

Axioms 6.2.2 **Separation Axioms**

vi) If $PQ//RS$, then P, Q, R, and S are distinct collinear points, $PQ//SR$ and $RS//PQ$.

vii) If P, Q, R, and S are distinct collinear points, then at least one of the following holds: $PQ//RS$, $PR//QS$, or $PS//QR$.

viii) If $PQ//RS$ and $PR//QT$, then $PQ//ST$.

ix) If $H(PQ, RS)$, then $PQ//RS$.

Theorem 6.2.2

i) If $PQ//RS$, then $QP//RS$, $QP//SR$, $RS//QP$, $SR//PQ$, and $SR//QP$.

ii) If A, B, C, and D are distinct collinear points, then exactly one of the following holds: $AB//CD$, $AC//BD$, or $AD//BC$.

Proof. See Problem 3. ■

Theorem 6.2.3 If X_p and X_q are determined from Definition 6.2.2 and Exercise 1 and $p \neq q$, then X_p and X_q are distinct.

Partial Proof. Show that, if $a < b < c$ for nonnegative integers (and so X_a, X_b, and X_c are determined from the definition), then $XX_b//X_aX_c$. Axiom (vi) then forces these points to be distinct. The remaining cases are similar.

Use induction to show that $XX_b//X_aX_c$, where $0 \leq a < b < c \leq k + 1$ holds for all integers $k \geq 1$. We have $H(XX_1, X_0X_2)$, so Axiom (ix) gives the case $k = 1$. Now assume that the case $0 \leq a < b < c \leq k + 1$ holds and let $0 \leq a < b < c \leq k + 2$. If $c < k + 2$, we are back at the induction hypothesis. Suppose that $c = k + 2$. Then, by definition, $H(XX_{k+1}, X_kX_{k+2})$. Axiom (ix) gives $XX_{k+1}//X_kX_{k+2}$. If $a = k$, then $b = k + 1$ and we are done. If $0 \leq a < k$, we have $XX_k//X_{k+1}X_a$ by the induction

JEAN VICTOR PONCELET

Jean Victor Poncelet (1788–1867) grew up in France during revolutionary times. He studied at the Ecole Polytechique under the influence of the legendary Monge. He then became an officer in Napoleon's army in the ill-fated campaign against Russia. He spent a year during 1813 and 1814 in a Russian prison. In prison he had the opportunity to reflect and write. He reconstructed his geometric education from memory and went on to discover many new results. Poncelet worked for various French governments after the fall of Napoleon and occasionally taught.

Projective geometry became a separate subject and moved to prominence with publication in 1822 of Poncelet's treatise on the subject, a revised and expanded version of his prison musings. Poncelet was an outspoken advocate of the synthetic approach, following the dictum of Lazare Carnot: " . . . to free geometry from the hieroglyphics of analysis." He realized the power of the general analytic approach compared to the isolated proofs of classical geometry. However, he felt that analytic geometry gave answers without giving insight. He developed projective geometry to provide general methods within the synthetic tradition. In the process he rediscovered many of the properties found previously by Desargues and others but then forgotten. He looked for properties of figures preserved by perspectivities. He was the first to realize the importance of duality.

The principle of continuity, which he used implicitly as an axiom, provided a powerful method for generalizing many results from understood cases to analogous cases. Thus a shape can pass continuously from circles and ellipses to parabolas to hyperbolas, much as the shadow of a lamp shade on a wall does as the lamp is tilted. The projective properties of tangents and other objects transfer along with the changing curves. For example, an asymptote of a hyperbola becomes just a special type of tangent. He even extended the principle of continuity to explore imaginary intersections of lines and conics that don't intersect in real points. Thus Poncelet initiated the study of complex projective geometry, although without using coordinates.

---

hypothesis. Axiom (viii), with $P = X$, $Q = X_{k+1}$, $R = X_k$, $S = X_{k+2}$, and $T = X_a$, gives $XX_{k+1}//X_{k+2}X_a$, which Theorem 6.2.2 converts to $XX_{k+1}//X_aX_{k+2}$. If $b = k + 1$, we are done. Problem 8 considers the remaining case, $a < b < k + 1$. ∎

The final axiom, the continuity axiom, extends the strategy of Theorem 6.2.3 to ensure that a projective line includes all points X_r, where r is a real number, together with the additional point X. Figure 6.7 illustrates how to match all but one of the points on a circle with the points on a line. In effect the final axiom ensures that the points on a projective line are arranged like the points on a circle. Visualizing lines in perspective drawings as circles is difficult because movement can be along the line in two directions, but only one seems to go toward the horizon. However, railroad tracks appear to intersect at the horizon in both directions. Theorem 6.2.1 forces each line to intersect the horizon (ideal line) in just one point, so these two directions must somehow meet at the same ideal point on the horizon, completing a circle. The (topological) arrangement of points of the entire projective plane is the same as the points in single elliptic geometry. (See Section 3.5.)

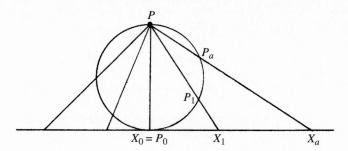

Figure 6.7

Axiom 6.2.3 **Continuity Axiom** Given any three points X, X_0 and X_1 on a projective line, there is a one-to-one correspondence between the real numbers r and all the points X_r on the line except X such that b is between a and c iff $X_a X_c // X_b X$.

Section 6.1 described the projective plane in terms of the familiar Euclidean plane together with an added "horizon," a line of ideal points. Based on the axioms presented we can now confirm that description. From Axiom (ii) we can start with four points O, X, Y, and U, no three of which are collinear. Think of $\overleftrightarrow{OX}$ and $\overleftrightarrow{OY}$ in Fig. 6.8 as the x- and y-axes and O as the origin for the points of the Euclidean plane. Axiom (x) and the discussion preceding Theorem 6.2.2 enable us to fill out the lines $\overleftrightarrow{OX}$ and $\overleftrightarrow{OY}$, provided we can find three points on each. For $\overleftrightarrow{OX}$ we use $O = X_0$, $\overleftrightarrow{OX} \cdot \overleftrightarrow{YU} = X_1$ and X. Similarly, for $\overleftrightarrow{OY}$ we use $O = Y_0$, $\overleftrightarrow{OY} \cdot \overleftrightarrow{XU} = Y_1$ and Y. Thus we have all the points X_a and Y_b on the "axes" $\overleftrightarrow{OX}$ and $\overleftrightarrow{OY}$. Next we include the points P_{ab} corresponding to the Euclidean plane, using two numbers, just as coordinates in analytic geometry. In Fig. 6.8, P_{ab} is the intersection of $\overleftrightarrow{X_a Y}$ and $\overleftrightarrow{XY_b}$. For any point P not on the line $\overleftrightarrow{XY}$, $\overleftrightarrow{PY}$ intersects $\overleftrightarrow{OX}$ at some point X_a and $\overleftrightarrow{PX}$ intersects $\overleftrightarrow{OY}$ at some point Y_b, which gives $P = P_{ab}$. Thus the points not on line $\overleftrightarrow{XY}$ look like the Euclidean plane. In effect, $\overleftrightarrow{XY}$ is the line of ideal points. However, the points on $\overleftrightarrow{XY}$ don't have natural coordinates in this procedure. Indeed, we used up all possible pairs (a, b) to label the points P_{ab} not on

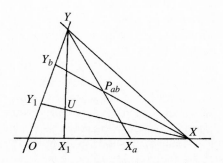

Figure 6.8

$\overleftrightarrow{XY}$. In Section 6.3 we present homogeneous coordinates, which are an elegant solution to this problem, involving the use of three coordinates for each point.

6.2.1 Duality

Poncelet noticed that, without parallel lines, lines have the same properties as points in projective plane geometry. That is, the words *point* and *line* in any axiom or theorem of projective geometry (and any definitions used in this statement) can be exchanged to get another theorem, called the *dual*. For example, the dual of "Every line has infinitely many points on it" is "Every point has infinitely many lines on it." Similarly, *collinear* and *concurrent* are dual concepts, as are *complete quadrangles* and *quadrilaterals*. People think about points differently from lines, so this structural similarity was (and is) hard to see. However, duality is an aesthetically pleasing property, which doubles the number of theorems available, often giving us theorems we might not have imagined.

Exercise 2 State the duals of Axioms (i), (ii), and (iii).

Theorem 6.2.4 The duals of Axioms (i), (ii), and (iii) hold.

Proof. See Problem 6. ∎

Next we develop the duals of harmonic sets of points and separation of points. The labeling of Fig. 6.9 provides the key to connecting a harmonic set of points $H(PQ, RS)$ and a harmonic set of lines $H(pq, rs)$. In particular, a harmonic set of points $H(PQ, RS)$ and a point A not on their line determine a harmonic set of the lines through A and these points. Conversely, a harmonic set of lines $H(pq, rs)$ and a line not on their common point determine a harmonic set of points. Similarly, Definition 6.2.3 uses the separation of points to define the separation of lines. (We assume that the definition is well-defined; that is, different choices of the line k give the same result of whether or not $pq//rs$ or $H(pq, rs)$. See Tuller [9] for more information.)

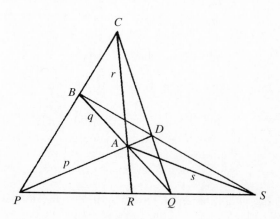

Figure 6.9

Definition 6.2.3 Let p, q, r, and s be concurrent lines on O, k be any line not on O, and $P = p \cdot k$, $Q = q \cdot k$, $R = r \cdot k$, and $S = s \cdot k$. Define $pq//rs$ iff $PQ//RS$ and $H(pq, rs)$ iff $H(PQ, RS)$.

Theorem 6.2.5 The duals of Axioms (iv)–(x) hold.

Proof. We prove the dual of Axiom (iv). (See Problem 6 for the others, which are similar.)

The dual of Axiom (iv) states, "Given three distinct lines p, q, and r on a point O, there is a unique line s on O, distinct from p, q, and r, such that $H(pq, rs)$." To prove this axiom, we start with the three lines p, q, and r on O. Let k be any line not on O. By Axiom (iii), k intersects the lines p, q, and r in the points P, Q, and R, respectively. By Theorem 6.2.1, P, Q, and R are distinct. Hence the hypothesis of Axiom (iv) holds, and there is a unique point S on k distinct from P, Q, and R such that $H(PQ, RS)$. By Theorem 6.2.1 $s = \overleftrightarrow{OS}$ is distinct from p, q, and r. Definition 6.2.3 gives $H(pq, rs)$. Finally, note that, as S is unique, s also is unique. ∎

Once we have the duals of the axioms, the duals of all the theorems follow immediately. Indeed, we could mechanically write the proof of a dual by switching the words *point* and *line* and so on throughout the original proof.

Exercise 3 Write the duals of Theorems 6.2.1, 6.2.2, and 6.2.3.

6.2.2 Perspectivities and projectivities

The notion of a perspectivity originates in perspective drawing. Recall that a perspectivity maps the points P_i of one line to the points Q_i of another line, using a point O such that for each i, O, P_i, and Q_i are collinear. Theorem 6.2.6 shows that harmonic sets and the relation of separation are preserved in perspectivities and so in perspective drawings.

Theorem 6.2.6 A perspectivity preserves harmonic sets of points and the relation of separation. That is, if a perspectivity from O maps the collinear points P, Q, R, and S to the collinear points P', Q', R', and S' and $H(PQ, RS)$, then $H(P'Q', R'S')$. Similarly, if $PQ//RS$, then $P'Q'//R'S'$.

Proof. See Problem 9. ∎

Definition 6.2.4 A composition of perspectivities is a *projectivity*.

Exercise 4 Explain why Theorem 6.2.6 applies to projectivities as well as perspectivities.

By Problem 2 of Section 6.1, there is a projectivity that maps any three collinear points to any three collinear points. Theorem 6.2.7 reveals that the images of these three points determine the images of all other points on the original line.

Theorem 6.2.7 **Fundamental Theorem of Projective Geometry** A projectivity of a line is completely determined by three points on the original line and their images.

Proof. WLOG call the three points on the original line X, X_0, and X_1 and suppose that the projectivity takes them to Y, Y_0, and Y_1, respectively. Theorem 6.2.5 ensures that any point X_p described in Theorem 6.2.3 must go to Y_p. Problem 2 and the continuity axiom extend this process, matching each X_r with Y_r. ∎

Exercise 5 Define and illustrate the concept of a line perspectivity taking concurrent lines to concurrent lines. Write the duals of Theorems 6.2.6 and 6.2.7.

PROBLEMS FOR SECTION 6.2

1. a) Prove the rest of Theorem 6.2.1.
   b) What other arrangements of P, Q, R, and S form harmonic sets for $H(PQ, RS)$? Prove your answer.

2. Describe all numbers q such that an X_q is constructed according to Definition 6.2.2. Explain why all positive real numbers r can be written as limits of these q.

3. Prove Theorem 6.2.2. [*Hint:* For part (ii) use Axiom (viii).]

4. Let r represent the Euclidean point $(r, 0)$.
   a) In Problem 8 of Section 6.1, you saw that $H(0\ 1, a\ \frac{a}{2a-1})$. Use similarity to show, for $k > 0$, $H(0\ k, ak\ \frac{ak}{2a-1})$.
   b) Show that $H(0\ \frac{1}{n+1}, \frac{1}{n}\ \frac{1}{n+2})$.
   c) Use part (b) to show how to construct poles that appear equally spaced in a perspective painting.

5. Art books give various methods of constructing equally spaced poles in a perspective painting. (See Powell [6].) Explain how the following construction blends harmonic sets and Euclidean ideas. Start with the horizon line, the pole nearest the viewer, and the base of the next pole drawn in (Fig. 6.10). Extend line b connecting the two bases to the horizon line to find the appropriate ideal point. Draw line t connecting this ideal point with the top of the first pole. All the poles will have their bases on b and tops on t. Draw the second pole parallel to the first pole. Draw line p through the ideal point parallel to the poles. Draw line k_1 from the top of the first pole through the base of the second pole to point P on p. Line k_2 connecting the top of the second pole with P intersects b at the base of the third pole. Continue as in Fig. 6.10 to determine the other poles.

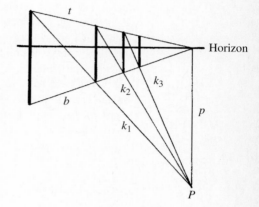

Figure 6.10

6. a) Prove Theorem 6.2.4.
   b) State the duals of Axioms (v)–(x).
   c) Prove the duals of Axioms (v)–(x).

7. a) In a complete quadrangle with vertices T_1, T_2, T_3, and T_4, the *diagonal points* are the three points of intersection of the "opposite" sides: $\overleftrightarrow{T_1T_2} \cdot \overleftrightarrow{T_3T_4}$, $\overleftrightarrow{T_1T_3} \cdot \overleftrightarrow{T_2T_4}$, and $\overleftrightarrow{T_1T_4} \cdot \overleftrightarrow{T_2T_3}$. Draw a picture to illustrate this situation and prove that the diagonal points aren't collinear.
   b) Define the dual concept of part (a) and illustrate the dual theorem.

8. Complete the proof of Theorem 6.2.3. [*Hint:* Use Axiom (viii) with $P = X$.]

9. Prove Theorem 6.2.6 [*Hint:* $H(\overleftrightarrow{OP}\ \overleftrightarrow{OQ}, \overleftrightarrow{OR}\ \overleftrightarrow{OS})$.]

10. For a fixed Euclidean point P, let k_i be the line on P having slope i and let k be the vertical line through P.
    a) Does this labeling of lines correspond to the labeling of points on a projective line?

In particular, do the duals of the separation axioms hold?

b) In analytic geometry, if m_1 and m_2 are the bisectors of the angles formed by m_3 and m_4, then $H(m_1m_2, m_3m_4)$. Investigate whether this relation holds for the labeling of the lines given, following the construction of a harmonic set of lines from Problem 7 of Section 6.1, with $m_1 = k$ and $m_2 = k_0$.

6.3 ANALYTIC PROJECTIVE GEOMETRY

We develop homogeneous coordinates to represent analytically all the points and lines in projective geometry and to emphasize their duality. In addition, homogeneous coordinates enable us to consider projective transformations in Section 6.4. In Section 6.2 we showed that two coordinates aren't sufficient to describe all points in the projective plane. In Chapter 4 we used three coordinates $(x, y, 1)$ for points, enabling transformations to move all points. In the process we demonstrated that row vectors $[a, b, c]$ represent lines, where $(x, y, 1)$ is on $[a, b, c]$ provided that $ax + by + c \cdot 1 = 0$. In addition, we showed that nonzero multiples $[\lambda a, \lambda b, \lambda c]$ of $[a, b, c]$ represent the same line. The duality of points and lines in projective geometry suggests using triples (x, y, z) for the enlarged set of projective points. As in Chapter 4 we use (x, y, z) for the column vector $\begin{bmatrix} x \\ y \\ z \end{bmatrix}$. Examples 1, 2, and 3 provide three ways to view these points. Example 3 is the most important representation. We leave it as an exercise to verify that the following interpretation satisfies the first three axioms of Section 6.2.

Interpretation By a *point* in the projective plane, we mean a nonzero column vector (x, y, z), where (x, y, z) and $(\lambda x, \lambda y, \lambda z)$ represent the same point for $\lambda \neq 0$. By a *line* in the projective plane, we mean a nonzero row vector $[a, b, c]$, where $[a, b, c]$ and $[\lambda a, \lambda b, \lambda c]$ represent the same line for $\lambda \neq 0$. A point (x, y, z) is *on* the line $[a, b, c]$ iff $ax + by + cz = 0$.

Exercise 1 Find the point on the lines $[2, -1, 3]$ and $[4, -2, 5]$.

Example 1 We can think of an ordinary Euclidean point (x, y) as the projective point $(x, y, 1) = (\lambda x, \lambda y, \lambda)$. Then "ideal points" have 0 for their third coordinate. For example, the points on the line $y = x$ (or $[1, -1, 0] = [\lambda, -\lambda, 0]$) are of the form (x, x, z). Note that the ideal point $(x, x, 0)$ is not only on $[1, -1, 0]$, but also on all the Euclidean lines parallel to it, which are of the form $[1, -1, c]$, or more familiarly, $y = x + c$. However, there is no reason in projective geometry to single out any line or point as "ideal" or different from any other; homogeneous coordinates make all points and lines equivalent. ●

Example 2 We can relate each point of the projective plane to two opposite points on a sphere. For each projective point (x, y, z), there are two scalars λ and $-\lambda$ such that $(\lambda x, \lambda y, \lambda z)$ and $(-\lambda x, -\lambda y, -\lambda z)$ have length 1 and so are on the sphere. Projective lines correspond to great circles. Example 3 explains why the great circle corresponding to $[a, b, c]$ is in the plane perpendicular to the point (a, b, c). The identification of opposite points on the sphere gives single elliptic geometry. (See Section 3.5.) The points and lines of single elliptic geometry match exactly the points and lines of projective geometry.

Single elliptic geometry has distance and angle measure in addition to the projective notions. ●

Example 3 Linear algebra in $\mathbf{R}^3$ matches projective geometry. A projective point is a one-dimensional subspace (a line through the origin). The scalar λ merely moves a point along this line. A projective line is a two-dimensional subspace (a plane through the origin). Recall that $[a, b, c]$ contains points (x, y, z), with $ax + by + cz = 0$ or $[a, b, c] \cdot (x, y, z) = 0$. The dot product of two vectors is 0 whenever the vectors are perpendicular. So $[a, b, c]$ must be the Euclidean plane through the origin perpendicular to the vector (a, b, c). Linear transformations take subspaces to subspaces, so projective transformations have a ready representation. ●

Exercise 2 Verify that the interpretation of Example 3 satisfies Axioms (i), (ii), and (iii) of a projective plane from Section 6.2.

Example 4 Three distinct points (p, q, r), (s, t, u), and (v, w, x) are collinear iff the determinant $\begin{vmatrix} p & s & v \\ q & t & w \\ r & u & x \end{vmatrix}$ is 0.

Solution. The three points are on the line $[a, b, c]$ iff $\begin{bmatrix} a & b & c \end{bmatrix} \begin{bmatrix} p & s & v \\ q & t & w \\ r & u & x \end{bmatrix} = \vec{0}$, which gives a system of three homogeneous equations. From linear algebra, this system has a nonzero solution $[a, b, c]$ (that is, the points are collinear) iff the matrix is singular and the determinant is 0. ●

Representing projectivities, harmonic sets, and separation analytically is easier if we use two homogeneous coordinates for collinear points. We represent the points on a line as (u, v), where $(u, v) = (\lambda u, \lambda v)$, for $\lambda \neq 0$. We postpone the treatment of projectivities to Section 6.4 because they are a type of transformation.

Example 5 The points on $[1, -1, 2]$ are of the form (x, y, z), where $x - y + 2z = 0$ or $y = x + 2z$. This last equation enables us to eliminate the y-value, so the two coordinates (x, z) are sufficient to describe which point on this line we are considering. From the Euclidean point of view, the point $(x, 1) = (\lambda x, \lambda)$ corresponds to the real number x and $(1, 0) = (\lambda, 0)$ corresponds to ∞ or the ideal point. By solving the equation $y = x + 2z$ for x or z, we could eliminate either of these variables instead. These methods of determining two homogeneous coordinates (and others) are compatible; they correspond in linear algebra to changes of coordinates. ●

6.3.1 Cross ratios

The concept of the cross ratio of four collinear points, initially explored by the ancient Greeks for Euclidean geometry, provides the analytic key for both harmonic sets and separation. We use the notationally easy form of the cross ratio given in Example 6

Chapter 6 Projective Geometry

whenever possible. Theorem 6.3.1 shows that our analytic interpretation matches Axiom (x), the most powerful separation axiom.

Example 6 For subscripts as in Axiom (x), the cross ratio of four distinct collinear points X_a, X_b, X_c, and X_d is $R(a,b,c,d) = \frac{a-c}{a-d} \div \frac{b-c}{b-d}$. Recall that $H(0\,1, x\frac{x}{2x-1})$. For example, $R(0, 1, 3, \frac{3}{5}) = \frac{-3}{-3/5} \div \frac{-2}{2/5} = -1$. Verify that for any $x \neq 0, \frac{1}{2}$, or 1 we have $R(0, 1, x, \frac{x}{2x-1}) = -1$. ●

Definition 6.3.1 The *cross ratio* of four collinear points $P = (p, q)$, $S = (s, t)$, $U = (u, v)$, and $W = (w, x)$ is

$$R(P, S, U, W) = \frac{\begin{vmatrix} p & u \\ q & v \end{vmatrix}}{\begin{vmatrix} p & w \\ q & x \end{vmatrix}} \div \frac{\begin{vmatrix} s & u \\ t & v \end{vmatrix}}{\begin{vmatrix} s & w \\ t & x \end{vmatrix}}.$$

Remark If $q = t = v = x = 1$, this formula reduces to the special case of Example 6.

Interpretation For four collinear points P, S, U, and W, $H(PS, UW)$ iff $R(P, S, U, W) = -1$ and $PS//UW$ iff $R(P, S, U, W) < 0$. Cross ratios, harmonic sets, and separation for lines are given dually.

Exercise 3 Suppose that $a < c < b < d$. Use Example 6 to verify that $R(a, b, c, d) < 0$ and so $X_aX_b//X_cX_d$.

Theorem 6.3.1 Axiom (x) holds in the analytic projective plane.

Proof. Let $a < b < c$ be real numbers. Then the homogeneous coordinates of the four points of the axiom are $X_a = (a, 1)$, $X_b = (b, 1)$, $X_c = (c, 1)$, and $X = (1, 0)$. Then

$$R(X_a, X_c, X_b, X) = \frac{\begin{vmatrix} a & b \\ 1 & 1 \end{vmatrix}}{\begin{vmatrix} a & 1 \\ 1 & 0 \end{vmatrix}} \div \frac{\begin{vmatrix} c & b \\ 1 & 1 \end{vmatrix}}{\begin{vmatrix} c & 1 \\ 1 & 0 \end{vmatrix}} = ((a-b)/-1) \div ((c-b)/-1) = (a-b)/(c-b).$$

As $a < b < c$, this fraction is negative and so $X_aX_c//X_bX$, as required. ∎

6.3.2 Conics

Projective geometry provides a unified way to study conics. The general equation of a conic in usual coordinates is $ax^2 + 2bxy + cy^2 + 2dx + 2ey + f = 0$. In homogeneous coordinates this equation becomes $ax^2 + 2bxy + cy^2 + 2dxz + 2eyz + fz^2 = 0$. Every term is now of second degree in the variables x, y, and z, so we say that it is a *homogeneous* second-degree equation. (Homogeneous first-degree equations, $ax + by + cz = 0$, represent lines.) Because a conic has a second-degree equation, it can intersect a line in zero, one, or two points. Linear algebra makes the homogeneous equation even more useful, as Exercise 4 illustrates. The matrix form explains why we choose to have the factors of 2 in the general equation. However, not every such matrix gives a conic, as Exercise 5 illustrates. Quadratic forms in linear algebra are closely tied to conics and their generalizations. (See Fraleigh and Beauregard [3, Chapter 7].)

6.3 Analytic Projective Geometry

Exercise 4 Show that a point $P = (x, y, z)$ is on the conic with equation $ax^2 + 2bxy + cy^2 + 2dxz + 2eyz + fz^2 = 0$ iff $P^T C P = 0$, where P^T is the transpose of P and $C = \begin{bmatrix} a & b & d \\ b & c & e \\ d & e & f \end{bmatrix}$.

Exercise 5 Verify that $C = \begin{bmatrix} 2 & -3/2 & 1 \\ -3/2 & 1 & -1/2 \\ 1 & -1/2 & 0 \end{bmatrix}$ has a determinant of 0 and represents the product of the two lines $2x - y = 0$ and $x - y + z = 0$.

Interpretation By a *conic* we mean a symmetric invertible 3×3 matrix. Two such matrices represent the same conic iff one is the multiple of the other by some real number $\lambda \neq 0$. A point P is on a conic C iff $P^T C P = 0$.

From a projective viewpoint, circles, ellipses, parabolas, and hyperbolas—the Euclidean types of conics—are indistinguishable. However, if, as in Fig. 6.11, we arbitrarily designate one line k as an ideal line, hyperbolas intersect the ideal line in two points, parabolas intersect it in one point, and circles and ellipses do not intersect it. (To distinguish circles from ellipses requires measures of angles or distances.) We define a *tangent* to a conic to be a line with only one point on the conic. (In Euclidean geometry this definition fails, for the parabola $y = x^2$ and the hyperbola $y = \frac{1}{x}$ each have only one point of intersection with vertical lines $x = b$, but vertical lines aren't tangents. For the hyperbola we require $b \neq 0$.) The asymptotes to a hyperbola in Euclidean geometry are simply tangents in projective geometry.

Exercise 6 Verify $x^2 - yz = 0$ and $xy - z^2 = 0$ are the homogeneous equations for the conics $y = x^2$ and $y = \frac{1}{x}$. Explain why in projective geometry the (Euclidean vertical) lines $[1, 0, b]$ are not tangent to these conics. (Assume that $b \neq 0$ for the second conic.)

Two distinct Euclidean circles intersect in at most two points, but Exercise 7 below shows that two conics can intersect in more than two points. Five points, no three collinear, are required to completely determine a conic.

Exercise 7 Graph $y = x^2$ and $x^2 + (y - 2)^2 = 2$ and find their intersections.

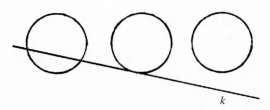

Figure 6.11

PROBLEMS FOR SECTION 6.3

1. a) Find the point of intersection of the lines $[m, -1, b]$ and $[m, -1, b']$. Interpret this situation in Euclidean geometry.

   b) Verify that the line $[a, b, c]$ on the points (p, q, r) and (s, t, u) is given by (the transpose of) their cross product, or equivalently by
   $$\left[\left|\begin{array}{cc} q & t \\ r & u \end{array}\right|, \left|\begin{array}{cc} p & s \\ r & u \end{array}\right|, \left|\begin{array}{cc} p & s \\ q & t \end{array}\right|\right], \text{ where } \left|\begin{array}{cc} v & y \\ w & z \end{array}\right|$$
   is the determinant of a 2×2 matrix. Describe how you can find the intersection of two lines.

2. Follow Van Staudt's coordinatization of Section 6.2, using $O = (0, 0, 1)$, $X = (1, 0, 0)$, $Y = (0, 1, 0)$, and $U = (1, 1, 1)$. Find homogeneous coordinates for X_a, Y_b, and P_{ab}. Find homogeneous coordinates for the point of intersection of $[0, 0, 1]$ and $\overleftrightarrow{OP_{ab}}$.

3. a) Verify that no three of points $(0, 0, 1)$, $(6, 0, 1)$, $(3, 3, 1)$, and $(8, 4, 1)$ are collinear. Find the diagonal points of the complete quadrangle they form and verify that these diagonal points are not collinear. (See Problem 7 of Section 6.2.) Use the interpretation of Example 1 to graph the complete quadrangle and its diagonal points.

   b) Pick two of the diagonal points of part (a), say, A and B, and find the coordinates of the line k on them. Find the intersections C and D of k with the other two sides of the complete quadrangle. Explain why the four points A, B, C, and D form a harmonic set. Verify that they form a harmonic set by using a cross ratio.

4. Use Example 6 to verify Axioms (v), (vi), and (vii) and Theorem 6.2.1, part (iii).

5. a) Illustrate Axiom (viii) by using five points on a horizontal line. If the coordinates of the points, from left to right, are a, b, c, d, and e, how do these letters match the letters of the points in the axiom? Use the form of the cross ratio in Example 6 to verify Axiom (viii) for the case in your drawing.

   b) What other orderings of the five points are compatible with the hypothesis of Axiom (viii)? Verify Axiom (viii) for these other cases.

6. Let $A = (1, 0)$, $B = (0, 1)$, $C = (1, 1)$, and $D = (r, 1)$.

   a) Verify that $R(A, B, C, D) = r$.

   b) There are 24 orderings of the four points A, B, C, and D, but there are only six different cross ratios. Use the coordinates in part (a) to find the five other values of the cross ratios besides r. List the orderings that have the same cross ratio, r, as the ordering A, B, C, D.

   c) Explore whether, for any four distinct collinear points, the relationship of the six different cross ratios of part (b) holds.

   d) Explore what values the cross ratio $R(A, B, C, D)$ can have for three distinct points, with one of them repeated.

7. a) In ordinary analytic geometry $y = x/(x + 1)$ is a hyperbola with asymptotes $y = 1$ and $x = -1$. Convert these equations to homogeneous equations. Find the ideal point of each asymptote and verify that it is the intersection of the asymptote and the conic.

   b) Find the asymptotes of $x^2 - y^2 = 1$ and repeat part (a) for this conic.

   c) Convert the equation of the parabola $y = x^2 - x$ to homogeneous form and verify that the ideal line $[0, 0, 1]$ has one point of intersection with it.

8. Explain why a general point on $x^2 + y^2 - z^2 = 0$ (the unit circle) is of the form $(\cos \alpha, \sin \alpha, 1)$. (In particular, why can you always choose $z = 1$?) Use analytic geometry to find the homogeneous coordinates of the line tangent to this circle at this point. [Hint: The coordinates of these tangents and their points of tangency have a remarkable property. (The set of tangents forms a *line conic*, the dual of a point conic, which is called a conic.)]

9. a) Show the general equation for the family of conics through the four points $(1, 1, 1)$, $(1, -1, 1)$, $(-1, 1, 1)$, and $(-1, -1, 1)$ to be $ax^2 + cy^2 + fz^2 = 0$.

   b) Use Example 1 to graph the three degenerate conics through these four points and find their equations. (Each is a pair of lines.)

   c) Let $f = 1$ and pick specific values for a and c. Graph the resulting conic.

   d) Show every point (p, q, r) to be on just one of the conics in the family of part (a), including the degenerate ones of part (b).

   e) What types of Euclidean conics are in the family of part (a)?

10. (Calculus) a) Write the equation $y = x^2$ in homogeneous coordinates.
   b) Show that the lines $[a, b, c]$ tangent to the equation in part (a) satisfy the equation of the line conic $a^2 - 4bc = 0$ as follows. First, pick x_0, a real number, and find the coordinates of the point $(x_0, y_0, 1)$ on the conic. Next use calculus to find the equation of the tangent line. Now convert this line to the form $[a, b, c]$ and verify that it satisfies the equation $a^2 - 4bc = 0$.
   c) Graph $y = x^2$ and selected tangents.

11. (Calculus) Repeat Problem 10 for the hyperbola $y = \frac{1}{x}$ and the corresponding line conic $4ab - c^2$. Explain the similarities in these problems from a projective viewpoint.

12. Suppose that P is a point on the conic C. Claim: The line $l = P^T C$ is tangent to C at P.
   a) Verify the claim by using the matrix form for conics in Problems 10 and 11.
   b) Show that P is on the line l of the claim.
   c) Suppose that P and Q are two points on the conic C and that l and k are their corresponding lines in the claim. Show that $l = k$ iff $P = Q$. Explain why this equation shows that the lines in the claim intersect the conic in only one point. Write a proof of the claim.
   d) Show that all tangents l satisfy the equation of the line conic $lC^{-1}l^T = 0$.

6.4 PROJECTIVE TRANSFORMATIONS

The usefulness of homogeneous coordinates becomes apparent in terms of transformations. Recall from Chapter 4 that a transformation is a one-to-one function from a space onto itself and that points are column vectors and lines are row vectors. Thus projectivities (transformations of collinear points) and collineations (transformations of all the points of the projective plane) correspond to invertible matrices. If we use two homogeneous coordinates for points on a line, a projectivity maps points (u, v) to points (s, t) and can be represented as an invertible 2×2 matrix. This matrix can be considered to map a line to itself.

Interpretation A *projectivity* is represented by an invertible 2×2 matrix. Two such matrices that differ by a nonzero constant represent the same projectivity.

Problem 2 of Section 6.1 shows that a projectivity can map any three collinear points to any three collinear points. Theorem 6.4.1 shows the surprising fact that 2×2 matrices can mimic this flexibility.

Theorem 6.4.1 For any distinct collinear points (u_1, v_1), (u_2, v_2), and (u_3, v_3) and distinct collinear points (s_1, t_1), (s_2, t_2), and (s_3, t_3), a unique projectivity maps (u_i, v_i) to (s_i, t_i).

Proof. First we show that a projectivity can map the three collinear points $(1, 0)$, $(0, 1)$, and $(1, 1)$ to any three distinct collinear points (s_1, t_1), (s_2, t_2), and (s_3, t_3). The matrix $\begin{bmatrix} s_1 & s_2 \\ t_1 & t_2 \end{bmatrix}$ maps $(1, 0)$ to (s_1, t_1) and $(0, 1)$ to (s_2, t_2). If we multiply the first column of the matrix by any scalar multiple λ, $(1, 0)$ still goes to $(s_1, t_1) = (\lambda s_1, \lambda t_1)$ and similarly for the second column, giving us the needed flexibility. For the matrix $\begin{bmatrix} \lambda s_1 & \mu s_2 \\ \lambda t_1 & \mu t_2 \end{bmatrix}$ the image of $(1, 1)$ is $(\lambda s_1 + \mu s_2, \lambda t_1 + \mu t_2)$, which we set equal to (s_3, t_3) and solve for λ and μ. Because (s_2, t_2) is not a multiple of (s_1, t_1) the determinant $\begin{vmatrix} s_1 & s_2 \\ t_1 & t_2 \end{vmatrix} \neq 0$. Thus there is a solution. In fact, the solutions form a family that are scalar multiples of one another. Thus a unique projectivity M takes $(1, 0)$, $(0, 1)$, and $(1, 1)$ to any three distinct points (s_1, t_1), (s_2, t_2), and (s_3, t_3). Similarly, a unique projectivity K takes $(1, 0)$, $(0, 1)$, and $(1, 1)$ to (u_1, v_1), (u_2, v_2), and (u_3, v_3). Then the projectivity represented by MK^{-1} uniquely satisfies the theorem's conditions. ∎

Theorem 6.4.2 Cross ratios, harmonic sets, and separation are preserved under projectivities.

Proof. Let (u_i, v_i) for $i = 1, 2, 3, 4$ be four collinear points and $M = \begin{bmatrix} a & c \\ b & d \end{bmatrix}$ be any projectivity. We first show the cross ratios $R(u_1, u_2, u_3, u_4)$ and $R(Mu_1, Mu_2, Mu_3, Mu_4)$ to be equal. A cross ratio involves four determinants of the coordinates of the points. For example, consider $\begin{vmatrix} u_1 & u_3 \\ v_1 & v_3 \end{vmatrix}$ and the corresponding determinant of the images $M(u_1, v_1)$ and $M(u_3, v_3)$: $\begin{vmatrix} Mu_1 & Mu_3 \\ Mv_1 & Mv_3 \end{vmatrix} = |M| \begin{vmatrix} u_1 & u_3 \\ v_1 & v_3 \end{vmatrix}$. The extra factor of $|M|$ appears in each of the four determinants in the cross ratio of the four images. As two of these factors of $|M|$ are in the numerator and two in the denominator, they cancel each other. Thus $R(u_1, u_2, u_3, u_4) = R(Mu_1, Mu_2, Mu_3, Mu_4)$. Harmonic sets and separation are defined in terms of the cross ratio, so they also are preserved under projectivities. ∎

Interpretation A *collineation* of the projective plane is represented by an invertible 3×3 matrix. Two matrices represent the same collineation iff one is a nonzero scalar multiple of the other. A point P is *fixed* by a collineation α iff $\alpha(P) = P$. A line k is *stable* under a collineation α iff every point on k is mapped by α to a point on k.

All the affine transformations from Chapter 4, including isometries and similarities, are collineations because all affine transformations have invertible matrices of the form $\begin{bmatrix} a & c & e \\ b & d & f \\ 0 & 0 & 1 \end{bmatrix}$. Theorem 6.4.3 describes how to find the images of lines and conics under collineations. Although points, lines, conics, and collineations have multiple representations, for simplicity we use just one such representation in the examples. In essence a collineation is a projectivity for each line, so Theorem 6.4.2 applies to both collineations and projectivities.

Theorem 6.4.3 The image of the line $[a, b, c]$ under the collineation M is $[a, b, c]M^{-1}$. The image of the conic C under M is $M^{-1T}CM^{-1}$.

Proof. See Problem 3. ∎

Theorem 6.4.4 The set of projectivities of a line to itself forms a group of transformations. The set of collineations forms a group of transformations.

Proof. See Problem 4. ∎

6.4 Projective Transformations

Example 1 Consider the Euclidean translation $T = \begin{bmatrix} 1 & 0 & 3 \\ 0 & 1 & -2 \\ 0 & 0 & 1 \end{bmatrix}$. In Chapter 4 we showed that a translation has no fixed affine points $(x, y, 1)$. However, we now have more points (x, y, z). Because $\lambda(x, y, z)$ represents the same point as (x, y, z), we need to consider the general eigenvector problem $T(x, y, z) = \lambda(x, y, z)$. Note that eigenvectors for any nonzero eigenvalue represent fixed points. We obtain three equations: $x + 3z = \lambda x$, $y - 2z = \lambda y$, and $z = \lambda z$. The last equation forces $\lambda = 1$ or $z = 0$. In turn, the first two equations force both $\lambda = 1$ and $z = 0$. Thus every "ideal point" $(x, y, 0)$ is fixed by a translation. Recall that ideal points are where Euclidean parallel lines meet and that translations take a line to a line parallel to itself. We also showed that all the stable lines of a translation in the affine plane are parallel, in this case $[-\frac{2}{3}, -1, c]$. As a collineation, there is one more stable line: $[0, 0, 1]$, the ideal line $z = 0$. All these stable lines can be written in the form $[\frac{-2}{3}a, -a, c]$ and go through the point $(3, -2, 0)$. ●

Example 2 The ideal line $[0, 0, 1]$ is stable for all similarities and affine transformations, for the bottom row of each of their inverses is $[0 \ 0 \ 1]$. Consider $A = \begin{bmatrix} 0 & 2 & 3 \\ 2 & 0 & 0 \\ 0 & 0 & 1 \end{bmatrix}$, which reflects over the line $y = x - z$ ($[1, -1, -1]$) and expands by a factor of 2 around the fixed point $(-1, -2, 1)$. There are two other fixed projective points, which we find by solving the general eigenvector problem $A(x, y, z) = \lambda(x, y, z)$. There are three values of λ such that the determinant of $A - \lambda I$ is zero: 1, 2, and -2. The three fixed points are $(-1, -2, 1)$, $(1, 1, 0)$, and $(1, -1, 0)$, respectively. To find the stable lines we need to solve the general eigenvector problem $[a, b, c]A^{-1} = \lambda[a, b, c]$. Verify that $A^{-1} = \begin{bmatrix} 0 & 0.5 & 0 \\ 0.5 & 0 & -1.5 \\ 0 & 0 & 1 \end{bmatrix}$ and that the three eigenvalues are 1, 0.5, and -0.5, the multiplicative inverses of the eigenvalues of A. The corresponding stable lines are $[0, 0, 1]$, $[-1, -1, -3]$, and $[1, -1, -1]$. Verify that these lines are stable and intersect in the three fixed points. ●

Because collineations are 3×3 matrices, we can send the points $X = (1, 0, 0)$, $Y = (0, 1, 0)$, and $Z = (0, 0, 1)$ to any three noncollinear points. However, as in Theorem 6.4.1, we can do even more. We can send X, Y, Z, and $U = (1, 1, 1)$ to any four points, provided no three of them are collinear.

Exercise 1 Verify that the collineation $\begin{bmatrix} 6 & -1 & 0 \\ 9 & 1 & 0 \\ 3 & -1 & 3 \end{bmatrix}$ takes X, Y, Z, and U to $(2, 3, 1)$, $(1, -1, 1)$, $(0, 0, 1)$, and $(1, 2, 1)$, respectively.

Theorem 6.4.5 Let P_1, P_2, P_3, and P_4 be four points, no three of which are collinear, and Q_1, Q_2, Q_3, and Q_4 be any four points. Then a unique collineation takes P_1, P_2, P_3, and P_4 to Q_1, Q_2, Q_3, and Q_4, respectively, iff no three of Q_1, Q_2, Q_3, and Q_4 are collinear.

Proof. See Problem 5. ∎

In Chapter 4 we were able to distinguish different types of isometries by the fixed points and stable lines of each. For example, rotations of any angle other than 180° and 0° have one fixed point and no stable lines. Translations have no fixed points and a family of parallel stable lines. Projective geometry has an additional line and the points on it, which radically alter the existence of fixed points and stable lines. As collineations, all isometries and, more generally, all affine transformations leave stable this added line. Translations fix every point on this added line. Theorem 6.4.6 relies on linear algebra to ensure the existence of fixed points and stable lines.

Theorem 6.4.6 Every collineation of the projective plane has at least one fixed point and at least one stable line.

Proof. To solve the general eigenvector problem $A(x, y, z) = \lambda(x, y, z)$ we find the values of λ for which the determinant of $A - \lambda I$ is zero. As $A - \lambda I$ is a 3×3 matrix, with λ appearing in the three diagonal entries, the determinant (characteristic polynomial) is a third-degree real polynomial in λ. Every third-degree polynomial crosses the x-axis and so has at least one real root. Thus every collineation has an eigenvalue and a nonzero eigenvector, which is a fixed point. The same argument applies to the inverse matrix, giving a stable line. ∎

Example 3 The matrix $\begin{bmatrix} 0 & -1 & 0 \\ 1 & 0 & 0 \\ 0 & 0 & -1 \end{bmatrix}$ is a rotatory reflection when thought of as a spherical isometry or three-dimensional Euclidean isometry. It rotates the sphere 90° around the z-axis and reflects it over the equator (xy-plane). Note that as it is a spherical isometry, no point is fixed. However, $(0, 0, 1)$ is mapped to $(0, 0, -1)$, which is the same projective point. This fixed point is for the only real eigenvalue, -1. The only stable line is $[0, 0, 1]$, the equator or the ideal line. •

Recall from Section 6.3 that a conic is determined by five points, no three collinear. Theorem 6.4.5 asserts that collineations are determined by four points and their images. Surprisingly, despite this disparity, we can map every conic to every other conic, although we can't always specify where various points on one conic map onto the other conic. Problem 9 illustrates this flexibility.

PROBLEMS FOR SECTION 6.4

1. **a)** Describe all 2×2 matrices that send $(1, 0)$ to itself. Repeat for the point $(0, 1)$ and the point $(1, 1)$. Use these results to explain why the only 2×2 matrices leaving these points fixed are of the form $\begin{bmatrix} \lambda & 0 \\ 0 & \lambda \end{bmatrix}$.
   b) Repeat part (a) with 3×3 matrices and the points $(1, 0, 0)$, $(0, 1, 0)$, $(0, 0, 1)$, and $(1, 1, 1)$.

2. Let $x = (x, 1)$ and $\infty = (1, 0)$. Recall from Section 6.1 that $H(0\,1, x\, \frac{x}{2x-1})$.

   a) Find the projectivity $\begin{bmatrix} a & c \\ b & d \end{bmatrix}$ that takes 0 to 0, 1 to 1, and 2 to $\frac{3}{2}$. Verify that the harmonic set $H(0\,1, 2\,\frac{2}{3})$ goes to a harmonic set.
   b) Find the projectivity $\begin{bmatrix} a & c \\ b & d \end{bmatrix}$ that takes 0 to 0, 1 to 1, and 2 to $\frac{1}{2}$. Why does the harmonic set $H(0\,1, 2\,\frac{2}{3})$ go to a harmonic set?
   c) Find the projectivity $\begin{bmatrix} a & c \\ b & d \end{bmatrix}$ that takes 0 to 0, 1

to 1, and 2 to $x \neq \frac{1}{2}$. Verify that the harmonic set $H(0\ 1, 2\ \frac{2}{3})$ goes to $H(0\ 1, x\ \frac{x}{2x-1})$. [*Hint:* Pick $a = x$.]

d) Find the projectivity $\begin{bmatrix} a & c \\ b & d \end{bmatrix}$ that takes 0 to 0, 1 to ∞, and 2 to k. Find the image of $\frac{2}{3}$. Explain why this image forms a harmonic set with 0, ∞, and k.

3. Prove Theorem 6.4.3. [*Hint:* See Theorem 4.3.1.]
4. Prove Theorem 6.4.4.
5. Prove Theorem 6.4.5. [*Hint:* See Theorem 6.4.1.]
6. Recall Desargues's theorem from Section 6.1: If $\triangle ABC$ and $\triangle A'B'C'$ are perspective from a point P, then they are perspective from a line. Let $A = (1, 0, 1)$, $B = (1, 1, 1)$, $C = (0, 1, 1)$, $A' = (2, 0, 1)$, $B' = (4, 4, 1)$, $C' = (0, 3, 1)$, and $P = (0, 0, 1)$.

    a) Graph these points. (See Example 1, Section 6.3.) Find the intersections $\overleftrightarrow{AB} \cdot \overleftrightarrow{A'B'}$, $\overleftrightarrow{AC} \cdot \overleftrightarrow{A'C'}$, and $\overleftrightarrow{BC} \cdot \overleftrightarrow{B'C'}$ and verify that these points are collinear.

    b) Find the collineation $\alpha = \begin{bmatrix} a & d & g \\ b & e & h \\ c & f & i \end{bmatrix}$ that fixes P and takes A to A', B to B', and C to C'. [*Hint:* pick $a = 12$.] Verify that the line you found in part (a) is stable under α. Show further that every point on this line is fixed by α.

    c) Make a conjecture generalizing part (b) and relate your conjecture to Desargues' theorem.

7. a) Let $M = \begin{bmatrix} 1/2 & 0 & 0 \\ 0 & -1 & -1 \\ 0 & 1/2 & -1/2 \end{bmatrix}$. Find M^{-1} and the image of the unit circle $x^2 + y^2 - z^2 = 0$ under M.

    b) Explain why $[\cos \theta, \sin \theta, -1]$ are lines tangent to the unit circle.

    c) Convert the image in part (a) to nonhomogeneous coordinates ($z = 1$). What type of Euclidean conic is this image?

    d) (Calculus) Use derivatives to show that $[4k, -1, -2k^2]$ is tangent to the conic in part (c). Show that the ideal line $[0, 0, 1]$ is also a tangent.

8. If line k is tangent to conic C and α is any collineation, prove that the image of k under α is tangent to the image of C under α. [*Hint:* See Problem 12 of Section 6.3.]

9. Investigate the effect of the family of collineations $M_w = \begin{bmatrix} 1 & 0 & 0 \\ 0 & 1 & 0 \\ w & 0 & 1 \end{bmatrix}$ on the conic $C = \begin{bmatrix} 1 & 0 & -1 \\ 0 & 1 & 0 \\ -1 & 0 & 0 \end{bmatrix}$ (the circle $x^2 + y^2 - 2xz = 0$).

    a) Find M_w^{-1} and M_w^{-1T} and the general image $M_w^{-1T} C M_w^{-1}$ of C under M_w.

    b) For the following values of w, convert $M_w^{-1T} C M_w^{-1}$ to the equation of a conic in nonhomogeneous coordinates ($z = 1$): $w = \frac{1}{2}$, $-\frac{1}{4}$, $-\frac{1}{2}$, and -1.

    c) Graph the original circle and the conics of part (b).

    d) Which values of w in part (a) take the circle to Euclidean ellipses? to hyperbolas? to parabolas?

    e) Verify that the lines $[1, 0, -2]$ and $[0, 1, -1]$ are tangent to the original circle by graphing them with the circle.

    f) Find the images of the lines of part (e) for the values of w in part (b) and graph them with the corresponding conics to verify that they are tangents.

6.5 SUBGEOMETRIES

Projective geometry originated as an extension of Euclidean geometry. In 1858 Arthur Cayley provided the historically important construction of Euclidean distance and angle measure within projective geometry. That is, he showed Euclidean geometry to be a subgeometry of projective geometry. He also related the distance in spherical geometry to projective geometry but was unaware of any other geometries at that time. Felix Klein in 1871 built on Cayley's construction and other insights to show that both hyperbolic and single elliptic geometries (see Chapter 3) were subgeometries of projective geometry. Klein's construction led him to pick the names hyperbolic and elliptic.

ARTHUR CAYLEY

The most fundamental mathematical contributions of Arthur Cayley (1821–1895) came during the 15 years he practiced law following his mathematical education at Cambridge University. Cayley was first in his class and started teaching there, but soon left because he didn't want to be ordained as a priest, then required of all Cambridge professors. Later, as a renowned mathematician, he taught at Cambridge (without ordination) from 1863 until his death, except for a year in the United States. Cayley published nearly two hundred papers while practicing law and hundreds more in his lifetime, largely on algebra and geometry.

At the age of 20, Cayley started publishing on algebraic invariants, a subject now displaced by topics in algebra and geometry that he and others developed from invariants. While in his early 20s he published some of the first works on n-dimensional geometry. In 1845, at age 24, he found an 8-dimensional algebra generalizing the complex numbers and Hamilton's recently published 4-dimensional quaternions. In 1849, he proposed the abstract definition of a group and later published the theorem on groups named after him.

Algebraists had studied determinants for some time, but Cayley's work on invariants led him to be the first, in 1855, to study matrices and their properties. He used matrices to represent systems of equations and transformations. He defined matrix multiplication so that it would correspond to the composition of transformations. His crowning achievement in what we now call linear algebra came in 1858 with publication of the Cayley–Hamilton theorem on the characteristic equation of a matrix. Invariants also led Cayley in the following year to his key derivation of Euclidean geometry within projective geometry relative to an absolute conic. He had long used projective geometry in his study of algebraic invariants, which included conics as second-degree invariants. His method of finding metrical properties inside projective geometry led to the unification of several geometries. Cayley learned of hyperbolic geometry after publishing this important paper, but he never accepted it as more than a logical curiosity.

(Under his classification Euclidean geometry was a "parabolic" geometry.) The unification of many geometries under projective geometry emphasized the usefulness of projective geometry and the importance of modern, abstract mathematics.

One geometry is a subgeometry of another in two regards. First, from Klein's Erlanger Programm (see Section 4.2), if all transformations of one geometry are transformations of a second geometry, the first is a subgeometry of the second. Second, the points, lines, and other terms of the subgeometry must be defined from the points, lines, and the like of the encompassing geometry. Fortunately, these two conditions reinforce each other. Figure 6.12 shows relationships of various geometries in terms of their transformation groups.

Recall that projective properties of points on a line must involve at least four points because any three points can be mapped to any three points by Theorem 6.2.7. Cross ratios, harmonic sets, and separation, involving four points, are preserved under all collineations by Theorem 6.4.2, as are projective properties. However, distance is a relation involving just two points. Cayley realized that he could define the distance between two points relative to two other points on a given line. To obtain these other

6.5 Subgeometries

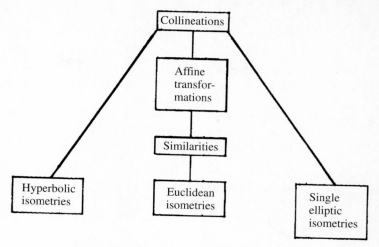

Figure 6.12 Some subgroups of collineations.

points he used the intersection of the line with a fixed conic, which he called the *absolute conic*. Cayley similarly defined the angle between two lines by using two other concurrent lines tangent to the absolute conic. We develop this idea fully only for distance in Klein's model of hyperbolic geometry. The other situations are more complicated, for they involve complex numbers and other advanced concepts. The transformations for various geometries are characterized as projective collineations that take the absolute conic to itself. (See Tuller [9, Chapter 7] for more information.)

6.5.1 Hyperbolic geometry as a subgeometry

From a projective point of view, in the Klein model of hyperbolic geometry in Chapter 3 we could use the points in the interior of any (real, nondegenerate) conic, but the unit circle is the simplest. In the following interpretation, this conic is the absolute conic in Cayley's terms. Note that any projective line through an interior point intersects the conic in two points. Indeed, one defines a point to be in the *interior* of a conic if every line on that point intersects the conic in two distinct real points. These two points of intersection of a line and the conic match the property that hyperbolas intersect the ideal line in two points. Klein used this analogy and other reasons to call this geometry hyperbolic. He based the formula for hyperbolic distance on the cross ratio of the two points and the intersections of the line they determine with the conic. Recall that all collineations preserve the cross ratio, so automatically Klein's distance formula is preserved under whichever collineations are hyperbolic. Poincaré defined hyperbolic distance in his model in the same way we do here. (See Sections 3.4 and 4.6.) Poincaré recognized that the cross ratio is preserved under both inversions and collineations. To simplify the distance formula we include a Euclidean way to compute it.

Interpretation for Hyperbolic Plane Geometry By *absolute conic* we mean $x^2 + y^2 - z^2 = 0$. By *point* we mean the points (x, y, z), with $x^2 + y^2 < z^2$, interior to the absolute conic. By *line* we mean the set of points interior to the absolute conic that are on a projective line $[a, b, c]$. The intersections of a line with the absolute conic are the *omega points* of that line. The *distance* between A and B is $d_H(A, B) = c \cdot |\log(R(A, B, \Omega, \Lambda))| = c \cdot \left|\log\left(\frac{A\Omega}{A\Lambda} \div \frac{B\Omega}{B\Lambda}\right)\right|$, where XY is the Euclidean distance between X and Y, c is some constant, and Ω and Λ are the two omega points of line $\overleftrightarrow{AB}$. By *hyperbolic isometry* we mean a collineation that leaves the absolute conic stable.

Example 1 Verify that the adjacent points P_i and P_{i+1} shown in Fig. 6.13 have the same distance between them. The x-coordinates of the points are $P_0 = 0$, $P_1 = \frac{1}{3}$, $P_2 = \frac{3}{5}$, $P_3 = \frac{7}{9}$, $P_4 = \frac{15}{17}$, $P_5 = \frac{31}{33}$, $P_{-i} = -P_i$, $\Omega = -1$, and $\Lambda = 1$.

Solution. The Euclidean distances between these points are simply the differences of their x-coordinates. Then $(P_0\Omega/P_0\Lambda) \div (P_1\Omega/P_1\Lambda) = (1/1) \div (\frac{4}{3}/\frac{2}{3}) = \frac{1}{2}$. Similarly, $(P_1\Omega/P_1\Lambda) \div (P_2\Omega/P_2\Lambda) = (\frac{4}{3}/\frac{2}{3}) \div (\frac{8}{5}/\frac{2}{5}) = \frac{1}{2}$. All the corresponding products equal $\frac{1}{2}$ or 2. In turn, the absolute values of their logarithms are all the same. Hence, whatever the constant c is, the distances all are the same. ●

Example 2 Euclidean rotations and mirror reflections fixing the origin map the unit circle to itself and so are hyperbolic isometries. A matrix X_x analogous to a translation shifting $(0, 0, 1)$ along the x-axis to $(x, 0, 1)$ has for its last column $(x, 0, 1)$, with $-1 < x < 1$. This translation leaves the omega points $(1, 0, 1)$ and $(-1, 0, 1)$ fixed. Then $X_x = \begin{bmatrix} a & d & x \\ b & e & 0 \\ c & f & 1 \end{bmatrix}$ must satisfy $X_x(1, 0, 1) = (a + x, b, c + 1) = \lambda(1, 0, 1)$ and $X_x(-1, 0, 1) = (-a + x, -b, -c + 1) = \lambda(-1, 0, 1)$. From these equations $a = 1$, $b = 0$, and $c = x$. Figure 6.14 suggests that X_x should shift $(0, 1, 1)$ and $(0, -1, 1)$ to the points on the unit circle directly above and below $(x, 0, 1)$. This condition forces $d = 0 = f$ and $e = \sqrt{1 - x^2}$. ●

Exercise 1 Let $X_x = \begin{bmatrix} 1 & 0 & x \\ 0 & \sqrt{1-x^2} & 0 \\ x & 0 & 1 \end{bmatrix}$. Verify that $C = \begin{bmatrix} 1 & 0 & 0 \\ 0 & 1 & 0 \\ 0 & 0 & -1 \end{bmatrix}$, the unit circle, is stable.

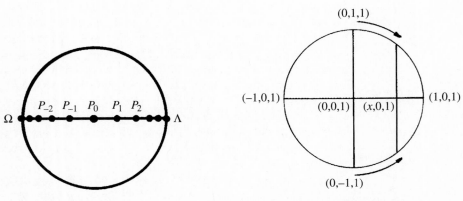

Figure 6.13   Figure 6.14

under any scalar multiple of X_x, and verify that the inverse of X_x is a scalar multiple of $\begin{bmatrix} 1 & 0 & -x \\ 0 & \sqrt{1-x^2} & 0 \\ -x & 0 & 1 \end{bmatrix}$. Explain why X_x is a hyperbolic isometry and why, in effect, X_{-x} is the inverse of X_x.

Theorem 6.5.1 The hyperbolic isometries form a group of transformations. The set $\{\lambda X_x : -1 < x < 1, \lambda \neq 0\}$ forms a group of transformations.

Proof. See Problem 3. ∎

The matrices X_x correspond to the transformations of velocities in the special theory of relativity. (See Section 5.5.) In Section 6.6 we explore this connection, including representing the geometry of relativity as a subgeometry of four-dimensional projective space. General hyperbolic isometries have a matrix form quite similar to spherical isometries. (See Section 4.5.) A hyperbolic isometry M must take C to itself. By Theorem 6.4.3, $M^{-1T}CM^{-1} = \lambda C$, for $\lambda \neq 0$. The special form of C simplifies this equation further. To emphasize the relationship of these isometries with spherical isometries, define the *h-inner product* of two vectors to be $(r, s, t) \cdot_h (u, v, w) = ru + sv - tw$. Only the minus sign distinguishes this h-inner product from the usual definition. Following this analogy, define the *h-length* of a vector (r, s, t) to be $(r, s, t) \cdot_h (r, s, t)$ and two vectors to be *h-orthogonal* iff their h-inner product is 0. (These definitions do not fulfill all the usual properties of inner products and lengths. For example, nonzero vectors can be h-orthogonal to themselves and so have zero h-lengths.)

Example 3 Find conditions on M so that M is a hyperbolic isometry.

Solution. For M to be a hyperbolic isometry, we must have $M^{-1T}CM^{-1} = \lambda C$ for some $\lambda \neq 0$. Multiply both sides of this equation by M^T on the left and M on the right to get $C = M^T \lambda C M$. We can factor out λ and move it to the other side to get $\frac{1}{\lambda}C = M^T C M$. If we write P for the first column, Q for the second column and R for the third column of M, then $M^T C M = M^T \begin{bmatrix} 1 & 0 & 0 \\ 0 & 1 & 0 \\ 0 & 0 & -1 \end{bmatrix} M = \begin{bmatrix} P \cdot_h P & P \cdot_h Q & P \cdot_h R \\ Q \cdot_h P & Q \cdot_h Q & Q \cdot_h R \\ R \cdot_h P & R \cdot_h Q & R \cdot_h R \end{bmatrix}$,

which is supposed to equal $\frac{1}{\lambda}C = \begin{bmatrix} 1/\lambda & 0 & 0 \\ 0 & 1/\lambda & 0 \\ 0 & 0 & -1/\lambda \end{bmatrix}$. Then the columns of M must be h-orthogonal to each other to give 0 off the main diagonal. Furthermore, the first two columns must have the same h-length, which must be the negative of the h-length of the third column. ●

Exercise 2 Verify that the X_x satisfy the conditions of Example 3.

6.5.2 Single elliptic geometry as a subgeometry

Single elliptic geometry (see Section 3.5) needs all projective points and lines. Thus the absolute conic must contain no (real) points. For the absolute conic we pick the imaginary conic $x^2 + y^2 + z^2 = 0$, whose only real solution is $(0, 0, 0)$, which isn't a projective point. Lines in single elliptic geometry therefore don't intersect the absolute conic, just as an ellipse has no intersection with the ideal line, suggesting one reason why Klein called this geometry elliptic. (He added single to the name to distinguish it from spherical geometry, in which lines intersect twice.) An isometry M for this geometry is a collineation that takes this degenerate conic $C = \begin{bmatrix} 1 & 0 & 0 \\ 0 & 1 & 0 \\ 0 & 0 & 1 \end{bmatrix}$ to itself. Because $C = I$, the identity, the equation $M^{-1T}CM^{-1} = \lambda C$ for $\lambda \neq 0$ reduces to $M^{-1T}M^{-1} = \lambda I$. Orthogonal matrices, the isometries for spherical geometry (see Section 4.5), satisfy the similar equation $M^T = M^{-1}$, or $M^T M = I$.

Theorem 6.5.2 A collineation M is an isometry of single elliptic geometry iff $M^T = \lambda M^{-1}$ for some nonzero real number λ. These isometries form a group of transformations.

Proof. See Problem 5. ∎

6.5.3 Affine and Euclidean geometries as subgeometries

Recall that the projective plane can be thought of as the affine or Euclidean plane and one extra "ideal" line, [0, 0, 1]. Cayley realized that [0, 0, 1], thought of as the degenerate conic $z^2 = 0$, functioned as the absolute conic. As affine geometry has no notions of distance or angle measure, we can directly give the interpretation of the affine plane as a subgeometry of projective geometry. Note that each affine line intersects the absolute conic in one point just as a parabola intersects the ideal line in one point. Klein considered affine geometry and its subgeometries to be parabolic.

Interpretation for Affine Plane Geometry By *absolute conic* we mean $z^2 = 0$. By *point* we mean a point not on the absolute conic: $(x, y, 1) = (\lambda x, \lambda y, \lambda)$. By *line* we mean a line other than the absolute conic: $[a, b, c] = [\lambda a, \lambda b, \lambda c]$, where not both a and b are zero. An *affine transformation* is a collineation that leaves the absolute conic stable.

Exercise 3 Verify that the affine transformations of Chapter 4 are the affine transformations of the preceding interpretation.

To derive Euclidean distance, Cayley needed to pick two specific points on the absolute conic $z^2 = 0$, the *circular points at infinity*, or $I = (1, i, 0)$ and $J = (1, -i, 0)$. Their name comes from the fact, noted by Poncelet, that these points are on every Euclidean circle. Example 4 reveals that all Euclidean isometries map I and J to themselves, showing the set $\{I, J\}$ to be stable under this group of transformations. The cross-ratio definition of distance (and angle measure) Klein employed in hyperbolic and single elliptic geometries works for Euclidean angle measure but not for Euclidean distance. Cayley utilized a more complicated method, which we omit.

Example 4 Show that all Euclidean isometries map I and J to themselves.

Solution. Recall from Section 4.3 that Euclidean isometries are of the form
$\begin{bmatrix} \cos\theta & -\sin\theta & c \\ \sin\theta & \cos\theta & f \\ 0 & 0 & 1 \end{bmatrix}$ or $\begin{bmatrix} \cos\theta & \sin\theta & c \\ \sin\theta & -\cos\theta & f \\ 0 & 0 & 1 \end{bmatrix}$. The first form, for direct isometries, takes $I = (1, i, 0)$ to $(\cos\theta - i\sin\theta, \sin\theta + i\cos\theta, 0) = (\cos\theta - i\sin\theta, i(\cos\theta - i\sin\theta), 0) = (1, i, 0)$. Similarly, J maps to itself. Matrices of the second form, for indirect isometries, switch I and J. ●

PROBLEMS FOR SECTION 6.5

1. **a)** Verify that the coordinate for each P_i in Example 1 is $(2^i - 1)/(2^i + 1)$.

   b) Verify that $X_{1/3}$, as defined in Exercise 1, takes P_i to P_{i+1}.

   c) Verify that $X_{1/3}$, as defined in Exercise 1, takes $-P_{i+1}$ to $-P_i$.

   d) Verify that $X_{1/3} \cdot X_{1/3}$ is a scalar multiple of $X_{3/5}$.

2. For the points $A = (a, 0, 1)$ and $B = (b, 0, 1)$, verify the identity $d_H(A, B) = d_H(X_x(A), X_x(B))$, where X_x is the hyperbolic translation of Exercise 1. [*Hint:* Simplify the cross-ratio for $X_x(A), X_x(B)$, $\Omega = (1, 0, 1)$, and $\Lambda = (-1, 0, 1)$ by factoring.]

3. **a)** Prove that the composition of the hyperbolic translations X_a and X_b is a scalar multiple of $X_{a \oplus b}$, where $a \oplus b = (a + b)/(1 + ab)$ is the addition of velocities in the special theory of relativity. (See Section 5.5.)

   b) Prove Theorem 6.5.1. [*Hint:* See Problem 7 of Section 5.5.]

4. Find the matrix for a hyperbolic translation Y_b along the y-axis. Verify that, in general, $X_a \cdot Y_b \neq Y_b \cdot X_a$. How are $X_a \cdot Y_b$ and $Y_b \cdot X_a$ related as matrices? Let $(c, d, 1)$ be any point inside the unit circle. Find hyperbolic translations X_a and Y_b such that $X_a \cdot Y_b(0, 0, 1) = (c, d, 1)$.

5. Prove Theorem 6.5.2. [*Hint:* See Theorem 4.5.1.]

6. **a)** Show that a conic $ax^2 + 2bxy + cy^2 + 2dxz + 2eyz + fz^2 = 0$ has the points $(1, i, 0)$ and $(1, -i, 0)$ on it iff it is a Euclidean circle; that is, $a = c$ and $b = 0$.

   b) Show that all similarities take the circular points at infinity to themselves.

7. Find the properties of an inner product and of length in a linear algebra text and explain which ones hold and which ones fail for the h-inner product and the h-length. (See, for example, Fraleigh and Beauregard [3].)

8. **a)** Show that every single elliptic isometry that is simultaneously an affine transformation must be a similarity. Are there any other restrictions on these transformations? [*Hint:* What are the images of the points $O = (0, 0, 1)$, $X_1 = (1, 0, 1)$, and $Y_1 = (0, 1, 1)$ under a similarity?]

   b) Describe all collineations that are simultaneously single elliptic isometries, hyperbolic isometries, and Euclidean isometries. Justify your answer and show that these collineations form a group.

6.6 PROJECTIVE SPACE

Homogeneous coordinates and collineations can be readily extended to higher dimensional projective spaces, which are significant models that are used for many purposes. For example, three-dimensional projective space provides perspective views for computer-aided design (CAD). And the Lorentz transformations in the special theory of relativity can be seen as isometries in a subgeometry of four-dimensional projective space related to hyperbolic geometry.

Interpretation By a *point* in n-dimensional projective space $\mathbf{P}^n$ we mean a one-dimensional subspace of the $n + 1$ dimensional vector space $\mathbf{R}^{n+1}$. By *line* we mean a two-dimensional subspace, by *plane* we mean a three-dimensional subspace, and so on. A point (line, and so on) is *on* a line (plane, and so on) iff one is a subset of the other. By a *collineation* in $\mathbf{P}^n$ we mean an invertible $(n+1) \times (n+1)$ matrix. Any nonzero scalar multiple of a matrix represents the same collineation.

Example 1 Two points in $\mathbf{P}^3$ (or any $\mathbf{P}^n$) have a unique line on them. In other words, two one-dimensional subspaces are spanned by a unique two-dimensional subspace. However in $\mathbf{P}^3$ two lines can fail to intersect, which corresponds to skew lines in Euclidean space. For example, the lines (two-dimensional subspaces) $\{(x, y, 0, 0) : x, y \in \mathbf{R}\}$ and $\{(0, 0, z, t) : z, t \in \mathbf{R}\}$ intersect only at the origin $(0, 0, 0, 0)$, the zero-dimensional subspace, which isn't a projective point. Two distinct planes (three-dimensional subspaces) in $\mathbf{P}^3$ must intersect in a line (a two-dimensional subspace). Each plane has three basis vectors, giving six possible vectors. The whole space needs only four basis vectors, so there must be an overlap of at least two. Because the planes are distinct, the overlap is exactly two—a line. In $\mathbf{P}^3$, point and planes are duals and a line is "self-dual." Thus the dual in $\mathbf{P}^3$ of "Two points determine a line" is "Two planes determine a line." In general, the dual of an i-dimensional subspace in $\mathbf{P}^n$ is an $(n + 1 - i)$-dimensional subspace. ●

6.6.1 Perspective and computer-aided design

A general collineation in $\mathbf{P}^3$ (a 4×4 matrix) can be broken into component parts, most of which appear in affine transformations. Recall that the rightmost column of an affine matrix describes how the origin moves and corresponds to a translation. The upper left 3×3 submatrix determines the type of affine transformation: rotation, reflection, shear, dilation, and so on. The bottom row of an affine matrix is $[\,0\ \ 0\ \ 0\ \ 1\,]$. The bottom row of a collineation provides flexibility lacking in affine transformations. In a CAD system the change of the lower right entry from a 1 to another value magnifies or shrinks the entire picture by the same factor—a much faster computer alteration than changing all the upper entries by the reciprocal factor. The first three entries of the bottom row determine perspective views in each dimension. The matrix $\begin{bmatrix} a & a & a & t_x \\ a & a & a & t_y \\ a & a & a & t_z \\ p_x & p_y & p_z & s \end{bmatrix}$ summarizes this discussion, where a stands for affine, p for perspective, s for scaling, and t for translation. (See Penna and Patterson [5] for more information on projective geometry and computer graphics.)

Example 2 We illustrate the effects of the perspective entries by using several related matrices to project a cube. (For convenience we ignore the translation and scaling entries.) By Theorem 4.5.4, we need only consider what the transformations do to the 4×8 matrix $V = \begin{bmatrix} 0 & 0 & 0 & 0 & 1 & 1 & 1 & 1 \\ 0 & 0 & 1 & 1 & 0 & 0 & 1 & 1 \\ 0 & 1 & 0 & 1 & 0 & 1 & 0 & 1 \\ 1 & 1 & 1 & 1 & 1 & 1 & 1 & 1 \end{bmatrix}$ containing the eight vertices of the cube.

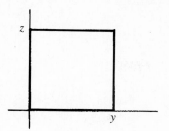

Figure 6.15 Image of a cube under PV.

The matrix $P = \begin{bmatrix} 0 & 0 & 0 & 0 \\ 0 & 1 & 0 & 0 \\ 0 & 0 & 1 & 0 \\ 0 & 0 & 0 & 1 \end{bmatrix}$ represents a projection parallel to the x-axis; it isn't a collineation because it collapses three-dimensional objects to two dimensions. On a computer screen, in nonhomogeneous coordinates the eight vertices in PV, together with their edges, would appear as a square (Fig. 6.15). To see the three-dimensional form of a cube we need to have a different viewing angle. Hence we rotate the cube $-30°$ around the z-axis and then $-30°$ around the (original) y-axis, using

the matrix $R = \begin{bmatrix} \cos -30 & 0 & -\sin -30 & 0 \\ 0 & 1 & 0 & 0 \\ \sin -30 & 0 & \cos -30 & 0 \\ 0 & 0 & 0 & 1 \end{bmatrix} \begin{bmatrix} \cos -30 & -\sin -30 & 0 & 0 \\ \sin -30 & \cos -30 & 0 & 0 \\ 0 & 0 & 1 & 0 \\ 0 & 0 & 0 & 1 \end{bmatrix} =$

$\begin{bmatrix} 3/4 & \sqrt{3}/4 & 1/2 & 0 \\ -1/2 & \sqrt{3}/2 & 0 & 0 \\ -\sqrt{3}/4 & -1/4 & \sqrt{3}/2 & 0 \\ 0 & 0 & 0 & 1 \end{bmatrix}$. Then PRV gives the familiar, nonperspective view of

the cube (Fig. 6.16). Note that all the faces in Fig. 6.16 are parallelograms.

Let $P_x = \begin{bmatrix} 1 & 0 & 0 & 0 \\ 0 & 1 & 0 & 0 \\ 0 & 0 & 1 & 0 \\ -1/5 & 0 & 0 & 1 \end{bmatrix}$, $P_{xy} = \begin{bmatrix} 1 & 0 & 0 & 0 \\ 0 & 1 & 0 & 0 \\ 0 & 0 & 1 & 0 \\ -1/5 & -1/5 & 0 & 1 \end{bmatrix}$, and $P_{xyz} =$

$\begin{bmatrix} 1 & 0 & 0 & 0 \\ 0 & 1 & 0 & 0 \\ 0 & 0 & 1 & 0 \\ -1/5 & -1/5 & -1/5 & 1 \end{bmatrix}$. Then PRP_xV gives a one-point perspective view (Fig.

6.17), $PRP_{xy}V$ gives a two-point perspective view (Fig. 6.18), and $PRP_{xyz}V$ gives a three-point perspective view (Fig. 6.19).

In Fig. 6.17, note that the sides of the cube parallel to the x-axis all meet at a *vanishing point* denoted V_x. The other sides remain parallel. We can compute the coordinates of this vanishing point by finding the projection of the ideal point along the x-axis, $(1, 0, 0, 0)$: $PRP_x(1, 0, 0, 0) = (0, -\frac{1}{2}, -\frac{\sqrt{3}}{4}, -\frac{1}{5}) = (0, 2.50, 2.17, 1)$, or, in the nonhomogeneous coordinates of the figure, $(2.50, 2.17)$. Figure 6.18 includes the vanishing points for both the x- and y-axes, V_x and V_y. Actually, P_{xy} automatically gives vanishing points for all directions in the xy-plane, which appear on the "horizon line"

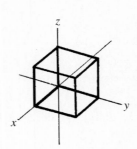

Figure 6.16 Image of a cube under PRV.

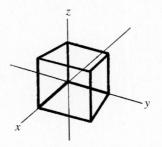

Figure 6.17 Image of a cube under PRP_xV.

$\overleftrightarrow{V_xV_y}$ shown in Fig. 6.18. For example, the diagonals connecting $(0, 0, 1, 1)$ to $(1, 1, 1, 1)$ and $(0, 0, 0, 1)$ to $(1, 1, 0, 1)$ meet at the vanishing point $(-0.92, 1.71)$, the projection of the ideal point $(1, 1, 0, 0)$. In three-point perspective (Fig. 6.19) every direction has a vanishing point. •

Projective geometry, like single elliptic geometry, is not oriented. (See Section 3.5.) In effect, there is no consistent way to define clockwise rotations around a point or direction on a line. One consequence of this nonorientability is that collineations can alter the appearance of a figure beyond what an artist or computer operator needs. For example, projectivities (and so collineations) can map any three collinear points to any three collinear points. Therefore these transformations can alter the betweenness relations of points on a line. However, no perspective view of a real object will ever turn it inside out. Stolfi [7] has created an oriented projective geometry for CAD by not allowing negative scalars. We place a tilde (~) over a point to indicate an oriented point. An *oriented point* is still a vector in $\mathbf{R}^4$, but two such vectors $\vec{v}$ and $\vec{w}$ represent the same point only if there is a positive real number k such that $\vec{w} = k \cdot \vec{v}$. In effect each

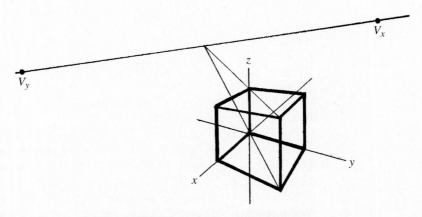

Figure 6.18 Image of a cube under $PRP_{xy}V$.

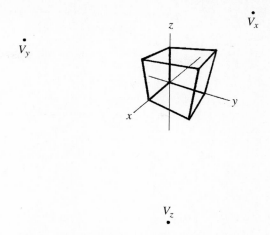

Figure 6.19 Image of a cube under $PRP_{xyz}V$.

projective point P is split into two opposite oriented points $\widetilde{P}$ and $-\widetilde{P}$. This corresponds exactly to the relationship between spherical and single elliptic geometry: Two opposite points in spherical geometry represent the same point in single elliptic geometry. All invertible matrices are still collineations in oriented projective geometry, although their effect changes and two matrices represent the same collineation when they differ by a positive scalar.

Definition 6.6.1 $\widetilde{R}$ is *between* $\widetilde{P}$ and $\widetilde{Q}$ iff there are positive real numbers a and b such that $\widetilde{R} = a\widetilde{P} + b\widetilde{Q}$. The line segment $\overline{\widetilde{P}\widetilde{Q}}$ is the set $\{\widetilde{R} : \widetilde{R}$ is between $\widetilde{P}$ and $\widetilde{Q}, \widetilde{R} = \widetilde{P}$ or $\widetilde{R} = \widetilde{Q}\}$. A set S of oriented points is *convex* iff, for all $\widetilde{P}$ and $\widetilde{Q}$ in S, $\overline{\widetilde{P}\widetilde{Q}}$ is a subset of S.

Example 3 The oriented points between $(0, \widetilde{0}, 1)$ and $(1, \widetilde{0}, 1)$ are $(x, \widetilde{0}, 1)$, where $0 < x < 1$.

Proof. Let $(u, \widetilde{v}, w)$ be between $(0, \widetilde{0}, 1)$ and $(1, \widetilde{0}, 1)$. Then there are positive reals a and b such that $(u, \widetilde{v}, w) = a(0, \widetilde{0}, 1) + b(1, \widetilde{0}, 1) = (b, 0, \widetilde{a+b}) = (\frac{b}{a+b}, 0, 1)$. Note that $x = b/(a+b)$ must be between 0 and 1 because both a and b are positive. WLOG we can pick a and b so that $a + b = 1$. Then (a, b) are the barycentric coordinates of $(x, \widetilde{0}, 1)$ in terms of $(0, \widetilde{0}, 1)$ and $(1, \widetilde{0}, 1)$. (See Section 2.3 for a discussion of barycentric coordinates.) ●

Remark If we used these same definitions with regular projective points P and Q and any nonzero scalars, the entire line $\overleftrightarrow{PQ}$ would be "between" P and Q and so would be a "line segment." Theorem 6.6.1 would still be provable (changing the word *positive* to *nonzero*), but that would not be particularly helpful for the only convex sets would be a single point, a (projective) line, a (projective) plane, and the like.

Theorem 6.6.1 Collineations in oriented projective geometry preserve betweenness and convexity.

Proof. Let γ be a collineation and $\widetilde{P}$ and $\widetilde{Q}$ be any two points. Then $\widetilde{R} = a\widetilde{P} + b\widetilde{Q}$, for a and b positive, is between $\widetilde{P}$ and $\widetilde{Q}$. Because γ is a linear transformation, $\gamma \widetilde{R} =$

Figure 6.20 Oriented points on a line.

$\gamma(a\widetilde{P} + b\widetilde{Q}) = a\gamma\widetilde{P} + b\gamma\widetilde{Q}$, which shows that $\gamma\widetilde{R}$ is between $\gamma\widetilde{P}$ and $\gamma\widetilde{Q}$. Convexity is defined in terms of betweenness, so collineations preserve convexity by the argument used in Theorem 4.4.6. ∎

Oriented projective geometry has some peculiarities. For example, two oriented projective lines intersect in two oriented points. Also an oriented line in effect has two copies of the real line on it (Fig. 6.20).

6.6.2 Subgeometries of projective space

Euclidean, hyperbolic, single elliptic, and other geometries of n dimensions are subgeometries of the projective geometry of the same number of dimensions. The transformation groups are entirely analogous to the corresponding groups in two dimensions. We briefly consider three-dimensional hyperbolic space to lead into Minkowski geometry, used in the special theory of relativity.

The points of three-dimensional hyperbolic space are the points in the interior of the unit sphere $x^2 + y^2 + z^2 - t^2 = 0$, which we take as the absolute quadratic surface. Hyperbolic isometries leave this surface stable.

Exercise 1 Extend the definitions from Section 6.5 of h-inner product, h-orthogonal, and h-length to vectors in $\mathbf{R}^4$.

Theorem 6.6.2 A 4×4 nonsingular matrix represents a hyperbolic isometry iff its columns are h-orthogonal, the first three have the same h-length, and the last column has the opposite h-length.

Proof. See Problem 5. ∎

In Section 5.5 we discussed the Lorentz transformations, which preserve $\Delta x_A^2 + \Delta y_A^2 + \Delta z_A^2 - \Delta t_A^2$, a value closely related to the equation of the quadratic surface for hyperbolic space. Recall that the equation $\Delta x_A^2 + \Delta y_A^2 + \Delta z_A^2 - \Delta t_A^2 = \Delta x_B^2 + \Delta y_B^2 + \Delta z_B^2 - \Delta t_B^2$ expressed the invariance of the distance and time measurements between two events from two frames of reference moving at a constant velocity relative to each other. Transformations preserving all values $k = \Delta x_A^2 + \Delta y_A^2 + \Delta z_A^2 - \Delta t_A^2$ clearly leave $x^2 + y^2 + z^2 - t^2 = 0$ stable and so are related to hyperbolic isometries. Actually, the four-dimensional geometry for relativity, called Minkowski geometry, needs all points (x, y, z, t). As with affine geometry, an extra coordinate is needed to permit movement of the origin. Hence the points are written $(x, y, z, t, 1)$. Thus Minkowski geometry is a subgeometry of $\mathbf{P}^4$, rather than $\mathbf{P}^3$. A Lorentz transformation is therefore a collineation of $\mathbf{P}^4$, but its upper left 4×4 submatrix is a hyperbolic isometry. As in affine transformations, the bottom row is $[\,0\ \ 0\ \ 0\ \ 0\ \ 1\,]$ and the right column represents translations of the origin.

Theorem 6.6.3 The Lorentz transformations are collineations of $\mathbf{P}^4$, where the bottom row is $[0\ 0\ 0\ 0\ 1]$ and the upper left 4×4 submatrix is a hyperbolic isometry, with the first three columns having h-length $= \pm 1$ and the fourth column having h-length $= \mp 1$.

Proof. See Problem 6. ∎

In Section 5.5 we showed that the value k above could be zero if the two events were "lightlike." For example, the differences in space and time coordinates of two points on the path of a light beam would have $k = 0$. In effect, the h-length of the difference of two lightlike events is zero. Similarly, two events can have k positive or negative depending on whether their relationship is "spacelike" or "timelike," respectively. (For further information on the special theory of relativity see Taylor and Wheeler [8].)

Projective geometry encompasses even more geometric ideas than Cayley and Klein envisioned. Even though projective geometry falls far short of "all geometry," as Cayley exclaimed in the quote opening this chapter, it has proven its worth in classical geometry, CAD systems, and many other applications.

PROBLEMS FOR SECTION 6.6

1. Find and prove a condition similar to that in Example 4 of Section 6.3, describing when four points of $\mathbf{P}^3$, projective space, are in the same plane.

2. Decide for which values of n the sets in parts (a)–(c) must intersect in $\mathbf{P}^n$. Justify your answers.
   a) A line and a plane
   b) Two planes
   c) A k-dimensional subspace and a j-dimensional subspace
   d) What happens to the intersection in part (c) as n decreases?

3. Draw a three-point perspective image of the cube in Example 2, using -0.25 for the first three entries in the bottom row. Find the vanishing points V_x, V_y, and V_z. What effect does changing the entries from -0.2 to -0.25 have?

4. Draw a three-point perspective image of the cube in Example 2, using $+0.2$ for the first three entries in the bottom row. Find the vanishing points V_x, V_y, and V_z. What effect does changing these entries from -0.2 to $+0.2$ have?

5. Prove Theorem 6.6.2. [*Hint:* See Example 3 of Section 6.5.]

6. Prove Theorem 6.6.3.

7. Prove that the intersection of any collection of convex sets in oriented projective geometry is again a convex set.

8. a) Graph the circle $x^2 + y^2 = 4$ and the hyperbola $x^2 - y^2 = 1$ and shade in the interiors of the circle $x^2 + y^2 < 4$ and of the hyperbola $x^2 - y^2 < 1$. Note that the intersection of their interiors has two separate regions, which therefore can't form a convex set.

   b) Explain why the oriented points $(x, \widetilde{y, z})$, with $x^2 + y^2 < 4z^2$ and $z > 0$, are interior to one oriented representation of the circle in part (a) and why these oriented points form a convex set.

   c) Verify that the collineation $\begin{bmatrix} 0 & 0 & 1 \\ 1/2 & 0 & 0 \\ 0 & 1/2 & 0 \end{bmatrix}$ takes the circle of part (a) to the hyperbola of part (a). Verify that points interior to the circle are taken to points interior to the hyperbola.

   d) Explain how to resolve the following seeming contradiction between Theorem 6.6.1, Problem 7, and the preceding parts of this problem. The oriented interior of the circle is convex, the circle is mapped to the hyperbola, and Theorem 6.6.1 implies that the hyperbola's oriented interior is convex. Problem 7 says that the intersection of two such convex sets must be convex, yet the intersection in part (a) clearly is not convex.

PROJECTS FOR CHAPTER 6

1. Investigate how artists make perspective drawings. (See Powell [6].)

2. Use parts (a) and (b) and Fig. 6.21 to explain why Desargues's theorem holds in three-dimensional Euclidean space. Desargues's theorem: If two triangles are perspective from a point, then they are perspective from a line.

   a) First let triangles $\triangle ABC$ and $\triangle DEF$ be in two nonparallel planes and perspective from P. Explain why the plane through P, A, and B must include D, E, and R. Explain why the two triangles must be perspective from the line on the intersection of the planes containing the triangles.

   b) Explain how to use part (a) twice to prove Desargues's theorem if the two triangles are in parallel planes or the same plane.

   c) What adjustments would be needed in this argument for three-dimensional projective space?

   d) State the converse of Desargues's theorem and explain how you could prove it without using duality.

3. Verify that the points on a Euclidean circle satisfy all the separation axioms except Axiom (ix), involving harmonic sets. Give an interpretation of harmonic sets of points on circle using harmonic sets of lines through the center. Does your interpretation satisfy Axiom (ix)?

4. Jacob Steiner defined conics using projectivities as follows. Let k_i be the family of lines on point P, let m_i be the family of lines on point Q, and assume that k_i is related to m_i by a projectivity of lines that isn't a perspectivity. Then the points R_i, which are the intersections of k_i and m_i, form a conic.

   a) Explore this method with graph paper. Let $P = (5, 12)$, $Q = (10, -7)$, lines k_i intersect the y-axis at $(0, i)$, and lines m_i intersect the x-axis at $(i, 0)$. Find various points R_i and sketch the conic.

   b) Repeat part (a) with $P = (-10, -6)$ and $Q = (6, 10)$.

   c) Identify the types of Euclidean conics you obtained in parts (a) and (b). Explain why points P and Q are always on the conic. Experiment with other placements of P and Q and other ways to relate the families of lines.

5. a) Use a dynamic geometry program to explore the following theorem in the special case of a circle. Pascal's theorem: Let A_1, A_2, A_3, A_4, A_5, and A_6 be any six points on a conic. Then the three points of intersections of the pairs of lines $\overleftrightarrow{A_1A_2} \cdot \overleftrightarrow{A_4A_5}$, $\overleftrightarrow{A_2A_3} \cdot \overleftrightarrow{A_5A_6}$, and $\overleftrightarrow{A_3A_4} \cdot \overleftrightarrow{A_6A_1}$ are collinear. (See Coxeter [2] for a proof.)

   b) Repeat part (a) where A_1, A_3, and A_5 are on one line and A_2, A_4, and A_6 are on another line.

   c) State and illustrate the dual of Pascal's theorem. (Pascal showed his theorem in 1640. Brianchon showed the dual in 1806. Only later did the

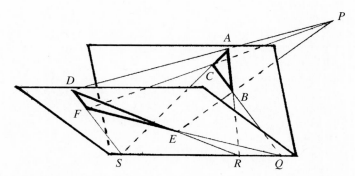

Figure 6.21 $\triangle ABC$ and $\triangle DEF$ are perspective from point P and from line $\overleftrightarrow{QR}$, where the planes containing the triangles intersect.

concept of duality reveal the connection between these theorems.)

d) State and illustrate the dual of part (b).

6. In Section 2.2 we defined an ellipse by using two foci. When we transform an ellipse to another ellipse by an affine transformation, do the old foci map to the new foci? Investigate with the ellipses $x^2/a^2 + y^2/b^2 = 1$ and affine transformations that map the x- and y-axes to themselves.

7. Relate homogeneous coordinates and barycentric coordinates. (See Section 2.3.)

8. A *correlation* is a transformation that maps points to lines and lines to points. Investigate correlations and how they relate to duality. (See Cederberg [1] or Coxeter [2].)

9. Investigate the possible types of eigenvalues and eigenvectors of a collineation. Find the possible sets of fixed points and stable lines. How does the set of fixed points relate to the set of stable lines for a collineation? [*Hint:* Use duality.] (See Fraleigh and Beauregard [3] for more on linear algebra.)

10. Investigate how computer-aided design uses projective geometry. (See Penna and Patterson [5].)

11. Investigate oriented projective geometry. (See Stolfi [7].)

12. Investigate the special theory of relativity. (See Taylor and Wheeler [8].)

13. Convert various ordinary equations, such as $y = x^2/(x^2 - 1)$, to homogeneous coordinates, in this case $x^2 y - x^2 z - yz^2 = 0$. Find the ideal points of these homogeneous equations and relate them to the graphs and asymptotes of the original equations. In algebraic geometry, usually a graduate-level subject, homogeneous coordinates are used to explore polynomial equations.

14. Write an essay discussing Klein's definition of geometry (see Section 4.2) in light of the various subgeometries of projective geometry.

15. Write an essay comparing the advantages and disadvantages of synthetic and analytic approaches to projective geometry. Do you agree with Poncelet that analytic geometry gives answers without insight? Explain.

Suggested Readings

[1] Cederberg, J. *A Course in Modern Geometries*. New York: Springer-Verlag, 1989.

[2] Coxeter, H. *Projective Geometry*. Toronto: University of Toronto Press, 1974.

[3] Fraleigh, J., and R. Beauregard. *Linear Algebra*. Reading, Mass.: Addison-Wesley, 1987.

[4] Kline, M. *Mathematical Thought from Ancient to Modern Times*. New York: Oxford University Press, 1972.

[5] Penna, M., and R. Patterson. *Projective Geometry and its Applications to Computer Graphics*. Englewood Cliffs, N.J.: Prentice Hall, 1986.

[6] Powell, W. *Perspective*. Tustin, Calif.: Walter Foster, 1989.

[7] Stolfi, J. *Oriented Projective Geometry*. Boston: Academic Press, 1991.

[8] Taylor, E., and J. Wheeler. *Spacetime Physics*. New York: W. H. Freeman, 1966.

[9] Tuller, A. *A Modern Introduction to Geometries*. New York: Van Nostrand Reinhold, 1967.

Suggested Media

1. "The Art of Renaissance Science: Galileo and Perspective," video, American Mathematical Society, Providence, 1991.
2. "Central Perspectivities," $13\frac{1}{2}$-minute film, International Film Bureau, Chicago, 1971.
3. "Conics in Perspective," 24-minute video, Films for the Humanities and Sciences, Princeton, N.J., 1996.

4. "Points of View: Perspective and Projection," 25-minute film, University Media, Solana Beach, Calif., 1975.
5. "Poles and Polars," $4\frac{1}{2}$-minute film, Educational Solutions, New York, 1979.
6. "Principles and Methods of Direct Perspective," 52-minute video, Pepper Publications, Tucson, 1987.
7. "Scientific Images in the Renaissance," 52-minute video, Churchill Films, Los Angeles, 1986.

7
Finite Geometries

Experimenters testing varieties of seed, fertilizers, or pesticides need to guard against other factors accidentally favoring one variety over another. For example, the drainage or quality of the soil will vary from one part of a field to another. Thus differences in the success of different varieties might be due to extraneous factors rather than the varieties themselves. Finite geometries provide a way to minimize such potential bias in experiments in areas ranging from agriculture to taste testing. Mathematicians have also used finite geometries to develop error correcting codes.

> The moving power of mathematical invention is not reasoning but imagination. —*Augustus DeMorgan*

7.1 Overview and History

Finite geometries as a subject arose from the investigations of geometric axioms at the end of the nineteenth century. The advent of hyperbolic and other geometries prompted a renewed interest in axioms. In particular, geometers sought models satisfying certain axioms but not others. Often these models had finitely many points, whence the term finite geometries. The first finite geometry was a three-dimensional example with 15 points developed in 1892 by Gino Fano. Geometers soon realized that finite geometries and their axiom systems were interesting for their own sake. Geometers incorporated into them various intriguing designs, results, and problems that had preceded this explicit development of finite geometry.

As so often happens in mathematics, significant applications arose from the blend of interesting problems and mathematical structure. In turn these applications have enriched mathematics by suggesting new areas to explore. Statistical design theory drew on finite geometries and expanded the types of geometric structures and theorems. More recently computers have greatly influenced the study of finite geometries, both by the ability to consider more complicated examples and to meet the need for sophisticated designs. Examples 1 and 2 illustrate that brute computing force may solve particular problems but that more general problems require mathematical insight.

Example 1 Euler posed the "36-officer problem" in 1782. He imagined six military regiments, each with the same six ranks and supposed one officer of each rank from each regiment. Euler sought a square array of these 36 officers so that no two officers of the same rank or from the same regiment were in the same row or column. He was convinced that this problem had no solution but was unable to give a proof. (The impossibility was finally proved in 1900 by an exhaustive search.) He was able to construct a solution to the corresponding problem for $5^2 = 25$ officers, among others (Fig. 7.1). He also conjectured that similar problems with n^2 officers, where $n = 4k + 2$, would be impossible. Euler's intuition was uncharacteristically flawed. In 1960 after extensive research, mathematicians showed that they could solve the n^2 officers problem for every value of n except 2 and 6. ●

$$
\begin{array}{ccccc}
A1 & B2 & C3 & D4 & E5 \\
D3 & E4 & A5 & B1 & C2 \\
B5 & C1 & D2 & E3 & A4 \\
E2 & A3 & B4 & C5 & D1 \\
C4 & D5 & E1 & A2 & B3 \\
\end{array}
$$

Figure 7.1 A solution to the 25-officer problem. The five ranks are A, B, C, D, and E, and the five regiments are 1, 2, 3, 4, and 5.

LEONHARD EULER

The Swiss mathematician Leonhard Euler (1707–1783) was the most prolific mathematician of all time—it took the St. Petersburg Academy 47 years to finish publishing the more than two hundred papers left when he died. Euler (pronounced "oiler") finished his university degree in theology at age 15. Meanwhile he had started studying mathematics with Johann Bernoulli and published his first paper at age 18. He followed Bernoulli to the St. Petersburg Academy in Russia and stayed there until he was 34. Euler then went to Berlin for 25 years at the request of Frederick the Great, who reportedly said, "The greatest mathematician should be at the court of the greatest king." Euler returned to Russia in 1766 blind in one eye and soon became completely blind. However, his production of mathematics didn't diminish. He merely dictated his papers and used his phenomenal memory to accomplish even the most complicated computations in his head. (Euler created the 36-officer problem, with its strong visual appeal, when he was blind.) Euler knew and made major contributions to all of mathematics and many of its applications in physics, although we discuss only some of his geometric work.

Euler, like mathematicians of his time, focused on particular problems which led to deeper, general insights. Euler's investigation of the Königsberg Bridge problem in 1735 marks the beginning of graph theory, an area of mathematics closely tied to combinatorics. Euler's formula (and its generalizations), discussed in Section 1.6, is a key theorem in graph theory and topology. Euler gave his argument for this formula in 1751. Euler's many textbooks established the importance of functions in analytic geometry and calculus. In his 1748 textbook he first introduced parametric equations (see Section 2.3). The problem of representing the curved surface of the earth on flat maps led to a variety of Euler's investigations and differential geometry.

Euler's fame as a mathematician provided one reason to tackle the general problem in Example 1, but a modern application provided another impetus. In the 1920s Sir Ronald A. Fisher developed statistical design theory. He and others wanted to test, for example, how the interactions of varieties of fertilizers and pesticides affected the growth of a type of plant. They needed to test each variety of fertilizer with each variety of pesticide. For n varieties of each, there are n^2 combinations. Previously, others had arranged different varieties of fertilizer in the rows and different varieties of pesticide in the columns. However, fields in nature, unlike mathematics, aren't uniform. Hence some rows or columns may provide better drainage, soil nutrients, or other growing conditions than others. To control for such "confounding variables," Fisher realized that he needed a design wherein each type of fertilizer and each type of pesticide appeared once in each row and column. Thus Euler's 36-officer problem became the prototype for Fisher's design problems. Fortunately, geometers had already solved many of these types of problems by using affine planes, which we develop in Sections 7.2 and 7.4. Fisher used these geometric solutions, starting a vital interaction between finite geometries and design theory.

Example 2 In 1850, the Rev. Thomas Kirkman posed and solved the "15-schoolgirl problem." He supposed that the girls took walks every day in an artificially regimented style of five rows of three each. He asked for daily arrangements of them so that any two girls walked in the same row just once a week. Computers have listed all 845 essentially different ways to solve this problem, but there is no easy way to generate such a solution. One such solution, from Berman and Fryer [3], follows.

Sunday	Monday	Tuesday	Wednesday	Thursday	Friday	Saturday
1 2 3	1 4 5	1 6 7	1 8 9	1 10 11	1 12 13	1 14 15
4 8 12	2 8 10	2 9 11	2 12 14	2 13 15	2 4 6	2 5 7
5 10 15	3 13 14	3 12 15	3 5 6	3 4 7	3 9 10	3 8 11
6 11 13	6 9 15	4 10 14	4 11 15	5 9 12	5 11 14	4 9 13
7 9 14	7 11 12	5 8 13	7 10 13	6 8 14	7 8 15	6 10 12

Which of the numerical conditions in this problem imply the others? The total number of girls (15) and the size of each row (3) clearly determine the number of rows (5). We use the condition that each pair of girls occurs just once to determine the number of days possible. For any girl, the 14 other girls must be put with this girl in pairs so that there can only be seven nonrepeating days. This reasoning is a simple example of combinatorics, a branch of mathematics that provides insight into numerical relationships. ●

Finite geometry is currently an active area of research benefiting from the cross-fertilization of geometry, algebra, and combinatorics. Transformational geometry and group theory gave powerful insights about finite geometries in the twentieth century just as they have since the nineteenth century for traditional geometries. Combinatorics provides essential insights into finite and other areas of geometry. For example, the consequences of Euler's formula (see Chapter 1) were identified through combinatorial reasoning. Error-correcting codes from coding theory, which we discuss in Section 7.3, are used in electronic data transmission and benefit from the interaction of these areas of mathematics.

PROBLEMS FOR SECTION 7.1

1. a) Find a solution to the "9-officer" problem.
   b) Find a solution to the "16-officer" problem.
   c) Explain why there can be no solution to the "4-officer" problem.

2. a) Describe how the As, Bs, and other letters in Fig. 7.1 relate to each other. Similarly describe the placement of the numbers.
   b) Find a different solution to the "25-officer" problem and describe any patterns you find in the placement of the letters and numbers.
   c) Does Fig. 7.1 avoid the risk of confounding variables that Fisher needed for his statistical work? Repeat for your solution in part (b). Discuss your answers.

3. Modify Example 2 so there are nine girls. Find daily arrangements of nine girls in rows of three so that each pair of girls appears in the same row just once. How many days can they go walking before they repeat partners?

4. Generalize Example 2 to a v-schoolgirl problem, wherein there are still three girls in each row and every two girls are in the same row just once.
   a) Show that v must be an odd multiple of 3, say, $v = 6n + 3$.
   b) For $v = 6n + 3$, find the number of days that these girls can go for walks without any girl walking with another girl twice. (Combinatorics won't guarantee such arrangements of these $6n + 3$ girls.)

5. Generalize Problem 4 so that there are four girls in each row but that two girls are still in the same row just once.

a) Show that v, the total number of girls, must satisfy $v = 12k + 4$.

b) Repeat Problem 4(b) for $v = 12k + 4$.

6. Generalize Problem 5, wherein there are n girls in each row.

7. A geometric figure **G** satisfies the following conditions.

   i) **G** is made of points, some pairs of which are adjacent.

   ii) Every point of **G** is adjacent to $k = 2$ other points.

   iii) No three points of **G** are mutually adjacent.

   iv) In **G** one point is adjacent to any two non-adjacent points.

   a) Find a figure satisfying all four conditions. How many points are there?

   b) Repeat part (a) for $k = 3$ in condition (ii). (This figure is called the Petersen graph.)

   c) For a general k in part (ii), find a formula for the total number of points in a figure satisfying all four conditions.

   d) For various values of k, find other figures that satisfy conditions (i), (ii), and (iv) but not (iii).

7.2 AFFINE AND PROJECTIVE PLANES

7.2.1 Affine planes

Focusing on a few of the undefined terms and axioms of Euclidean or other geometries helps us understand these geometries better. This restriction also leads naturally to geometries having finitely many points. With affine planes we explore relations of points and lines, especially parallelism, but can't use them to consider betweenness, distance, and congruence. (Of course, mathematicians have explored these concepts in other geometric systems.) The axiomatic system of an *affine plane* has the undefined terms *point*, *line*, and *on* and the following axioms.

Axioms 7.2.1

i) Every two distinct points have exactly one line on them both.

ii) There are at least four points with no three on the same line.

iii) For every line and point not on that line, there is a unique line on that point that has no point in common with the given line.

Exercise 1 Verify that the Euclidean plane is an affine plane.

Definition 7.2.1 Two lines, k and l, are *parallel*, written $k \| l$, iff either $k = l$ or no point is on both k and l.

Definition 7.2.2 An affine plane with n points on each line is of *order n*.

Example 1 Figure 7.2 depicts a model of an affine plane of order 2 and Fig. 7.3 depicts a model of order 3. In each figure, lines parallel to one another are drawn with the same kind of line for clarity. We can use Fig. 7.3 to obtain solutions to the nine-schoolgirl problem in Problem 3 of Section 7.1. For example, for the first day the rows could be the horizontal lines, for the second day the rows could be the vertical lines, and so on. ●

After proving some initial, elementary properties of all affine planes in Theorems 7.2.1 and 7.2.2, in Theorem 7.2.3 we show that all finite affine planes have an order, as stated in Definition 7.2.2. In Theorem 7.2.4 we use a combinatorial argument

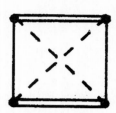
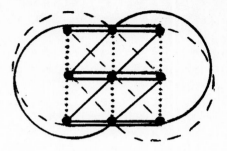

Figure 7.2 An affine plane of order 2.

Figure 7.3 An affine plane of order 3.

to reveal necessary relationships between the numbers of points and lines. However, such counting arguments can't guarantee the existence of affine planes.

Theorem 7.2.1 In an affine plane

  i) two distinct lines have at most one point in common,
  ii) for every point there is a line not on that point,
  iii) for every line there is a point not on that line,
  iv) every line has at least two points on it, and
  v) every point is on at least three lines.

Proof. For parts (iii) and (iv), let k be any line. Let A, A_1, A_2, and A_3 be the four points guaranteed by Axiom (ii). At least one of these four points, say, A, is not on k, showing part (iii). Consider the lines k_i, which are on A and A_i. Axiom (ii) ensures that these lines are distinct. At most, one of them is parallel to k, by Axiom (iii) and hence the other two intersect k. These two lines already have A in common, so they must intersect k in different points by part (i). See Problem 1 for the remaining parts. ∎

Theorem 7.2.2 Parallelism is an equivalence relation for lines in an affine plane. That is, for all lines k, l, and m, three properties hold: reflexive, $k \| k$; symmetric, if $k \| l$, then $l \| k$; and transitive, if $k \| l$ and $l \| m$, then $k \| m$.

Proof. See Problem 2. ∎

Theorem 7.2.3 If some line of an affine plane has n points on it, then each line has n points on it and each point has $n + 1$ lines on it.

Exercise 2 Draw diagrams to illustrate the proof of Theorem 7.2.3.

Proof. Let k be a line with n points on it, say, $P_1, \ldots, P_n$, where $n \geq 2$. First, let l be any line not parallel to k. By Axiom (ii), there is a point Q on neither l nor k. By Axiom (i), Q has n lines on it that intersect k plus a parallel to k, by Axiom (iii), for a total of $n + 1$ lines. Theorem 7.2.1 guarantees all to be distinct lines. In turn, Axiom (iii) forces all but one line to be on l, so l has n points on it. Now suppose that a line m is parallel to k. Use the axioms to find a line j not parallel to k. We can use the transitive property of parallelism to show that m is not parallel to j. (See Problem 2(b).) Now, from the first part of this proof, we know that j must have the same number of points as k and as m. Hence m has the same number of points as k. The preceding argument for the point Q shows that any point not on line k has $n + 1$ lines on it. For a point on line k, the same reasoning holds with a line not on that point. ∎

Exercise 3 What happens in Theorem 7.2.3 for a line with infinitely many points on it?

Theorem 7.2.4 In an affine plane if some line has n points on it, then there are n^2 points and $n^2 + n$ lines, and each line has n lines parallel to it, including itself.

Proof. Let P be any point. Then there are $n + 1$ lines on P and each of them has $n - 1$ points on it other than P by Theorem 7.2.3. Thus there are $(n + 1)(n - 1)$ points besides P, giving a total of $(n + 1)(n - 1) + 1 = n^2$ points. See Problem 3 for the rest of this proof. ∎

Example 2 There is no affine plane of order six.

Solution. We show that an affine plane with six points on a line would provide a solution to Euler's 36-officer problem, which we assume to be impossible. So, for a contradiction, suppose that there were an affine plane of order 6. We choose any two nonparallel lines to determine the rows and columns of the officers and fix any one point P (Fig. 7.4). Pick a line k through P other than a row or column. Put all the officers from the first regiment on k and from each other regiment on one of the parallels of k. Then no two officers from the same regiment would be in the same row or column by Theorem 7.2.1. Next, pick a line m through P other than k or a row or a column. Put the officers of each rank on one of the parallels of m. Again, we would have one officer from each rank in each row and column. This outcome contradicts the fact that there is no solution to Euler's problem. Hence there can be no affine plane of order 6. For a direct proof, see Anderson [1, 91]. ●

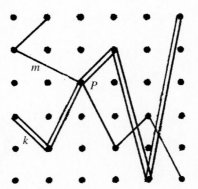

Figure 7.4 Using a hypothetical affine plane of order 6 to solve the 36-officer problem.

270   Chapter 7   Finite Geometries

The idea behind Example 2 enabled Fisher to turn affine planes into a solution of the statistical design problems discussed in Section 7.1. But for which orders are there actually affine planes? By 1896, geometers had shown algebraically that affine planes (and projective planes) of order n exist if $n = p^k$, where p is any prime number. In Section 7.4 we develop analytic geometries with this approach. No other orders of n are known to give planes, and many geometers believe that no other finite orders are possible. Theoretical results have eliminated infinitely many other values of n. An extensive computer search in 1989 revealed that there is no affine or projective plane of order 10, the smallest order not previously decided theoretically. Proving which orders of n give planes is one of the key open problems in finite geometry. (See Mullen [8].)

7.2.2 Projective planes

Axiom (iii) for affine planes characterizes parallelism. Projective geometry focuses on a contrasting idea from perspective drawings in art, whereby parallel lines appear to meet at a point on the horizon. Axiom (iii) characterizes projective planes and is the only change from the axioms of affine planes. (The axioms here are the first three axioms of the real projective plane in Chapter 6.) The axiomatic system for projective planes has the undefined terms *point*, *line*, and *on* and the following axioms.

Axioms 7.2.2
  i) Every two distinct points have exactly one line on them both.
  ii) There are at least four points with no three on the same line.
  iii) Every two lines have at least one point on them both.

Exercise 4 Explain why spherical geometry isn't a model of these axioms but single elliptic geometry is.

Example 3 Figure 7.5 gives a model of a projective plane with three points on each line.   ●

Theorem 7.2.5 In a projective plane
  i) two distinct lines have one point on them both,
  ii) there are at least four lines with no three on the same point,

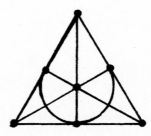

Figure 7.5 A projective plane of order 2.
(The curve is the seventh line.)

iii) every line is on at least three points and
iv) every point is on at least three lines.

Proof. See Problem 5. ∎

Projective planes have an important and mathematically aesthetic property called *duality*. Every theorem about points and lines remains a theorem when we systematically switch the words *point* and *line*. The original theorem and the switched theorem are called *duals* of each other. The first two parts of Theorem 7.2.5 contain the duals of Axioms (i) and (ii). [The dual of Axiom (iii) is weaker than that of Axiom (i), so this dual holds.] Because the duals of the axioms are all theorems, whenever we prove a theorem in projective geometry we immediately obtain its dual by switching the words *point* and *line* in the proof.

Theorem 7.2.6 If one line of a projective plane has $n + 1$ points on it, then all lines have $n + 1$ points on them and all points have $n + 1$ lines on them.

Proof. See Problem 6. ∎

Theorem 7.2.7 If one line of a projective plane has $n + 1$ points on it, there are $n^2 + n + 1$ points and $n^2 + n + 1$ lines.

Proof. See Problem 7. ∎

Definition 7.2.3 A projective plane of *order n* has $n + 1$ points on a line.

A natural connection exists between affine and projective planes of the same order, explaining Definition 7.2.3. Theorems 7.2.4 and 7.2.7 show that a projective plane of order n has one more line and $n + 1$ more points than an affine plane of the same order. The following construction shows that we can convert an affine plane into a projective plane of the same order by adding one line and its points. This result corresponds to the addition of a horizon line in perspective drawing. For an affine plane of order n, we collect the lines parallel to each other in a class. For each of the $n + 1$ classes of parallel lines, we add a new point that we define to be on each of those parallel lines. We also define these $n + 1$ new points all to be on the same new line. We leave it as an exercise to verify that the affine points and lines, together with the new points and line, satisfy the axioms of a projective plane. The other direction works also. Take a projective plane and delete any one line and the points on it. The remaining points and lines form an affine plane, where lines become parallel if they formerly intersected on the deleted line. (For more on affine and projective planes see Karteszi [7].)

Exercise 5 Convert the affine planes in Figs. 7.2 and 7.3 to projective planes using the construction given above.

PROBLEMS FOR SECTION 7.2

1. Prove the rest of Theorem 7.2.1.
2. a) Prove Theorem 7.2.2.
   b) If a line in an affine plane intersects one of two parallel lines, prove that it intersects the other.

3. Prove the rest of Theorem 7.2.4.

4. a) In the models illustrated in Figs. 7.2 and 7.3 find the maximum number of points so that no three are on the same line.

   b) In an affine plane of order n, prove there can be at most $n + 2$ points such that no three are on the same line.

   c) Repeat part (b) for a projective plane of order n.

5. Prove Theorem 7.2.5.

6. Prove Theorem 7.2.6. [*Hint:* Modify the proof of Theorem 7.2.3 and use duality.]

7. Prove Theorem 7.2.7. [*Hint:* Modify the proof of Theorem 7.2.4 and use duality.]

8. a) For which axiom of affine planes is its dual provable? Prove it.

   b) For the other axioms, show that their duals must be false.

   c) For which parts of theorems of affine planes are the duals provable? Prove them.

9. A *weak projective plane* satisfies Projective Axioms (i) and (iii) and the following replacements for Axiom (ii).

   ii′) There exist a point and a line not on that point.

   ii″) Every line has at least two points on it.

   a) Find a model of a weak projective plane of order 1.

   b) Prove that duality holds in this revised axiomatic system.

   c) Find a model of a weak projective plane with two lines having different numbers of points on them.

   d) Suppose that we replace Axiom (ii) for affine planes with the Axioms (ii′) and (ii″). Show that every such "weak affine plane" actually is an affine plane.

7.3 DESIGN THEORY

Design theory encompasses many notions besides affine and projective geometries. Design theory developed from mathematicians' natural inclination to generalize and explore and to meet the needs of applications, especially designs for statistical experiments. We concentrate on balanced incomplete block designs.

Some sampling trials require comparison of each of a number of varieties of something with all the other varieties available. Often one individual can't be expected to judge fairly among a large number of varieties. For example, taste comparisons by one person need to be done in a relatively short session. The organizer of the trials must guard against biases caused by which varieties are tasted together. Fisher and others developed designs in which each individual tests the same number of varieties (called the *size of the block*) and all pairs of varieties are compared the same number of times. Such designs are called balanced incomplete block designs (BIBD). *Balanced* refers to the uniformity of the arrangement, and *incomplete* refers to the fact that no block includes every variety. Fisher proved a number of results about BIBD. (See Anderson [1, Chapter 6].)

Remark Solutions to the n^2-officer problem are not BIBDs, although they are related, as in Example 2 of Section 7.2.

Definition 7.3.1 A *balanced incomplete block design* (BIBD) is an arrangement of v varieties in b blocks (subsets), each of size k, so that each pair of varieties appear in λ blocks. We write (v, k, λ) to describe the numerical type of a BIBD.

Example 1 An affine plane of order n is a BIBD, with the lines as the blocks: $v = n^2$, $b = n^2 + n$, $k = n$, and $\lambda = 1$. Note that each point (variety) is on the same number, $r = n + 1$, of lines (blocks). ●

SIR RONALD A. FISHER

Sir Ronald A. Fisher (1890–1962) was one of the key people in establishing statistics as a mathematical discipline. Throughout his career he blended his interests in biology and statistics, building on his strong mathematical ability and insight. In 1911, as an undergraduate at Cambridge University, he gave a talk marking him as one of the first to see how to combine Darwin's theory of natural selection with Mendel's recently rediscovered ideas in genetics. He strongly advocated eugenics, the movement to improve people's genetic inheritance. (His involvement was innocent of and prior to the use of eugenics ideas for political and racist ends in various countries, most notoriously Nazi Germany.)

From an early age his eyesight was poor, which required active compensation. For example, in high school he studied spherical trigonometry entirely orally. His extraordinary visualization skills enabled him to perform all the three-dimensional thinking and computing without either a text or paper and pencil. His geometric intuition also characterized his statistical thinking. His first result, as well as many others, depended on considering a statistical sample of n points as a vector in n-dimensional Euclidean space. Many other statisticians, lacking Fisher's geometric abilities, found his reasoning difficult to follow and felt he depended too much on intuition. He made fundamental contributions in many areas of statistics, including tests of hypotheses and analysis of variance.

He combined his statistical, mathematical, and biological interests throughout his career. In 1919, he became the statistician at Rothamsted Experimental Station, where daily practical problems led him to a wide variety of theoretical discoveries in statistics. It was here that he realized the need both for randomized sampling and the careful design of experiments to study the interaction of variables. His biological experiments led to development of statistical design theory.

Example 2 A projective plane of order n is a BIBD, with the lines as the blocks: $v = n^2 + n + 1$, $b = n^2 + n + 1$, $k = n + 1$, and $\lambda = 1$. Again, each point is on the same number, $r = n + 1$, of lines. ●

Example 3 Given a BIBD $(v, k, 1)$ we can make a BIBD (v, k, λ) simply by repeating each block of the original BIBD λ times. However, experimenters often need to place a pair of varieties with different varieties in different blocks to gain additional information. The following list gives blocks for a BIBD (7, 3, 2) with no repetitions of blocks. Note that each line is a BIBD (7, 3, 1), so there are other ways to form a (7, 3, 2) BIBD by using repetition.

$$\begin{array}{ccccccc} 124 & 235 & 346 & 457 & 156 & 267 & 137 \\ 126 & 237 & 134 & 245 & 356 & 467 & 157 \end{array}$$ ●

Checking even the short list in Example 3 to verify that it is a (7, 3, 2) BIBD is tedious. It would be much worse with a larger design and harder still to find such a design by trial and error. Indeed, not every set of values (v, k, λ)—even those values that satisfy the combinatorial relations of Theorem 7.3.1—has a BIBD.

274   **Chapter 7**   **Finite Geometries**

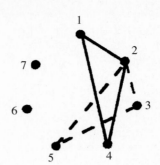

Figure 7.6   A cyclic way to represent a projective plane with seven points.

Exercise 1   Explain why there can be no BIBD (6, 3, 1).

Mathematicians and statisticians have found efficient ways to construct BIBDs and other designs. In Section 7.4 we use finite analytic geometries to construct many affine and projective spaces in any number of dimensions. Although other designs don't have such representations, symmetry often aids their construction.

Example 4   Figure 7.6 and symmetry provide the key to the BIBD of Example 3. In a regular seven-sided polygon, there are three different distances between vertices. Note that $\triangle 124$ has each of these three distances. When we rotate $\triangle 124$ to each of the seven possible positions, we get seven different blocks that satisfy a (7, 3, 1) BIBD. In fact, this design is a projective plane of order 2. Now $\triangle 157$ is the mirror image of $\triangle 124$, so the same reasoning applies to it. The 14 blocks of Example 3 are simply the rotations and mirror reflections of $\triangle 124$. Because of the different lengths of the sides, we can be sure that a given edge (pair of vertices) appears in exactly two triangles. ●

Before attempting to construct a BIBD with certain values, you should determine whether the given values are compatible with the conditions of a BIBD. Theorem 7.3.1 gives the necessary relations among the values for a BIBD. However, there is no guarantee, as Exercise 2 indicates, that values satisfying these relations always have a corresponding BIBD.

Theorem 7.3.1   In a BIBD (v, k, λ), each variety is in the same number r of blocks, $r(k-1) = \lambda(v-1)$ and $v \cdot r = b \cdot k$.

Exercise 2   Verify that the values in Example 1 for an affine plane of order n satisfy Theorem 7.3.1. However, there is no affine plane of order 6.

Proof.   Let V be any variety of a BIBD and r_V be the number of blocks containing V. Each such block has $k - 1$ other varieties. So $r_V(k-1)$ counts all appearances of these other varieties in blocks containing V. The value λ counts the number of times each of these other varieties appears with V. Thus $\lambda(v-1)$ also counts all appearances of other varieties in blocks containing V. Hence $r_V(k-1) = \lambda(v-1)$ for every V. As v, k and

λ are constant, all varieties must be in the same number $r_V = r$ of blocks, and the first equation of the theorem holds. Similarly, $vr = bk$ counts the number of times that any variety appears in any block in two ways: Each of the v varieties appears r times, and each of the b blocks contains k varieties. ∎

If we think of varieties as points and blocks as lines, a BIBD with $\lambda = 1$ satisfies the geometric axiom "two distinct points are on a unique line." In the 1840s and 1850s, Rev. Thomas Kirkman and Jacob Steiner (1796–1863) investigated such systems, with the added restriction that each line had three points on it, years before the more general notion of a BIBD was defined. A *Steiner triple system* is a BIBD with $k = 3$ and $\lambda = 1$. (See Anderson [1], Berman and Fryer [3] and Karteszi [7] for more on BIBD and Steiner systems.)

Theorem 7.3.2 There is a Steiner triple system $(v, 3, 1)$ iff $v = 6n + 1$ or $v = 6n + 3$.

Proof. Problem 3 shows v must equal either $6n + 1$ or $6n + 3$. (See Anderson [1,112] for the existence of such systems.) ∎

Example 5 Is there a finite "hyperbolic" plane with three points on a line?

Solution. The key axiom of hyperbolic geometry is that every point P not on a line m has at least two lines on it that do not intersect m. Such a plane with three points per line would be a Steiner triple system in which each point (variety) has at least two more lines than there are points on a line. That is $r \geq k + 2$. Note that a projective plane of order 2 gives a Steiner triple system with $v = 7$ and that an affine plane of order 3 gives a Steiner triple system with $v = 9$. By Theorem 7.3.2 the smallest possible hyperbolic plane with three points on a line would have $v = 13$ points. Theorem 7.3.1 implies that there must be $r = 6$ lines on every point and a total of $b = 26$ lines. However, combinatorics give no clue about how to construct such a design. Consider the vertices of a regular 13-gon (Fig. 7.7). Thirteen rotations of this polygon are symmetries of it. We could rotate two candidates for lines to find all 26 lines. Note the six different distances between vertices of the polygon. We can think of each line of three points as a triangle and look

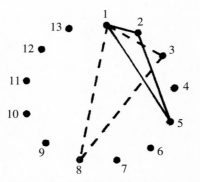

Figure 7.7 A cyclic way to represent a hyperbolic plane with 13 points.

for two triangles whose six sides include all the different lengths. After a reasonably short application of trial and error, we find that $\triangle 125$ and $\triangle 138$ satisfy this relationship. Hence their rotated images give the lines for a finite hyperbolic plane. ●

7.3.1 Error-correcting codes

The advances in computers and their availability have vastly increased the amount of data transmitted as strings of 1s and 0s. Although computers are far more precise than human beings, errors in these transmissions occasionally occur because of static or for other reasons. With the use of words in real language, we can often correct errors because the context of the message or the word itself helps us recognize where the error must be and what the correction should be. However, a string such as 0010110111 provides no clue by itself about possible errors. Hence we need to build into the transmission a code to aid in finding and correcting errors; BIBD have important connections with such codes.

One way to obtain error correction is to send each individual 0 or 1 three times. The preceding message then would be sent as 000000111000111111000111111111. If just one of the three repetitions is accidentally altered, the receiving computer can correct the error by majority rule. If we assume that two or three errors in a triple are extremely unlikely, compared to none or one, this code allows us to correct most errors. However, the price of error correcting with this code is utilization of three times as much data as the actual message requires.

We can convert a BIBD into a matrix of 0s and 1s to construct a list of code words. The columns of the matrix, called an *incidence matrix*, are the varieties of the BIBD and the rows are the blocks. We put a 1 in an entry if the corresponding variety is in the corresponding block. Otherwise we put in a 0. For example, the top row of Example 3 has the incidence matrix
$$\begin{pmatrix} 1 & 1 & 0 & 1 & 0 & 0 & 0 \\ 0 & 1 & 1 & 0 & 1 & 0 & 0 \\ 0 & 0 & 1 & 1 & 0 & 1 & 0 \\ 0 & 0 & 0 & 1 & 1 & 0 & 1 \\ 1 & 0 & 0 & 0 & 1 & 1 & 0 \\ 0 & 1 & 0 & 0 & 0 & 1 & 1 \\ 1 & 0 & 1 & 0 & 0 & 0 & 1 \end{pmatrix}.$$
If we used the seven rows as code words, we would be able to detect any single error among the seven digits received. For example, 0010010 differs from 0011010 only in the fourth digit, but it differs from all the others in more digits. In the language of Definition 7.3.2, the Hamming distance between any two of the seven code words is 4, whereas 0010010 and 0011010 have a Hamming distance of 1.

Definition 7.3.2 A *code* is a set of n-tuples of 0s and 1s, each of which is a *code word*. The *Hamming distance* between two n-tuples is the number of places in which they differ.

Theorem 7.3.3 A code can detect as many as k errors if the Hamming distance between any two code words is at least $k + 1$. A code can correct as many as k errors if the Hamming distance between any two code words is at least $2k + 1$.

Proof. Suppose that all the code words are n-tuples. The number of errors in a received n-tuple is the number of places that this n-tuple differs from the code word sent, or the Hamming distance between them. Suppose that the Hamming distance between any two code words is at least $k + 1$. If between 1 and k errors occur, then the received n-tuple is not a code word and so will be detected as an error. Similarly, suppose that the Hamming distance between code words is at least $2k + 1$. Then a received n-tuple with between 1 and k errors will have a smaller Hamming distance from the original code word than from any other code word, enabling us to correct it. ∎

Coding theory combines abstract algebra with design theory to develop more efficient codes than the incidence matrices of BIBDs. However, many of these codes are related to BIBDs. (See Anderson [1, Chapters 6 and 7] and Gallian [6, Chapter 31] for more information.)

PROBLEMS FOR SECTION 7.3

1. **a)** Find a formula for the number of blocks b in terms of v, k, and λ in a BIBD (v, k, λ).
   b) Verify that all the conditions of Theorem 7.3.1 are satisfied with a BIBD $(v, k, 1)$ when r is any multiple of k.

2. Fisher showed that $b \geq v$ in any BIBD, with $v > k$. (See Anderson [1, 85].)
   a) If you assume that $b \geq v$, what else does Theorem 7.3.1 tell you?
   b) If $b = v$ and $\lambda = 1$, find v, b, and r in terms of k. Now set $n + 1 = k$ and find v in terms of n. Relate this result to projective planes. (BIBD with $b = v$ are called *symmetric*.)
   c) Prove Axioms (i) and (ii) of a projective plane for a symmetric BIBD $(v, k, 1)$, with $k \geq 3$. [*Hint:* For Axiom (ii), given two varieties (points), count the number of varieties not on the block (line) they determine. Do the same when you add the third point.]
   d) Prove Axiom (iii) of a projective plane for a symmetric BIBD $(v, k, 1)$, with $k \geq 3$. [*Hint:* Count the number of blocks that meet a given block in one of its varieties and the total number of blocks that meet it.]

3. **a)** In Theorem 7.3.2 prove that $v = 6n + 1$ or $v = 6n + 3$. [*Hint:* Use factoring and Theorem 7.3.1 twice, first to show that v must be odd and then to eliminate $v = 6n + 5$.]
   b) Show that $v = 6n + 1$ and $v = 6n + 3$ are compatible with all conditions of Theorem 7.3.1.

4. Define a *Steiner double system* and prove that, for every $v \geq 2$, there is a Steiner double system.

5. **a)** Find a design for a projective plane of order 3 by using symmetries of the vertices of a regular 13-gon.
   b) Program a computer to search for a projective plane of order 4 using symmetries of the vertices of a regular 21-gon.

6. Modify Example 5 to find a Steiner triple system for $v = 19$.

7. In a round-robin tournament each player (or team) plays every other player (or team). The organizer of such a tournament wants to schedule as many matches at once as possible. If the number of players is even, say, $2n$, conceivably n matches at a time can be scheduled and thus the tournament arranged so that no one has to wait while others play.
   a) Find a schedule for a round-robin tournament with 4 players.
   b) Find the number of rounds in a round-robin tournament with $2n$ players if there are n matches at a time.
   c) Show for all even numbers $2n$ that there is a tournament schedule with no waiting as follows. Use symmetry to find a "near" schedule for the vertices of a regular polygon with $2n - 1$ vertices: Just one vertex sits out each round. Then place the last team (the $2n$th) at the center of the polygon and convert the near schedule to a schedule for all $2n$ teams.

8. A *Steiner quadruple system* is a BIBD (v, k, λ), with $k = 4$ and $\lambda = 1$. Show that $v = 12n + 1$ or $12n + 4$. Generalize. [*Hint:* See Problem 3.]

9. **a)** Write the code words in the code based on the $(7, 3, 2)$ BIBD of Example 3. How many errors can this code detect? How many can it correct?

   b) Repeat part (a) for the code based on the affine plane of order 3, a BIBD $(9, 3, 1)$.

10. You can add rows to the incidence matrix given in the text in a way that doesn't give a BIBD. For example, you can add rows that have four or more 1s. By Theorem 7.3.3, if the Hamming distance between any two (new and old) code words is at least 3, you will still be able to correct a single error while having more code words.

    a) Explain why, if you want to be able to correct single errors, there is no point in adding code words with one, two, or three 1s.

    b) Find all seven code words with four 1s that have a Hamming distance of at least 4 from each other and from the seven code words in the original matrix. How do these new code words relate to the code words?

    c) Find the two other code words that are a Hamming distance of at least 3 from each other and the 14 already found.

    d) Suppose that you have a set of n code words, each a seven-dimensional vector of 0s and 1s, and that each code word is a Hamming distance of at least 3 from every other code word. Find the total number of vectors possible (including code words and noncode words). Explain why every code word has seven vectors at a Hamming distance of 1 from it. Explain why n can be no larger than 16.

7.4 FINITE ANALYTIC GEOMETRY

We can imitate the analytic geometry and transformations presented in Chapters 4 and 6 for finite geometries whenever we have algebraic structures corresponding to the arithmetic of the real numbers. Finite fields provide the analog that we use, although a more general approach is possible. (See Blumenthal [4].) Fields are number systems having many of the familiar properties of the four usual operations of addition, subtraction, multiplication, and division. Although there are other fields, we concentrate on the fields of integers modulo a prime number. (See Gallian [6] for more information on fields.)

Example 1 Let $\mathbf{Z}_3$ be the set $\{0, 1, 2\}$ together with addition and multiplication modulo 3. That is, after doing the usual arithmetic, we subtract multiples of 3 until we get back to a number in the given set. For example, $2 + 2 = 4$ becomes 1 (*mod* 3) because $4 - 3 = 1$. We write $2 + 2 \equiv 1$ (*mod* 3). Think of the three numbers in the set placed around a circle (Fig. 7.8). The following tables give all additions and multiplications. $\mathbf{Z}_3$ is a field.

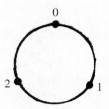

Figure 7.8

7.4 Finite Analytic Geometry

+	0	1	2		×	0	1	2
0	0	1	2		0	0	0	0
1	1	2	0		1	0	1	2
2	2	0	1		2	0	2	1

●

Example 2 Let $\mathbf{Z}_5$ be the set $\{0, 1, 2, 3, 4\}$ together with addition and multiplication modulo 5, given in the following tables. For example, $3 \times 4 \equiv 2 \ (mod \ 5)$ because $3 \times 4 = 12 \equiv 12 - 5 - 5 = 2 \ (mod \ 5)$. $\mathbf{Z}_5$ is a field.

+	0	1	2	3	4		×	0	1	2	3	4
0	0	1	2	3	4		0	0	0	0	0	0
1	1	2	3	4	0		1	0	1	2	3	4
2	2	3	4	0	1		2	0	2	4	1	3
3	3	4	0	1	2		3	0	3	1	4	2
4	4	0	1	2	3		4	0	4	3	2	1

●

Definition 7.4.1 By $\mathbf{Z}_n$ we mean the set $\{0, 1, \ldots, n-1\}$ together with addition and multiplication modulo n. That is, after doing the usual arithmetic, we subtract multiples of n until we get back to a number in the given set.

Example 3 Although $\mathbf{Z}_4$ satisfies many familiar algebraic properties, it misses one of the defining properties to qualify as a field. The number 2 doesn't have a multiplicative inverse. That is, none of the numbers in the set $\{0, 1, 2, 3\}$ when multiplied by 2 and reduced modulo 4 give us 1, the multiplicative identity. Thus we can't divide by 2. We leave the addition and multiplication tables for $\mathbf{Z}_4$ as an exercise. ●

Theorem 7.4.1 $\mathbf{Z}_p$ is a field iff p is a prime. There is, up to isomorphism, exactly one field with p^k elements, where p is a prime number.

Proof. See Gallian [6, 213 and 328]. ■

We can use any field to form an analytic model of affine or projective geometries, just as we did with the real numbers in Chapters 4 and 6. The number of elements in a field is its *order* and equals the order of the corresponding affine or projective plane. Using algebra, we can develop matrices with entries from any field to describe the affine transformations and collineations of the corresponding geometries. As in Chapter 4, we need three coordinates for points so that the transformations can move the origin. For a field $\mathbf{F}$, $\mathbf{F}^3$ is the three-dimensional vector space over $\mathbf{F}$. If you are familiar with abstract algebra, you can show that the affine planes defined (and projective planes defined later) do indeed satisfy the axioms of Section 7.2. The proofs that these systems satisfy the axioms follow the proofs for the usual analytic models using the real numbers. (See Karteszi [7].)

Definition 7.4.2 Given a field $\mathbf{F}$, define $\mathbf{AF}^2$, the *affine plane over* $\mathbf{F}$, as follows. The *points* of $\mathbf{AF}^2$ are column vectors $(x, y, 1)$ of $\mathbf{F}^3$, and the *lines* are row vectors $[a, b, c]$ from $\mathbf{F}^3$, where a and b are not both 0. Two row vectors represent the same line iff one is a scalar multiple of the other by a nonzero element of **F**. A point $(x, y, 1)$ is *on* a line $[a, b, c]$ iff $ax + by + c1 \equiv 0 \ (mod\ 5)$.

Example 4   The 25 vectors of $\mathbf{AZ}_5^2$ form an affine plane of order 5. To find the line through points $(2, 2, 1)$ and $(1, 4, 1)$, we need to solve the system of two equations $a \cdot 2 + b \cdot 2 + c \cdot 1 \equiv 0 \ (mod\ 5)$ and $a \cdot 1 + b \cdot 4 + c \cdot 1 \equiv 0 \ (mod\ 5)$. The solution is $[3, 4, 1]$ or any of its multiples, modulo 5, such as $[2 \cdot 3, 2 \cdot 4, 2 \cdot 1] \equiv [1, 3, 2] \ (mod\ 5)$. (Note that the Euclidean line through $(2, 2)$ and $(1, 4)$ is $y = -2x + 6$, which becomes $[2, 1, -6]$ and, by multiplying by 4, is equivalent to $[3, 4, 1]$ modulo 5.) Figure 7.9 illustrates the five lines $[m, 4, 0]$ through the origin, more familiarly known as $y = mx$. (In $\mathbf{Z}_5$, 4 acts as -1 does in ordinary arithmetic.) Each line can be considered to cycle both horizontally and vertically, as the numbers in $\mathbf{Z}_5$ cycle. For example, line $[2, 4, 0]$ goes over 1 and up 2 from one point to the next: $(0, 0, 1), (1, 2, 1), (2, 4, 1), (3, 1, 1), (4, 2, 1)$, and back to $(0, 0, 1)$. The "vertical" line $[1, 0, 0]$ provides the sixth line through the origin guaranteed by Theorem 7.2.3. Every other point $(x, y, 1)$ also has six lines on it, each parallel to one of these six. ●

Exercise 1   Give algebraic conditions for two lines $[a, b, c]$ and $[a', b', c']$ to be parallel. Verify that two such lines are equal or don't intersect.

As in Chapter 4, we represent affine transformations by 3×3 invertible matrices whose bottom row is $[0\ 0\ 1]$. For a field of order n, there are at most n^6 such matrices because only six entries aren't fixed, but not all are invertible. We can use the determinant of a matrix $(mod\ n)$ to determine whether that matrix is invertible. We use a combinatorial argument to find the number of invertible matrices with the bottom row of $[0\ 0\ 1]$. Again, as in Chapter 4, the images of points $(0, 0, 1), (1, 0, 1)$, and $(0, 1, 1)$ determine an affine transformation. The condition that the matrix is invertible implies that these points must be mapped to three distinct points not all on the same line. For the field of order n, the affine plane has n^2 points and $(0, 0, 1)$ can be mapped to any of them. Once we know where $(0, 0, 1)$ goes, $(1, 0, 1)$ has $n^2 - 1$ places to go. For $(0, 1, 1)$ there remain $n^2 - n$ places to go because it can't be mapped to any of the n points on the line

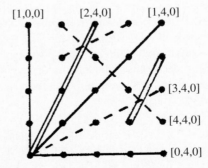

Figure 7.9   The lines through $(0, 0, 1)$ in $\mathbf{AZ}_5^2$.

7.4 Finite Analytic Geometry

through the other two points. Thus there are $n^2(n^2 - 1)(n^2 - n) = n^6 - n^5 - n^4 + n^3$ affine plane transformations over the field of order n.

Definition 7.4.3 For a field $\mathbf{F}$, define $\mathbf{PF}^2$, the *projective plane over* $\mathbf{F}$, as follows. The *points* of $\mathbf{PF}^2$ are the nonzero column vectors (x, y, z) of $\mathbf{F}^3$, where two vectors represent the same point iff one is a scalar multiple of the other by a nonzero element of $\mathbf{F}$. The *lines* of $\mathbf{PF}^2$ are nonzero row vectors $[a, b, c]$ from $\mathbf{F}^3$, where two row vectors represent the same line iff one is a scalar multiple of the other by a nonzero element of $\mathbf{F}$. A point (x, y, z) is *on* a line $[a, b, c]$ iff $ax + by + cz = 0$.

Example 5 Consider the projective plane $\mathbf{PZ}_3^2$ (Fig. 7.10). $\mathbf{Z}_3^3$ has $3^3 = 27$ vectors. The 26 nonzero vectors pair up to give 13 points because of the two nonzero scalars, 1 and 2. The figure emphasizes that $\mathbf{PZ}_3^2$ adds four points and one line to $\mathbf{AZ}_3^2$, which is represented in Fig. 7.3. As there are four points on each line, we can define both $H(AB, CD)$ and $AB//CD$ from Chapter 6 if A, B, C, and D are distinct collinear points. These definitions satisfy Axioms (i)–(vii) and (ix) of Chapter 6. ●

Exercise 2 Verify that Axioms (viii) and (x) of Chapter 6 fail in $\mathbf{PZ}_3^2$.

Recall that the transformations of the projective plane, called collineations, can be represented by invertible 3×3 matrices. As before, two matrices represent the same collineation if one is a nonzero scalar multiple of the other. As in Chapter 6, a collineation is determined by where it sends four points, no three of which are collinear. Problem 7 shows that $(n^2 + n + 1)(n^2 + n)(n^2)(n - 1)^2 = n^8 - n^6 - n^5 + n^3$ collineations exist for the projective plane over a field of order n.

Theorem 7.4.2 The set of affine transformations for $\mathbf{AF}^2$ and the collineations for $\mathbf{PF}^2$ each form a group.

Proof. We replace the specific field $\mathbf{R}$ with the general field $\mathbf{F}$ in the proofs of Theorem 4.4.5 and Theorem 6.4.3. ∎

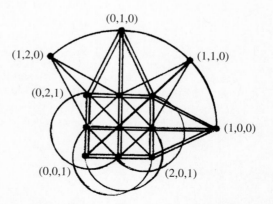

Figure 7.10 A representation of $\mathbf{PZ}_3^2$.

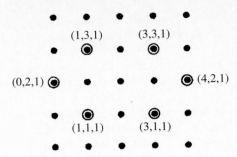

Figure 7.11 The oval $4x^2 + 2y^2 + 4x + 2y + 3 = 0$ in $\mathbf{AZ}_5^2$.

7.4.1 Ovals in finite projective planes

The preceding paragraphs indicate how well finite planes mimic many familiar geometric concepts. Mathematicians also have explored other traditional geometric concepts in a finite setting. In a finite plane a simple non-algebraic way to describe the analog to conics is to use ovals. In Euclidean geometry, no three points on a conic are collinear. For only a finite number of points, not very many sets can fulfill this property. Figure 7.11 shows a set of six points in $\mathbf{AZ}_5^2$ with no three on the same line. Some exploring will reveal that no other point can be added to this set. Indeed, in $\mathbf{AZ}_p^2$ and in $\mathbf{PZ}_p^2$, with $p > 2$, we can never find more than $p + 1$ points with no three points collinear. (When $p = 2$, all four points in $\mathbf{AZ}_2^2$ form a set with no three points collinear. We omit this special case.) Conveniently, these sets in $\mathbf{PZ}_p^2$ correspond to second-degree equations. A second-degree equation in an affine plane might give fewer than the expected number of points because an affine plane can be seen as the corresponding projective plane minus a line and its points. Therefore ovals, the analog to conics, are usually defined for projective planes. (See Beck et al. [2] and Karteszi [7, 110–119] for more information on ovals in projective planes and finite analytic geometry.)

Definition 7.4.4 In a projective plane of order n, an *oval* is a set of $n + 1$ points, no three of which are collinear. A line is *tangent* to an oval iff the line and the oval have just one point in common.

Theorem 7.4.3 Every point on an oval has exactly one tangent to the oval.

Proof. Let P be any point on an oval and $P_1, \ldots, P_n$ be the other points on the oval. The definition of an oval guarantees that the lines PP_i for $i > 0$ are all distinct, accounting for n of the $n + 1$ lines through P. The remaining line cannot intersect the oval except at P, so it must be a tangent. ∎

PROBLEMS FOR SECTION 7.4

1. Use $\mathbf{AZ}_3^2$ in this problem.
   a) Find the line on $(1, 2, 1)$ and $(0, 1, 1)$.
   b) Find the intersection of the line in part (a) with $[2, 2, 2]$.

c) Draw a picture to illustrate the effect of the affine transformation $\begin{bmatrix} 0 & 1 & 0 \\ 1 & 0 & 0 \\ 0 & 0 & 1 \end{bmatrix}$ on the affine plane $\mathbf{AZ}_3^2$. Is the effect in $\mathbf{AZ}_3^2$ similar to the effect of that matrix on the real affine plane? Explain.

d) Repeat part (c) for $\begin{bmatrix} 0 & 2 & 2 \\ 1 & 0 & 0 \\ 0 & 0 & 1 \end{bmatrix}$, $\begin{bmatrix} 2 & 0 & 0 \\ 0 & 2 & 0 \\ 0 & 0 & 1 \end{bmatrix}$, and $\begin{bmatrix} 1 & 0 & 1 \\ 0 & 1 & 2 \\ 0 & 0 & 1 \end{bmatrix}$.

2. Repeat Problem 1, using the affine plane $\mathbf{AZ}_5^2$. Generalize.

3. Suppose that you try to form an affine plane using $\mathbf{Z}_4$, which isn't a field.

   a) Find a line that doesn't have four points on it.
   b) Find two distinct lines with more than one point of intersection.
   c) Find two lines with different slopes but no points of intersection.
   d) Which of the axioms of an affine plane fail when you try to use $\mathbf{Z}_4$?
   e) Repeat parts (a)–(d) for $\mathbf{Z}_6$, replacing *four* with *six* in part (a).

4. Use the projective plane $\mathbf{PZ}_7^2$ in this problem.

   a) Find the point on lines [2, 3, 1] and [3, 1, 4].
   b) Find a complete quadrangle, as defined in Section 6.1.
   c) For the three collinear points $P = (0, 0, 1)$, $Q = (1, 0, 1)$, and $R = (1, 0, 0)$, find a point S such that $H(PQ, RS)$, as defined in Section 6.1. Is S unique?

5. Repeat Problem 4, using the projective plane $\mathbf{PZ}_5^2$.

6. Define a projectivity in $\mathbf{PF}^2$. (See Section 6.4.) Count the number of projectivities if the field has order n. [*Hint:* See Theorem 6.4.1.]

7. Show that $(n^2 + n + 1)(n^2 + n)(n^2)(n - 1)^2 = n^8 - n^6 - n^5 + n^3$ collineations exist in a projective plane over a field of order n.

8. a) Find the points on the oval $x^2 + y^2 - z^2 = 0$ in $\mathbf{PZ}_3^2$. (For simplicity, you may assume throughout that $z = 0$ or $z = 1$.)
   b) Repeat for $x^2 + y^2 - 2z^2 = 0$ in $\mathbf{PZ}_3^2$.
   c) Describe the points not on either of the ovals in parts (a) and (b).
   d) Repeat for $x^2 + 2y^2 - kz^2 = 0$ in $\mathbf{PZ}_5^2$, where $k = 1, \ldots, 4$.

9. Find the points in the affine plane $\mathbf{AF}_5^2$ on ovals $x^2 + y^2 = 1$, $x^2 + 4y = 0$, and $x^2 + 3y^2 = 1$. Explain the difference in numbers of points.

PROJECTS FOR CHAPTER 7

1. a) In a BIBD explain why, if $\lambda > 0$, then $k \geq 2$. Explain why BIBDs with $k = 2$ are easy to construct but not particularly interesting.

   b) Use Theorem 7.3.1 to find all triples (v, k, λ) with $v \leq 25$ for which there could be a BIBD (v, k, λ) with $v > k > 2$ and $\lambda = 1$. [*Hint:* Explain why $k \leq 5$.]

   c) For each triple (v, k, λ) in part (b), try to construct a BIBD. Explore if a different (nonisomorphic) BIBD can have the same values for v, k, and λ.

   d) Try to construct BIBDs with $(v, k, 2)$ for $v \leq 19$ that aren't built from BIBDs with $(v, k, 1)$ by simple repetition.

2. Explore the affine and projective planes $\mathbf{AF}_4^2$ and $\mathbf{PF}_4^2$, where $\mathbf{F}_4$ is the field of order four with the following addition and multiplication.

+	0	1	a	b		×	0	1	a	b
0	0	1	a	b		0	0	0	0	0
1	1	0	b	a		1	0	1	a	b
a	a	b	0	1		a	0	a	b	1
b	b	a	1	0		b	0	b	1	a

3. a) Define points and planes for the affine space $\mathbf{AF}^3$, where $\mathbf{F}$ is a field of order n.
   b) Count the number of points, lines, and planes in $\mathbf{AF}^3$.
   c) Count the number of affine transformations in $\mathbf{AF}^3$.
   d) Count the number of points and lines in $\mathbf{AF}^k$, the k-dimensional affine space over the field $\mathbf{F}$. [*Hint:* How many lines are there on each point?]
   e) Explain why the number of affine transformations in $\mathbf{AF}^k$ is $n^k \prod_{j=0}^{k-1}(n^k - n^j)$ if $\mathbf{F}$ is of order n.

4. a) Repeat Project 3, parts (a), (b), and (d) for projective spaces $\mathbf{PF}^3$ and $\mathbf{PF}^k$.

   b) Explain why $\mathbf{PF}^3$ has $(n^3 + n^2 + n + 1)(n^3 + n^2 + n)(n^3 + n^2)(n^3)(n-1)^3$ collineations.

5. Program a computer to search for projective planes of orders 4 and 5, using the symmetries of a regular polygon as in Section 7.3.

6. Define a point to be *exterior* to an oval if two tangents to the oval are on that point. Define a point to be *interior* to an oval if no tangents to the oval are on that point. Choose different ovals and find their interior and exterior points. Look for formulas counting the number of exterior and interior points for an oval in PZ_p^2. Prove your formulas.

7. Explore Desargues's theorem (see Section 6.1) in $\mathbf{PZ}_p^2$ for various $p > 2$. (Explain why Desargues's theorem isn't interesting in $\mathbf{PZ}_2^2$.) (Smart [9] has an axiomatization of Desargues's configuration, which is shown in Fig. 6.4.)

8. The axiomatic system for a *weak hyperbolic plane* has the undefined terms *point*, *line*, and *on* and the following axioms.

   i) Every two distinct points have exactly one line on them both.

   ii) There are at least four points with no three on the same line.

   iii) Given a line and a point not on that line, there are at least two lines on that point with no point in common with the given line.

   a) Show that any set of $n \geq 5$ points is a weak hyperbolic plane if you take the lines to be all subsets of two elements.

   b) Find a model of a weak hyperbolic plane with some lines having only one point on them.

   c) Find a model of a weak hyperbolic plane with six points, every line with at least two points on it, and two lines with different numbers of points on them.

   Define a *hyperbolic plane* to be a weak hyperbolic plane satisfying this stronger version of Axiom (iii):

   iii′) Given a line with n points on it and a point not on that line, there are exactly n lines through that point which do not have any point in common with the given line.

   c) Prove: If one line of a hyperbolic plane has n points on it, then all do. Find the number of points and lines in terms of n and prove your answers.

   d) Are your results in part (c) consistent with Theorem 7.3.1? Explain.

   e) Use a computer to find a model of a hyperbolic plane similar to that in Example 5 of Section 7.3, but with $k = 5$.

9. You can define a notion of distance in finite affine planes and so investigate isometries.

   a) For a prime p with $p = 4n + 3$, define the distance in $\mathbf{AZ}_p^2$ between $(s, t, 1)$ and $(u, v, 1)$ to be $(s - u)^2 + (t - v)^2 \pmod{p}$. For $p = 3$ and $p = 7$, verify that two distinct points have a nonzero distance between them. How many points are at each possible distance from a given point? Do the points at a given distance from a point form an oval ("circle")? If so, what is the equation of that oval?

   b) Determine which affine transformations of $\mathbf{AZ}_p^2$ are isometries; that is, they preserve the distances of part (a). Relate these isometries to the isometries of Section 4.3. [*Hint:* In Section 4.5, an affine matrix was defined to be an isometry iff its upper left submatrix was orthogonal.] Count the number of isometries in $\mathbf{AZ}_p^2$. Prove that the isometries form a group.

   c) For a prime p not of the form $4n + 3$, the definition of distance in part (a) has the following curious property. There are distinct points with a distance of 0 between them. Verify this property for the primes 2, 5, 13, and 17. You can modify the distance formula for $p = 4n + 1$ by multiplying the term $(t - v)^2$ by some nonzero scalar of $\mathbf{Z}_p$. Experiment with different scalars. Do the points at a specific distance from a given point form an oval? If so, what is the equation of that oval? How does the equation relate to the scalar? Investigate isometries for these planes with these distances. Count the number of isometries.

   d) Use different definitions of distance to explore parts (a) and (b). Do they change the number of isometries? Is there a common formula for the number of isometries for $\mathbf{AZ}_p^2$?

   e) Define perpendicular in $\mathbf{AZ}_p^2$ for odd primes p. Do you need different definitions for different primes?

f) Explore similarities in $\mathbf{AZ}_p^2$, for $p = 4n + 3$ and $p = 4n + 1$. Count the number of similarities for $\mathbf{AZ}_p^2$ and show that they form a group.

10. Let potential code words be vectors of 0s and 1s of length n.

    a) Find the total number of these vectors.

    b) How many vectors are at a Hamming distance of 1 from a given vector?

    c) For a code to be able to correct one error, each pair of code words must be a Hamming distance of at least 3 apart. Use parts (a) and (b) to determine the maximum number of code words possible if each pair must be a distance of at least 3 apart. For various values of n, look for codes (sets of code words) that can correct one error and have as many code words as possible.

    d) Redo part (c) with codes that can correct two (or more) errors. [*Hint:* The binomial theorem may be helpful.]

    e) Redo part (c) with codes that can detect two or more errors.

11. Explore design theory. (See Anderson [1], Berman and Fryer [3] and Karteszi [7].)

12. Explore coding theory. (See Anderson [1] and Gallian [6].)

13. Explore finite geometries. (See Blumenthal [4] and Dembowski [5].)

14. Write an essay discussing the analogies between finite geometries and the familiar Euclidean (and projective) geometry. What insights can finite geometries provide?

Suggested Readings

[1] Anderson, I. *A First Course in Combinatorial Mathematics*. New York: Oxford University Press, 1989.

[2] Beck, A., M. Bleicher, and D. Crowe. *Excursions into Mathematics*. New York: Worth, 1969.

[3] Berman, G., and K. Fryer. *Introduction to Combinatorics*. New York: Academic Press, 1972.

[4] Blumenthal, L. *A Modern View of Geometry*. Mineola, N.Y.: Dover, 1961.

[5] Dembowski, P. *Finite Geometries*. New York: Springer-Verlag, 1968.

[6] Gallian, J. *Contemporary Abstract Algebra*. Lexington, Mass.: D. C. Heath, 1994.

[7] Karteszi, F. *Introduction to Finite Geometries*. Amsterdam: North Holland, 1976.

[8] Mullen, G. A candidate for the "next Fermat problem." *The Mathematical Intelligencer*, 1995, 17(3):18–22.

[9] Smart, J. *Modern Geometries*. Monterey, Calif.: Brooks/Cole, 1988.

A

Definitions, Postulates, Common Notions, and Propositions from Book I of Euclid's *Elements*

DEFINITIONS

1. A *point* is that which has no part.
2. A *line* is breadthless length.
3. The extremities of a line are points.
4. A *straight line* is a line which lies evenly with the points on itself.
5. A *surface* is that which has length and breadth only.
6. The extremities of a surface are lines.
7. A *plane surface* is a surface which lies evenly with the straight lines on itself.
8. A *plane angle* is the inclination to one another of two lines in a plane which meet one another and do not lie in a straight line.
9. And when the lines containing the angle are straight, the angle is called *rectilineal*.
10. When a straight line set up on a straight line makes the adjacent angles equal to one another, each of the equal angles is *right*, and the straight line standing on the other is called *perpendicular* to that on which it stands.
11. An *obtuse angle* is an angle greater than a right angle.
12. An *acute angle* is an angle less than a right angle.
13. A *boundary* is that which is the extremity of anything.
14. A *figure* is that which is contained by any boundary or boundaries.
15. A *circle* is a plane figure contained by one line such that all the straight lines falling upon it from one point among those lying within the figure are equal to one another.
16. And the point is called the *center* of the circle.

17. A *diameter* of the circle is any straight line drawn through the center and terminated in both directions by the circumference of the circle, and such a straight line also bisects the circle.
18. A *semicircle* is the figure contained by the diameter and the circumference cut off by it. And the center of the semicircle is the same as that of the circle.
19. *Rectilineal figures* are those which are contained by straight lines, *trilateral* figures being those contained by three, *quadrilateral* those contained by four, and *multilateral* those contained by more than four straight lines.
20. Of trilateral figures, an *equilateral triangle* is that which has three sides equal, an *isosceles triangle* that which has two of its sides alone equal, and a *scalene triangle* that which has its three sides unequal.
21. Further, of trilateral figures, a *right-angled triangle* is that which has a right angle, an *obtuse-angled triangle* that which has an obtuse angle, and an *acute-angled triangle* that which has its three angles acute.
22. Of quadrilateral figures, a *square* is that which is both equilateral and right-angled; an *oblong* [*rectangle*] that which is right-angled but not equilateral; a *rhombus* that which is equilateral but not right-angled; and a *rhomboid* [*parallelogram*] that which has its opposite sides and angles equal to one another but is neither equilateral nor right-angled. And let quadrilaterals other than these be called *trapezia*.
23. *Parallel* straight lines are straight lines which, being in the same plane and being produced indefinitely in both directions, do not meet one another in either direction.

THE POSTULATES

Let the following be postulated:

1. To draw a straight line from any point to any point.
2. To produce a finite straight line continuously in a straight line.
3. To describe a circle with any center and distance.
4. That all right angles are equal to one another.
5. That, if a straight line falling on two straight lines makes the interior angles on the same side less than two right angles, the two straight lines, if produced indefinitely, meet on that side on which are the angles less than the two right angles.

COMMON NOTIONS

1. Things which are equal to the same thing are also equal to one another.
2. If equals be added to equals, the wholes are equal.
3. If equals be subtracted from equals, the remainders are equal.

4. Things which coincide with one another are equal to one another.
5. The whole is greater than the part.

THE PROPOSITIONS OF BOOK I

(Modern paraphrases are shown in brackets)

I-1. On a given finite straight line, to construct an equilateral triangle.

I-2. To place at a given point (as an extremity) a straight line equal to a given straight line. [Construct a line segment of a given length.]

I-3. Given two unequal straight lines, to cut off from the greater a straight line equal to the less. [Construct a shorter line segment on a larger one.]

I-4. [SAS] If two triangles have two sides equal to two sides, respectively, and have the angles contained by the equal straight lines equal, they will also have the base equal to the base, the triangle will be equal to the triangle, and the remaining angles will be equal to the remaining angles, respectively, namely, those which the equal sides subtend.

I-5. In isosceles triangles, the angles at the base are equal to one another, and, if the equal straight lines be produced further, the angles under the base will be equal to one another. [Triangles with two congruent sides have two congruent angles.]

I-6. If in a triangle two angles be equal to one another, the sides which subtend the equal angles will also be equal to one another. [Triangles with two congruent angles have two congruent sides.]

I-7. Given two straight lines constructed on a straight line (from its extremities) and meeting in a point, there cannot be constructed on the same line (from its extremities), and on the same side of it, two other straight lines meeting in another point and equal to the former two, respectively, namely, each to that which has the same extremity with it. [Given the lengths of the three sides, only one triangle can be formed in a given position.]

I-8. [SSS] If two triangles have the two sides equal to two sides, respectively, and have also the base equal to the base, they will also have the angles equal which are contained by the equal straight lines.

I-9. To bisect a given rectilineal angle.

I-10. To bisect a given finite straight line.

I-11. To draw a straight line at right angles to a given straight line from a given point on it. [Construct the perpendicular to a line from a point on a line.]

I-12. To a given infinite straight line, from a given point which is not on it, to draw a perpendicular straight line. [Construct the perpendicular to a line from a point not on a line.]

I-13. If a straight line set up on a straight line make angles, it will make either two right angles or angles equal to two right angles. [The measures of supplementary angles add to 180°.]

I-14. If with any straight line, and at a point on it, two straight lines not lying on the same side make the adjacent angles equal to two right angles, the two straight lines will be in a straight line with one another. [If the measures of two adjacent angles add to 180°, then the two rays not shared by the angles form a line.]

I-15. If two straight lines cut one another, they make the vertical angles equal to one another. [Vertical angles are congruent.]

I-16. In any triangle if one of the sides be produced, the exterior angle is greater than either of the interior and opposite angles. [An exterior angle of a triangle is greater than either remote interior angle.]

I-17. In any triangle two angles taken together in any manner are less than two right angles. [The measures of any two angles of a triangle add to less than 180°.]

I-18. In any triangle the greater side subtends the greater angle. [The larger side is opposite the larger angle.]

I-19. In any triangle the greater angle is subtended by the greater side. [The greater angle is opposite the greater side.]

I-20. In any triangle two sides taken together in any manner are greater than the remaining one. [Triangle inequality: The lengths of any two sides of a triangle add to more than the length of the third side.]

I-21. If on one of the sides of a triangle, from its extremities, there be constructed two straight lines meeting within the triangle, the straight lines so constructed will be less than the remaining two sides of the triangle, but will contain a greater angle.

I-22. Out of three straight lines, which are equal to three given straight lines, to construct a triangle: thus it is necessary that two of the straight lines taken together in any manner should be greater than the remaining one.

I-23. On a given straight line and at a point on it, to construct a rectilineal angle equal to a given rectilineal angle. [Construct a given angle.]

I-24. If two triangles have the two sides equal to two sides, respectively, but have the one of the angles contained by the equal straight lines greater than the other, they will also have the base greater than the base.

I-25. If two triangles have the two sides equal to two sides, respectively, but have the base greater than the base, they will also have the one of the angles contained by the equal straight lines greater than the other.

I-26. [AAS and ASA] If two triangles have the two angles equal to two angles, respectively, and one side equal to one side, namely, either the side adjoining the equal angles, or that subtending one of the equal angles, they will also have the remaining sides equal to the remaining sides and the remaining angle to the remaining angle.

I-27. If a straight line falling on two straight lines makes the alternate angles equal to one another, the straight lines will be parallel to one another. [For two lines cut by a transversal if alternate interior angles are congruent, the lines are parallel.]

I-28. If a straight line falling on two straight lines makes the exterior angle equal to the interior and opposite angle on the same side, or the interior angles on the same side equal to two right angles, the straight lines will be parallel to one another. [For two lines cut by a transversal if corresponding angles are congruent, the lines are parallel. If the measures of the two interior angles on the same side of the transversal add to 180°, the lines are parallel.]

> *Note: The Parallel Postulate is needed to prove many propositions starting with I-29.*

I-29. A straight line falling on parallel straight lines makes the alternate angles equal to one another, the exterior angle equal to the interior and opposite angle, and the interior angles on the same side equal to two right angles. [Two parallel lines cut by a transversal have alternate interior (and exterior) angles congruent and corresponding angles congruent, and the measures of the two interior angles on the same side of the transversal add to 180°.]

I-30. Straight lines parallel to the same straight line are also parallel to one another. [If k is parallel to l and l is parallel to m, then k is parallel to m.]

I-31. Through a given point to draw a straight line parallel to a given straight line. [Construct a parallel to a given line through a given point.]

I-32. In any triangle, if one of the sides be produced, the exterior angle is equal to the two interior and opposite angles, and the three interior angles of the triangle are equal to two right angles. [The angle sum of a triangle is 180°. The sum of the measures of two angles equals the opposite exterior angle.]

I-33. The straight lines joining equal and parallel straight lines (at the extremities which are) in the same directions (respectively) are themselves also equal and parallel. [If $\overline{AB}$ is congruent and parallel to $\overline{CD}$ in quadrilateral $ABCD$, then $\overline{AD}$ is congruent and parallel to $\overline{BC}$.]

I-34. In parallelogrammic areas the opposite sides and angles are equal to one another, and the diameter bisects the areas. [In a parallelogram opposite sides are congruent, opposite angles are congruent, and the diagonals divide the parallelogram into congruent triangles.]

I-35. Parallelograms which are on the same base and in the same parallels are equal to one another. [Parallelograms with the same base and the same height have the same area.]

I-36. Parallelograms which are on equal bases and in the same parallels are equal to one another. [Parallelograms with congruent bases and heights have the same area.]

I-37. Triangles which are on the same base and in the same parallels are equal to one another. [I-35 for triangles.]

I-38. Triangles which are on equal bases and in the same parallels are equal to one another. [I-36 for triangles.]

I-39. Equal triangles which are on the same base and on the same side are also in the same parallels.

I-40. Equal triangles which are on equal bases and on the same side are also in the same parallels.

I-41. If a parallelogram have the same base with a triangle and be in the same parallels, the parallelogram is double of the triangle. [In effect, the area formula $A = \frac{1}{2} b \cdot h$ for triangles.]

I-42. To construct, in a given rectilineal angle, a parallelogram equal to a given triangle. [Construct a parallelogram with the same area as a triangle and with a given angle.]

I-43. In any parallelogram the complements of the parallelogram about the diameter are equal to one another.

I-44. To a given straight line to apply, in a given rectilineal angle, a parallelogram equal to a given triangle. [Construct a parallelogram with the same area as a triangle, with a given side and a given angle.]

I-45. To construct, in a given rectilineal angle, a parallelogram equal to a given rectilineal figure. [Generalizes I-44 to any polygon.]

I-46. On a given straight line to describe a square. [Construct a square with a given side.]

I-47. [The Pythagorean theorem.] In right-angled triangles the square on the side subtending the right angle is equal to the squares on the sides containing the right angle.

I-48. [The converse of the Pythagorean theorem.] If in a triangle the square on one of the sides be equal to the squares on the remaining two sides of the triangle, the angle contained by the remaining two sides of the triangle is right.

B

Hilbert's Axioms for Euclidean Plane Geometry

Undefined Terms   *Point, line, on, between,* and *congruence*

GROUP I: AXIOMS OF CONNECTION

I-1.   Through any two distinct points A, B, there is always a line m.

I-2.   Through any two distinct points A, B, there is not more than one line m.

I-3.   On every line there exist at least two distinct points. There exist at least three points which are not on the same line.

GROUP II: AXIOMS OF ORDER

II-1.   If point B is between points A and C, then A, B, C are distinct points on the same line, and B is between C and A.

II-2.   For any two distinct points A and C, there is at least one point B on the line $\overleftrightarrow{AC}$ such that C is between A and B.

II-3.   If A, B, C are three distinct points on the same line, then only one of the points is between the other two.

Definition   By the *segment* $\overline{AB}$ is meant the set of all points which are between A and B. Points A and B are called the *endpoints* of the segment. The segment $\overline{AB}$ is the same as segment $\overline{BA}$.

II-4.   (Pasch's axiom.) Let A, B, C be three points not on the same line and let m be a line which does not pass through any of the points A, B, C. Then if m passes through a point of the segment $\overline{AB}$, it will also pass through a point of segment $\overline{AC}$ or a point of segment $\overline{BC}$.

> *Note: This postulate may be replaced by the* separation axiom.

II-4′. A line m separates the points which are not on m into two sets such that if two points X and Y are in the same set, the segment $\overline{XY}$ does not intersect m, and if X and Y are in different sets, the segment $\overline{XY}$ does intersect m. In the first case X and Y are said to be on the *same side* of m; in the second case, X and Y are said to be on *opposite sides* of m.

Definition By the *ray* $\overrightarrow{AB}$ is meant the set of points consisting of those which are between A and B, the point B itself, and all points C such that B is between A and C. The ray $\overrightarrow{AB}$ is said to *emanate from* point A. A point A, on a given line m, divides m into two rays such that two points are on the same ray if and only if A is not between them.

Definition If A, B, and C are three points not on the same line, then the system of three segments $\overline{AB}$, $\overline{BC}$, $\overline{CA}$, and their endpoints is called the *triangle* $\triangle ABC$. The three segments are called the *sides* of the triangle, and the three points are called the *vertices*.

GROUP III: AXIOMS OF CONGRUENCE

III-1. If A and B are distinct points on line m, and if A' is a point on line m' (not necessarily distinct from m), then there is one and only one point B' on each ray of m' emanating from A' such that the segment $\overline{A'B'}$ is congruent to the segment $\overline{AB}$, written $\overline{AB} \cong \overline{A'B'}$.

III-2. If two segments are each congruent to a third, then they are congruent to each other. (From this it can be shown that congruence of segments is an equivalence relation; i.e., $\overline{AB} \cong \overline{AB}$; if $\overline{AB} \cong \overline{A'B'}$, then $\overline{A'B'} \cong \overline{AB}$; and if $\overline{AB} \cong \overline{CD}$ and $\overline{CD} \cong \overline{EF}$, then $\overline{AB} \cong \overline{EF}$.)

III-3. If point C is between A and B, and C' is between A' and B', and if $\overline{AC} \cong \overline{A'C'}$ and $\overline{CB} \cong \overline{C'B'}$, then $\overline{AB} \cong \overline{A'B'}$.

Definition By an *angle* is meant a point (called the *vertex* of the angle) and two rays (called the *sides* of the angle) emanating from the point.

If the vertex of the angle is point A and if B and C are any two points other than A on the two sides of the angle, we speak of the angle $\angle BAC$ or $\angle CAB$ or simply of angle $\angle A$.

III-4. If $\angle BAC$ is an angle whose sides do not lie on the same line and if $\overrightarrow{A'B'}$ is a ray emanating from A', then there is one and only one ray $\overrightarrow{A'C'}$ on a given side of line $\overleftrightarrow{A'B'}$, such that $\angle B'A'C' \cong \angle BAC$. In short, a given angle can be laid off on a given side of a given ray in one and only one way. Every angle is congruent to itself.

Definition If $\triangle ABC$ is a triangle then the three angles $\angle BAC$, $\angle CBA$, and $\angle ACB$ are called the *angles of the triangle*. Angle $\angle BAC$ is said to be *included by* the sides $\overline{AB}$ and $\overline{AC}$ of the triangle.

III-5. [SAS] If two sides and the included angle of the one triangle are congruent, respectively, to two sides and the included angle of another triangle, then each of the remaining angles of the first triangle is congruent to the corresponding angle of the second triangle.

GROUP IV: AXIOM OF PARALLELS

IV-1. (Playfair's axiom.) Through a given point A not on a given line m there passes at most one line which does not intersect m.

GROUP V: AXIOMS OF CONTINUITY

V-1. (Archimedean axiom.) If $\overline{AB}$ and $\overline{CD}$ are arbitrary segments, then there exists a number n such that if the segment $\overline{CD}$ is laid off n times on the ray $\overrightarrow{AB}$ starting from A, then a point E is reached, where $n \cdot \overline{CD} = \overline{AE}$, and where B is between A and E.

V-2. (Axiom of linear completeness.) The system of points on a line with its order and congruence relations cannot be extended in such a way that the relations existing among its elements as well as the basic properties of linear order and congruence resulting from axiom groups I, II and III and axiom V-1 remain valid.

> *Note: These axioms may be replaced by* Dedekind's axiom of continuity. *For every partition of the points on a line into two nonempty sets such that no point of either lies between two points of the other, there is a point of one set which lies between every other point of that set and every point of the other set.*

Selected Answers

CHAPTER 1

Section 1.1
1. 0.6%, $\pi \approx 3.1604938$. **3.** $1.4146296 \approx \sqrt{2}$, $42.42638 \approx 30\sqrt{2}$. **5a)** 72°, 54°, 108°, 36°, 108°, 72°, 72°, 36°. **5c)** $1 + x = (1 + \sqrt{5})/2$. **6a)** 360°, 540°, 720°. **6b)** $(n-2)180°$. **6c)** Same result. **7a)** 3×6, 4×4. **7b)** For sides x and y, $y = \frac{2x}{x-2}$. **7c)** $3 \times 7 \times 42$, $3 \times 8 \times 24$, $3 \times 9 \times 18$, $3 \times 10 \times 15$, $3 \times 12 \times 12$, $4 \times 5 \times 20$, $4 \times 6 \times 12$, $4 \times 8 \times 8$, $5 \times 5 \times 10$, $6 \times 6 \times 6$. **9.** Area of semicircle: $(AB/2)^2 \pi/2 = AB^2\pi/8$. Area of quarter circle: $(AB/\sqrt{2})^2 \pi/4 = AB^2\pi/8$. **11.** Use angles to show similar triangles. By similarity $\frac{c}{a} = \frac{a}{y}$, $\frac{c}{b} = \frac{b}{x}$. **13b)** Cone: $\int_0^h \pi(xr/h)^2 dx = \pi h r^2/3$, cylinder: $\int_0^h \pi r^2 dx = \pi h r^2$. **13c)** Radius: $r(i) = (i/n)(r/h)$, volume: $\pi(h/n)(r(i))^2$.

Section 1.2
1a) 1, 2, 3, 9, 10, 11, 12, 22, 23, 31, 42, 44, 45, 46. **1b)** 4, 8, 26, 34. **1c)** 5, 6, 13, 15, 32–38, 43, 47, 48. **1d)** 27–31, 39, 40. **3a)** As $\triangle ABD$ is similar to $\triangle CDB$, $\frac{1}{CD} = \frac{CD}{x}$. So $CD = \sqrt{x}$. **3d)** Use part (a) with $DB = \sqrt{x}, \sqrt{\sqrt{x}}$. **3e)** Use part (a) to construct length $\sqrt{2}$ and then length $\sqrt{1 + \sqrt{2}}$. **5a)** Find the intersections of the perpendicular bisectors of the sides. **5b)** Find the intersections of the angle bisectors. **11a)** A square and a rhombus. **11b)** Prove that both sets of two triangles formed by the diagonals are congruent using SSS. **11c)** AAAS does not work: Consider two rectangles with the same height and different widths. **11d)** SASAA and SASSS are sufficient, SAASA is not. **13b)** Use addition formulas for $\sin(a+b)$ and $\cos(a+b)$.

Section 1.3
1a) Three points, three lines. **1c)** Four points and six lines. **5a)** I-1, I-2, I-3 (first part), II-1, II-2, II-3, III-1, III-2, III-3, V-1, V-2. **5e)** V-2 (or I-1, I-2, I-3, II-1, II-2, II-4, and IV-1).

Section 1.4
1a) III-1, III-2, III-3, III-4. **1b)** Squares whose diagonals are vertical and horizontal; zero, one, or two points or one or two line segments. **1c)** no. **3a)** Three points, two lines; both lines are on all the points. **6a)** An infinite checkerboard. **6b)** A cube. **6c)** A checkerboard on a doughnut.

Section 1.5
1. $AD = AB^2/AC$. **2a)** $x/y = \sqrt{2}$. **2b)** $x/y = \sqrt{n}$. **2c)** $x/y = (1 + \sqrt{5})/2$.
4d) $y = \sin(x) \sim y = \frac{1}{k}\sin(kx)$. **13b)** Use integration by substitution where $u = \frac{x}{k}$.
13c) A similar volume is k^3 times the original volume.

Section 1.6
3a) $0.5, \sqrt{2}/2, \sqrt{3}/2$. **3b)** $2/(\pi\sqrt{3}) \approx 37\%$. **3c)** A tetrahedron, volume $1/3$. **5b)** The distance between $(1, b, 0)$ and $(-1, b, 0)$ is 2. By Problem 2, the distance between $(1, b, 0)$ and $(0, 1, b)$ is $\sqrt{1 + (b-1)^2 + b^2}$. Set these equal to find $b = (1 + \sqrt{5})/2$. **6a)** $AC = (1 + \sqrt{5})/2$, $BC = \sqrt{5 + \sqrt{5}}/\sqrt{2} \approx 1.902$. **6b)** $2/BC \approx 1.05$. **9b)** $195°$. **9c)** $\triangle ABN$ is $1/8$ of sphere, so $\triangle ACF$ is $1/48$ of sphere. **13a)** 42. **13b)** 92. **13d)** $720/(10n^2 + 2)°$, $n = 11$.
15a) $2\pi R \sin(r/R)$. **15b)** Use $f(x) = \sqrt{R^2 - x^2}$, $a = R\cos(r/R)$, $b = R$.

CHAPTER 2

Section 2.1
2. If the vertices are $(0, 0)$, $(0, a)$, (b, c), and (d, e), then the midpoints are $(0, \frac{a}{2})$, $(\frac{b}{2}, \frac{a+c}{2})$, $(\frac{b+d}{2}, \frac{c+e}{2})$, and $(\frac{d}{2}, 0)$. Find the lengths of opposite sides. **3.** If the vertices are $C = (0, 0)$, $(0, d)$, and (e, f), then $\cos C = e/\sqrt{e^2 + f^2}$. **6.** Distance formula: $\sqrt{k^2(x_1 - x_2)^2 + j^2(y_1 - y_2)^2}$, circle: $((x-a)/k)^2 + ((y-b)/j)^2 = r^2$. **9a)** $y = x^2$, $y = x - x^2$, $y = x^4$, $y = e^x$, $y = \ln x$.
9b) convex functions: $y = x^2$, $y = x^4$, $y = e^x$, concave functions: $y = x - x^2$, $y = \ln x$.
9c) Second derivative of convex (concave) function is positive (negative). **10b)** They are reflections in the real axis. **10c)** The distance a point is from the origin. **11a)** The product is the square of the modulus of that number. **11b)** Change scale (similarity). **11c)** Rotation around the origin.

Section 2.2
1a) The line that is the perpendicular bisector of the two points. **1b)** A circle on the sphere.
1c) The plane that is the perpendicular bisector of the two points. Parts (a) and (b) are the intersections of this plane with their respective domains. **1d)** A point, two points, and a line perpendicular to the plane determined by the three points. **3a)** Parabola. **4b)** $xy - 1 = 0$.
7a) Parabola. **7b)** Ellipse. **7c)** Degenerate ellipse (one point). **7d)** Hyperbola. **7e)** Ellipse.

Section 2.3
1a) $\cos^2 t + \sin^2 t = 1$. C and C^* are the same set of points—C^* moves around the circle twice as fast. **1d)** They give the same graph. **3a)** $(0, 0)$, $(2, 0)$. No instantaneous velocity at the bottom of the wheel, top of the wheel is going twice as fast as the train. **3b)** $(t - k\sin t, 1 - k\cos t)$, $(1 - k\cos t, k\sin t)$, $(1 - k, 0)$. **5a)** $x = r\cos\theta$, $y = r\sin\theta$, $r = \sqrt{x^2 + y^2}$, $\theta = \tan^{-1}(y/x)$.
5b) $x = f(\theta)\cos\theta$, $y = f(\theta)\sin\theta$. **5c)** $x^2 + y^2 - x = 0$. **5d)** $y = 2 - x$, $r = 2\sec(\theta - 210°)$.
9. All hold except II-4, III-1, V-1, and V-2. **11a)** $x + y + 1 = 0$, $\sqrt{2}x + y = 0$. **11b)** $nx + y = 0$, $\sqrt{2}x + y = 0$. **11c)** I-1, I-3, and IV-1.

Section 2.4
1. $-0.15 + 3.161x + 11.227x^2 + 2.804x^3$, $-97.4x^2 - 616x^3$. **3a)** $x = t - 1$, $y = -2 + 2t + 4t^2 - 3t^3$. **3b)** $x = t + 1$, $y = -1 + 4t^2 - 3t^3$. **3c)** $x = t$, $y = 1 + t - 8t^2 + 5t^3$.
3d) $x = t - 1$, $y = 2 - 2t - 4t^2 + 3t^3$. **3e)** $x = -t + 1$, $y = 2 - 2t - 4t^2 + 3t^3$.
5. Let $f(t) = 1 + t^2$, $g(t) = 2t + 4t^2 - 4t^3$. Use eight Hermite curves $(x_i(t), y_i(t))$, where $f = x_1 = y_2 = y_3 = x_8$, $g = y_1 = x_2 = y_4 = x_7$, $-f = x_4 = x_5 = y_6 = y_7$, and $-g = x_3 = y_5 = x_6 = y_8$. (Other f and g will work.) **7a)** Fifth degree. **7b)** $1 + x + 0.5x^3 - 0.5x^5$.

7c) $2n - 1$ degree.   **9a)** $x - x^3/3! + x^5/5! - x^7/7!$.   **9b)** $8/3(x/\pi) - (8/3)(x/\pi)^3$.   **9c)** From left to right: $y = 0.858 + 2.639x + 1.101x^2 + 0.111x^3$, $y = x + 0.057x^2 - 0.111x^3$, $y = -x - 0.057x^2 + 0.111x^3$, and $y = -0.858 - 2.639x - 1.101x^2 - 0.111x^3$.

Section 2.5
1a) $\alpha(1, 2, 4) + \beta(2, 0, 0)$, $2y - z = 0$.   **1b)** Use the dot product of $(2, 3, -4)$ and any (x, y, z) in the plane. Yes.   **3b)** Neither.   **3c)** Cut off one corner of a cube.   **9a)** Let e/v represent edges per vertex, and so on.

Dimension	1	2	3	4
e/v	1	2	3	4
f/v	0	1	3	6
c/v	0	0	1	4
v	2	4	8	16
e	1	4	12	32
f	0	1	6	24
c	0	0	1	8

CHAPTER 3

Section 3.1
2a) The centers of the two circles and one of their points of intersection, $(0.8, 0.6)$, form a right triangle.   **2b)** $(0.65, \pm 0.45)$.   **2d)** Angle at $(0, 0)$ is $69.4°$, at $(0.65, \pm 0.45)$ is $18.4°$.   **3.** II-4, III-4, and IV-1.   **5.** $(1, 1)$, $(4, 2)$, and $(4, \sqrt{10})$; $8.13°$, $63.43°$, and $90°$.

Section 3.2
1. All can occur.   **8.** Suppose in omega triangles $\triangle AB\Omega$ and $\triangle CD\Lambda$ that $\angle AB\Omega \cong \angle CD\Lambda$ and $\angle BA\Omega \cong \angle DC\Lambda$. For a contradiction suppose that AB is longer than CD. Construct B' between A and B with $AB' \cong CD$. Use Theorem 3.2.7 on $\triangle AB'\Omega$ and $\triangle CD\Lambda$ and then use Theorem 3.2.6 for $\triangle BB'\Omega$.

Section 3.3
2. Assume that C' is on $\overrightarrow{DC}$ with $DC' \cong BA$ so that $ABDC'$ is a Saccheri quadrilateral. Theorem 3.3.1 and Corollary 3.2.1 show AC' is ultraparallel to BD. Show C is between D and C'.

Section 3.4
3. The maximum defect a triangle can have is $180°$, $K = k \times 180°$.   **4.** The area of $\triangle AB_1B_i$ is finite even as $i \to \infty$, so the area of $\triangle AB_iB_{i+1}$ approaches 0 as $i \to \infty$.   **6.** $K_n = (n - 2)K$.

Section 3.5
2a) Consider two doubly right triangles with different included sides.   **3.** In spherical geometry, all lines perpendicular to a given line intersect in two opposite points.   **6a)** Half of a sphere, $2\pi r^2$.

CHAPTER 4

Section 4.1
1a) Both formulas clearly give functions. To show one-to-one and onto, solve $y = \alpha(x)$ and $y = \beta(x)$ for x to see that the choice of x in each is unique.   **1b)** $(2x - 1)^3$ and $2x^3 - 1$.

1c) $\sqrt[3]{x}$ and $0.5x + 0.5$. **1d)** $\sqrt[3]{0.5x + 0.5}$ and $0.5\sqrt[3]{x} + 0.5$, $\beta^{-1} \circ \alpha^{-1}$. **2a)** Rotation of 180° around the fixed point $(1, 2)$. All lines through $(1, 2)$ are stable. **2b)** Mirror reflection over the line $y = x + 1$, which are the fixed points. The other stable lines are $y = -x + c$. **2c)** Dilation by a factor of 2 about the fixed point $(0, 0)$. All lines through $(0, 0)$ are stable. **2d)** Dilation by a factor of 0.5 about the fixed point $(2, -2)$. All lines through $(2, -2)$ are stable.

Section 4.2
1. $(-1, 1)$, $(1, -1)$, $(y - 1, -x + 1)$. **7.** Note that $d(P, Q) + d(Q, R) = d(P, R)$ just when Q is between P and R. **8.** Use Theorem 4.2.7 and the fact that the composition is direct. It is a translation when the angles of rotation are opposite. **11.** $\mathbf{V}$ consists of translations, 180° rotations, vertical and horizontal mirror and glide reflections. Proof. Let l be a vertical line. $\iota(l) = l$ is vertical, so $\iota \in \mathbf{V}$. If $\alpha, \beta \in \mathbf{V}$, $\beta(l)$ and so $\alpha(\beta(l))$ are vertical, so $\alpha \circ \beta \in \mathbf{V}$. Suppose $\alpha \in \mathbf{V}$ and k the vertical line so that $\alpha(k) = l$. Then $\alpha^{-1}(l) = k$, so $\alpha^{-1} \in \mathbf{V}$.

Section 4.3
1. A is a rotation around $(1, 3, 1)$ with no stable lines. B is a translation with no fixed points and stable lines $y = -3.5x + c$. C is a mirror reflection over $y = x/3 + 5/9$, the fixed points. The other stable lines are $y = -3x + c$. D is a glide reflection over the stable line $y = x/3 + 1/6$. E is a rotation around $(1, 1 + \sqrt{2}, 1)$ with no stable lines.

3a) $\begin{bmatrix} \cos 30 & -\sin 30 & 3.5 - \sqrt{3} \\ \sin 30 & \cos 30 & 2 - 1.5\sqrt{3} \\ 0 & 0 & 1 \end{bmatrix}$. **3b)** $\begin{bmatrix} -0.6 & 0.8 & 0 \\ 0.8 & 0.6 & 0 \\ 0 & 0 & 1 \end{bmatrix}$. **3c)** $\begin{bmatrix} -0.6 & 0.8 & -0.8 \\ 0.8 & 0.6 & 0.4 \\ 0 & 0 & 1 \end{bmatrix}$.

3d) $\begin{bmatrix} -0.6 & 0.8 & c \\ 0.8 & 0.6 & 2c \\ 0 & 0 & 1 \end{bmatrix}$. **8a)** $\begin{bmatrix} -1 & 0 & 0 \\ 0 & -1 & 0 \\ 0 & 0 & 1 \end{bmatrix}$. **8b)** $\begin{bmatrix} -1 & 0 & 2u \\ 0 & -1 & 2v \\ 0 & 0 & 1 \end{bmatrix}$. **8c)** Direct isometry, rotation of 180°. **8d)** Lines through $(u, v, 1)$. **8e)** Translations twice the distance between the centers. You get the inverse translation.

Section 4.4
1a) $(1, 0, 1)$, $(0, 2, 1)$, $(-4, 0, 1)$, and $(0, -8, 1)$. Spiral. **1b)** $(1, 1, 1)$, $(-2, 2, 1)$, and $(-4, -4, 1)$. Yes. **1c)** M is a rotation by θ and a scaling by r, S is a rotation by $\theta/2$ and a scaling by $\sqrt{r}$. **1d)** For C use a rotation by $\theta/3$ and a scaling by $\sqrt[3]{r}$ and for N use a rotation by θ/n and a scaling by $\sqrt[n]{r}$. **11c)** This IFS fractal is the part of Fig. 4.25 on the x-axis. **12.** The four corners of the square, $(0, 0, 1)$, $(1, 0, 1)$, $(0, 1, 1)$, and $(1, 1, 1)$, must go to points in that square. For $\begin{bmatrix} a & b & c \\ d & e & f \\ 0 & 0 & 1 \end{bmatrix}$, the following must be between 0 and 1: $c, a + c, b + c, a + b + c, f, d + f, e + f$, and $d + e + f$.

Section 4.5
1b) $\begin{bmatrix} 1 & 0 & 0 & 0 \\ 0 & -1 & 0 & 0 \\ 0 & 0 & -1 & 0 \\ 0 & 0 & 0 & 1 \end{bmatrix}$, $\begin{bmatrix} -1 & 0 & 0 & 0 \\ 0 & 1 & 0 & 0 \\ 0 & 0 & -1 & 0 \\ 0 & 0 & 0 & 1 \end{bmatrix}$, rotation of 180° around the z-axis. The other of these two matrices. **1d)** Rotations of 120° around opposite vertices of the cube. **1e)** For example, $\begin{bmatrix} 1 & 0 & 0 & 0 \\ 0 & 0 & -1 & 0 \\ 0 & 1 & 0 & 0 \\ 0 & 0 & 0 & 1 \end{bmatrix}$ is a 90° rotation around the x-axis and $\begin{bmatrix} 0 & 1 & 0 & 0 \\ 0 & 0 & 1 & 0 \\ 1 & 0 & 0 & 0 \\ 0 & 0 & 0 & 1 \end{bmatrix}$ is a 120° rotation

around the axis through $(1, 1, 1, 1)$ and $(-1, -1, -1, 1)$. **1f)** $\begin{bmatrix} 1 & 0 & 0 & 0 \\ 0 & \cos\theta & -\sin\theta & 0 \\ 0 & \sin\theta & \cos\theta & 0 \\ 0 & 0 & 0 & 1 \end{bmatrix}$.

2a) $\begin{bmatrix} -1 & 0 & 0 & a \\ 0 & -1 & 0 & b \\ 0 & 0 & -1 & c \\ 0 & 0 & 0 & 1 \end{bmatrix}$, rotatory reflection. **2b)** Translation twice the distance between the centers. **3.** $\begin{bmatrix} \cos\theta & 0 & -\sin\theta & 0 \\ 0 & 1 & 0 & k \\ \sin\theta & 0 & \cos\theta & 0 \\ 0 & 0 & 0 & 1 \end{bmatrix}$ (or with the $-$ on the other $\sin\theta$, depending on the orientation of the axes).

Section 4.6
1a) Center $(0, 0)$ and radius 2, center $(-1, 0)$ and radius $\sqrt{10}$. **1b)** All x-axis except points between $x = 1$ and $x = 4$. **1c)** $r = \sqrt{c^2 - 5c + 4}$, where c is the x-coordinate of the center. **5.** No, because the inversion of at least some of the sides are circles. **6a)** Translation. **6b)** If $a > 0$, a dilation by a ratio of a. If $a < 0$, also a rotation of $180°$. A rotation. A composition of a dilation and a rotation. **6c)** A mirror reflection over the real axis. **7a)** $\overline{2/(z + 2i)} - 2i$. **7b)** $\overline{4/(z + 3i)} - 3i$. **7c)** E has center $-1.5i$ and radius 0.5. **7e)** $-4i/3$. **7f)** $v_C \circ v_D$. **9b)** The equator. Mirror reflection over the equator.

CHAPTER 5

Section 5.1
3. Iranian: rotations of $0°$, $120°$, and $240°$ and three mirror reflections. Byzantine: rotations of $0°$, $90°$, $180°$, and $270°$ and four mirror reflections. Afghani: rotations of $0°$, $60°$, $120°$, $180°$, $240°$, and $300°$. **5a)** 25%, 25%. **5b)** 33.3%, 33.3%. **5c)** 22.2%, 11.1%. **7a)** Mexican: translations, $180°$ rotations, vertical mirror reflections, and horizontal glide reflections. Chinese: translations, $180°$ rotations, vertical and horizontal mirror reflections, and horizontal glide reflections.

Section 5.2
1a) Gothic: C_3; Islamic: D_{10}; Gothic: C_2. **1b)** General: C_1; parallelogram: C_2; kite and isosceles trapezoid: D_1; rectangle and rhombus: D_2; and square: D_4. **3a)** Rotations of $0°$, $90°$, $180°$, and $270°$. Yes, C_4. **3b)** Rotations of $45°$, $135°$, $225°$, and $315°$. No. **3c)** Yes, C_8. **3e)** Color preserving: Rotations of $0°$, $120°$, and $240°$, yes, C_3; Color switching: Rotations of $40°$, $80°$, $160°$, $200°$, $280°$, and $320°$, no; Union: Yes, C_9. **6a)** Cube: 48; tetrahedron: 24; octahedron: 48; dodecahedron: 120; and icosahedron: 120. **6b)** Triangular prism: 12; square prism: 16. **6c)** $4n$. **6d)** Yes, four of the cube's vertices form a tetrahedron.

Section 5.3
1a) pmm2. **1b)** p112. **1c)** p1g1. **1d)** pmg2. **1e)** p1m1. **1f)** pm11. **3a)** pg. **3b)** p2. **3c)** p3. **3d)** cm. **3e)** p6m. **3f)** cmm. **11.** p4g, pgg.

Section 5.4
1. 8, 16, 48. Rotations of $180°$ around the x-, y-, and z-axes, mirror reflections over the xy-, xz-, and yz-planes, the identity and the central symmetry. (See Problem 2 of Section 4.5.) **6a)** Rotations of $90°$, $180°$, $270°$, and $0°$ around the centers of opposite faces, rotations of $120°$

Selected Answers

and 240° around opposite vertices, and rotations of 180° around centers of opposite edges; mirror reflections over six planes through opposite edges and three planes between opposite faces; and 15 rotatory reflections. **6b)** Rotations of 120°, 240°, and 0° around a vertex and the center of the opposite face and rotations of 180° around centers of opposite edges; mirror reflections over six planes through an edge and the center of the opposite edge; and six rotatory reflections. **6c)** For icosahedron: rotations of multiples of 72° around opposite vertices, rotations of 120° and 240° around centers of opposite faces, and rotations of 180° around centers of opposite edges; mirror reflections over 15 planes through opposite edges; 45 rotatory reflections.

Section 5.5
1a) D_2. **1b)** D_1, D_1, and C_2. **2a)** D_6. **2b)** D_1, D_1, and D_2. **2c)** D_2/D_1, D_1/C_1, and D_6/D_3. **5.** p31m. **6a)** 0.9945. **6b)** 0.2273. **6c)** 3, 5, and 7. **9a)** Hyperbola. **9b)** Hyperbola. **9c)** Asymptotes for parts (a) and (b).

Section 5.6
2a) $\ln 2/\ln 2 = 1$. **2b)** $\ln 3/\ln 2 \approx 1.585$ **2c)** $\ln 2/\ln 3 \approx 0.631$. **2d)** $\ln 5/\ln 3 \approx 1.465$. **2e)** $\ln 8/\ln 4 = 1.5$. **2f)** $\ln 3/\ln 2 \approx 1.585$. **4a)** $\ln 6/\ln 2 \approx 2.585$. **4b)** $\ln 13/\ln 3 \approx 2.335$. **4c)** $\ln 26/\ln 3 \approx 2.966$. **5a)** $1/9 + 5/(9^2) + 5^2/(9^3) + \ldots = 1/4$. **5b)** $1/27 + 13/(27^2) + 13^2/(27^3) + \ldots = 1/14$. **5c)** 0.

CHAPTER 6

Section 6.1
5. 0.6. **11a)** V, E, and D. **11b)** $\triangle BEV$, $\triangle CFU$, W and P, A, and D.

Section 6.2
1b) $H(PQ, SR)$, $H(QP, RS)$, $H(QP, SR)$, $H(RS, PQ)$, $H(RS, QP)$, $H(SR, PQ)$, and $H(SR, QP)$. **2.** $j/2^k$, where j and k are positive integers. **6b)** For example, (vi) becomes: If $pq//rs$, then p, q, r, and s are distinct, concurrent lines, $pq//sr$ and $rs//pq$. **10a)** Yes, yes. **10b)** Yes.

Section 6.3
1a) $(1, m, 0)$, parallel lines "meet" at infinity. **2.** $X_a = (a, 0, 1)$, $Y_b = (0, b, 1)$, $P_{ab} = (a, b, 1)$, and $(a, b, 0)$. **5a)** One of the orderings in (b). **5b)** The possible orderings of P, Q, R, S, and T are P, S, Q, R, T; P, T, R, Q, S; Q, R, T, P, S; Q, S, P, T, R; R, Q, S, P, T; and the reverse of these. **6b)** $1/r$, $r/(r-1)$, $(r-1)/r$, $1-r$, and $1/(1-r)$. $R(A, B, C, D) = R(B, A, D, C) = R(C, D, A, B) = R(D, C, B, A)$. **6c)** Yes. **6d)** 0, 1, and ∞. **7a)** $-xz + xy + yz = 0$, $y - z = 0$, and $x + z = 0$. $(1, 0, 0)$ and $(0, 1, 0)$. **7b)** $x + y = 0$, $x - y = 0$, and $x^2 - y^2 - z^2 = 0$. $(1, -1, 0)$ and $(1, 1, 0)$. **7c)** $x^2 - xz - yz = 0$. $(0, 1, 0)$. **9b)** In nonhomogeneous coordinates: $y = 1$, $y = -1$, $x = 1$, $x = -1$, $y = x$, and $y = -x$. **9e)** Unit circle, ellipses, and hyperbolas.

Section 6.4
1a) $\begin{bmatrix} a & b \\ 0 & d \end{bmatrix}$, where $a \neq 0$; $\begin{bmatrix} a & 0 \\ c & d \end{bmatrix}$, where $d \neq 0$; and $\begin{bmatrix} a & b \\ c & d \end{bmatrix}$, where $a + b = c + d$.

1b) $\begin{bmatrix} a & b & c \\ 0 & e & f \\ 0 & h & i \end{bmatrix}$, where $a \neq 0$; $\begin{bmatrix} a & 0 & c \\ d & e & f \\ g & 0 & i \end{bmatrix}$, where $e \neq 0$; $\begin{bmatrix} a & b & 0 \\ d & e & 0 \\ g & h & i \end{bmatrix}$, where $i \neq 0$; and

$\begin{bmatrix} a & b & c \\ d & e & f \\ g & h & i \end{bmatrix}$, where $a+b+c = d+e+f = g+h+i$. **2a)** $\begin{bmatrix} 3 & 0 \\ 1 & 2 \end{bmatrix}$. **2b)** $\begin{bmatrix} -1 & 0 \\ -3 & 2 \end{bmatrix}$.

6a) $(1,-2,1)$, $(4,-3,1)$, and $(-8,1,1)$ are on $3y + x + 5z = 0$. **6b)** $\begin{bmatrix} 12 & 0 & 0 \\ 0 & 12 & 0 \\ -1 & -3 & 7 \end{bmatrix}$.

7a) $\begin{bmatrix} 2 & 0 & 0 \\ 0 & -0.5 & 1 \\ 0 & -0.5 & -1 \end{bmatrix}$, $4x^2 - 2yz = 0$. **7c)** $y = 2x^2$, a parabola. **9a)** $\begin{bmatrix} 1 & 0 & 0 \\ 0 & 1 & 0 \\ -w & 0 & 1 \end{bmatrix}$,

$\begin{bmatrix} 1 & 0 & -w \\ 0 & 1 & 0 \\ 0 & 0 & 1 \end{bmatrix}$, and $(1+2w)x^2 + y^2 - 2xz = 0$. **9b)** $2x^2 + y^2 - 2x = 0$, $0.5x^2 + y^2 - 2x = 0$,

$2x = y^2$, and $-x^2 + y^2 - 2x = 0$. **9d)** For $w > -0.5$ and $w \neq 0$, the image is an ellipse. For $w < -0.5$, the image is a hyperbola. For $w = -0.5$, the image is a parabola. **9f)** For $w = 0.5$, $[2, 0, -2]$ and $[0.5, 1, -1]$. For $w = -0.25$, $[0.5, 0, -2]$ and $[-0.25, 1, -1]$. For $w = -0.5$, $[0, 0, -2]$ and $[-0.5, 1, -1]$. For $w = -1$, $[-1, 0, -2]$ and $[-1, 1, -1]$.

Section 6.5

4. $Y_b = \begin{bmatrix} \sqrt{1-b^2} & 0 & 0 \\ 0 & 1 & -b \\ 0 & -b & 1 \end{bmatrix}$. Transposes of each other. Use $a = -c$ and $b = -d/\sqrt{1-c^2}$.

Section 6.6

1. Four distinct points (a, b, c, d), (e, f, g, h), (i, j, k, l), and (m, n, o, p), in $\mathbf{P}^3$ are coplanar iff the determinant $\begin{vmatrix} a & e & i & m \\ b & f & j & n \\ c & g & k & o \\ d & h & l & p \end{vmatrix}$ is 0. **2a)** $n \leq 3$. **2b)** $n \leq 4$. **2c)** $n \leq k + j - 2$. **2d)** Its minimum dimension increases.

CHAPTER 7

Section 7.1

1a) A1 B2 C3 / C2 A3 B1 / B3 C1 A2   **1b)** A1 B2 C3 D4 / C2 D1 A4 B3 / D3 C4 B1 A2 / B4 A3 D2 C1   **2a)** Shift 2 to the right and 1 down. Shift 1 right and 2 down.   **3.** 1 2 3, 4 5 6, 7 8 9; 1 4 7, 2 5 8, 3 6 9; 1 6 8, 2 4 9, 3 5 7; and 1 5 9, 2 6 7, 3 4 8.   **5b)** $4k + 1$.   **7a)** A pentagon, five points.   **7c)** $k^2 + 1$.

Section 7.2

4a) 4, 4.   **8a)** (ii).   **8c)** Theorem 7.2.1 (i), (ii), (iii), and (iv).   **9a)** A triangle.   **9c)** Four points with three points on one line and all other lines having two points.

Section 7.3

1a) $b = \lambda(v/k)(v-1)/(k-1)$.   **2a)** $r \geq k$.   **2b)** $r = k = n + 1$; $v = b = k^2 - k + 1 = n^2 + n + 1$; $n = k - 1$ is the order of the projective plane.   **5a)** One line is 1, 2, 5, and 7. Rotate these numbers for the other lines.   **7a)** 1 vs. 2 and 3 vs. 4; 1 vs. 3 and 2 vs. 4; and 1 vs. 4 and 2 vs. 3.   **7b)** $2n - 1$.   **7c)** In the first round pair players that add to $2n$, n sits out.

Now rotate. With $2n$ players, $2n$ plays the person sitting out. **9a)**
$$\begin{pmatrix} 1 & 1 & 0 & 1 & 0 & 0 & 0 \\ 0 & 1 & 1 & 0 & 1 & 0 & 0 \\ 0 & 0 & 1 & 1 & 0 & 1 & 0 \\ 0 & 0 & 0 & 1 & 1 & 0 & 1 \\ 1 & 0 & 0 & 0 & 1 & 1 & 0 \\ 0 & 1 & 0 & 0 & 0 & 1 & 1 \\ 1 & 0 & 1 & 0 & 0 & 0 & 1 \end{pmatrix}$$
and
$$\begin{pmatrix} 1 & 1 & 0 & 0 & 0 & 1 & 0 \\ 0 & 1 & 1 & 0 & 0 & 0 & 1 \\ 1 & 0 & 1 & 1 & 0 & 0 & 0 \\ 0 & 1 & 0 & 1 & 1 & 0 & 0 \\ 0 & 0 & 1 & 0 & 1 & 1 & 0 \\ 0 & 0 & 0 & 1 & 0 & 1 & 1 \\ 1 & 0 & 0 & 0 & 1 & 0 & 1 \end{pmatrix}.$$
1 and 0. **9b)**
$$\begin{pmatrix} 1 & 1 & 1 & 0 & 0 & 0 & 0 & 0 & 0 \\ 0 & 0 & 0 & 1 & 1 & 1 & 0 & 0 & 0 \\ 0 & 0 & 0 & 0 & 0 & 0 & 1 & 1 & 1 \\ 1 & 0 & 0 & 1 & 0 & 0 & 1 & 0 & 0 \\ 0 & 1 & 0 & 0 & 1 & 0 & 0 & 1 & 0 \\ 0 & 0 & 1 & 0 & 0 & 1 & 0 & 0 & 1 \\ 1 & 0 & 0 & 0 & 1 & 0 & 0 & 0 & 1 \\ 0 & 1 & 0 & 0 & 0 & 1 & 1 & 0 & 0 \\ 0 & 0 & 1 & 1 & 0 & 0 & 0 & 1 & 0 \\ 1 & 0 & 0 & 0 & 0 & 1 & 0 & 1 & 0 \\ 0 & 1 & 0 & 1 & 0 & 0 & 0 & 0 & 1 \\ 0 & 0 & 1 & 0 & 1 & 0 & 1 & 0 & 0 \end{pmatrix}.$$
3 and 1.

Section 7.4

1a) $[1, 2, 1]$. **1b)** $(2, 0, 1)$. **1c)** Switches x- and y-coordinates. Similar to mirror reflection. **1d)** Rotation of $90°$ around $(1, 1, 1)$. Rotation of $180°$ around $(0, 0, 1)$. Translation of 1 to the right and 1 down. **3.** Many answers possible for each part. **3a)** $[2, 2, 2]$ has eight points. **3b)** $[2, 1, 0]$ and $[0, 1, 0]$ have both $(0, 0, 1)$ and $(2, 0, 1)$ on them. **3c)** $[1, 1, 0]$ and $[1, 3, 1]$. **3d)** (i) and (iii). **3e)** For part (a), $[2, 4, 2]$ has 12 points. For (b), $[3, 1, 0]$ and $[0, 1, 0]$. For (c), $[1, 1, 0]$ and $[1, 4, 1]$. For (d), (i) and (iii). **5a)** $(2, 0, 1)$. **5b)** Four points: $(0, 0, 1)$, $(0, 1, 1)$, $(1, 0, 1)$, and $(1, 1, 1)$. Six lines: $[1, 0, 0]$, $[0, 1, 0]$, $[1, 1, 4]$, $[1, 4, 0]$, $[0, 1, 4]$, and $[1, 0, 4]$. **5c)** $(3, 0, 1)$ is the only choice for S. **9.** $x^2 + y^2 = 1$ has $(0, 1, 1)$, $(0, 4, 1)$, $(1, 0, 1)$, and $(4, 0, 1)$. $x^2 + 4y = 0$ has $(1, 1, 1)$, $(4, 1, 1)$, $(2, 4, 1)$, $(3, 4, 1)$, and $(0, 0, 1)$. $x^2 + 3y^2 = 1$ has $(1, 0, 1)$, $(4, 0, 1)$, $(2, 2, 1)$, $(2, 3, 1)$, $(3, 2, 1)$, and $(3, 3, 1)$. In $\mathbf{PZ}_5^2$, they all have six points. The first has $(1, 2, 0)$ and $(1, 3, 0)$, the second has $(0, 1, 0)$, and the third has no other points.

Index

Notation

$\widehat{AB}$ Arc of a circle from A to B, 49
$\overleftrightarrow{AB}$ Length of segment $\overline{AB}$, 3
$\overleftrightarrow{AB}$ Line on A and B, 3
$\overline{AB}$ Line segment between A and B, 3, 293
$\overrightarrow{AB}$ Ray from A through B, 25, 294
$\angle ABC$ Angle ABC with vertex at B, 3, 294
$\triangle ABC$ Triangle with vertices A, B, and C, 3, 294
$AB//CD$ A and B separate C and D, 232
$\mathbf{AF}^2$ Affine plane over the field $\mathbf{F}$, 279
α^{-1} The inverse of α, 135
$\mathbf{C}$ The complex numbers, 65
$\mathbf{C}^{\#}$ The extended complex numbers, 170
$\mathbf{C}_n$ The cyclic group with n elements, 185
$d(A, B)$ Distance from A to B, 62
d_H Hyperbolic distance, 121, 250
d_T Taxicab distance, 31, 78
$\mathbf{D}_n$ The dihedral group with n rotations, 185
$\overline{\mathbf{D}_n}$ The symmetry group of a rectangular prism, 203
$\mathbf{F}$ A field, 279
$<g_1, g_2, \ldots, g_n>$ Generators of a group, 191
$H(PQ, RS)$ Harmonic set, 228
$k \cdot l$ The point of intersection of lines k and l, 231
$m\angle ABC$ Measure of angle $\angle ABC$, 14
$n!$ n factorial, product of first n positive integers, 205
$\overline{\mathbf{P}}$ The icosahedral group, 203
$\widetilde{P}$ Oriented point, 250
$\mathbf{PF}^2$ The projective plane over the field $\mathbf{F}$, 281
$\mathbf{Q}$ The rational numbers, 76
$\mathbf{R}$ The real numbers, 62
$\mathbf{S}_n$ The symmetric group on n elements, 205
$\overline{\mathbf{T}}$ The tetrahedral group, 203
$\vec{v}$ Vector, 85
$\overline{\mathbf{W}}$ The octahedral group, 203
$\|(x, y)\|$ Length of vector (x, y), 85
$\mathbf{Z}$ The integers, 77
$\mathbf{Z}_n$ The integers modulo n, 279
$\cong$ Congruent, 3
$\|$ Parallel, 12

Terms

AAS, 12
    Euclid I–26, 290
Absolute conic, 249
Achilles and the Tortoise, 4
Affine matrix, 147, 164
Affine plane, 252, 267, 279
Affine transformation, 147, 252, 282
Analytic,
    geometry, 59ff
    Greek meaning, 2
    model, 61ff, 238ff, 279ff
Angle of parallelism, 108

Index

Angle sum, 3
Antiprism, 206
Archimedean solids, 55
Area axioms, 117
Asymptote, 71
Attracting fixed point, 136
Axiomatic system, 24
Axioms, 25

Babylonian mathematics, 2
Balanced incomplete block design (BIBD), 272
Barycentric coordinates, 74
Base, 40
Between, betweenness, 23, 35, 257, 293
BIBD, 272
Bilateral symmetry, 180
Block, 272

CAD, 78, 165, 254
Cavalieri's Principle, 50
Center of inversion, 167
Central,
    angle, 19
    symmetry, 152, 166
Characteristic axiom of hyperbolic geometry, 99
Chemistry, 206ff
Circular points at infinity, 252
Circumscribed, 18
Closure, 136
Code, 276
Code word, 276
Collinear, 231
Collineation, 244, 254, 281
Color symmetry, 189, 199
Color-preserving symmetry, 198
Color-switching symmetry, 199
Combinatorics, 266
Complete, 31
    quadrangle, 228
    quadrilateral, 229
Complex numbers, conjugates, modulus, 65
Composition, 135
Computer-aided design (CAD), 78, 165, 254ff
Concurrent, 231
Congruence, 12, 111
Conic, 65, 240ff, 260
Conjugate, 65, 152
Consistent, 31

Constructible, 20
Construction, 9ff
Contraction mapping, 158
Converse, 13
Convex, 26, 257
    function, 65
Cross polytope, 89
Cross-ratio, 239ff
Crystal, 181, 204, 207ff
Crystallographic restriction, 194, 204
Cyclic group, 185

Decomposition, 16
Defect, 118
Degenerate conic, 69
Deltahedron, 56
Desargues's theorem, 229, 247, 260
Descartes's formula, 42
Design theory, 272ff
Diagonal point, 237
Differential geometry, 102
Dihedral group, 185
Dilation, dilatation, 159
Direct isometry, 140
Directrix, 66
Discrete, 190
Dodecahedron, 41
Doubling a cube, 14
Dual, 56, 92, 235, 271
Duality, 235, 271

Edge, 39
Egyptian Mathematics, 2
Eigenvalue, 150
Eigenvector, 150
Ellipse, 66
Equivalence relation, 176
Equivalent,
    logically, 13
    polygons, 117
Erlanger Programm, 142
Error correcting code, 276
Euclid's *Elements,* 2, 9ff, 287ff
Euler's formula, 42
Excess, 48, 125

Face, 39
Field, 278
Fifth postulate, 13, 288
Fixed point, 133, 244

Flow chart for wallpaper patterns, 196
Foci, focus, 66
Fractal, 156, 214ff
    curve, 220
    dimension, 220
    surface, 220
Frequency, 43
Frieze pattern, 190

Generator, 191
Geodesic, 102
    dome, 43ff
Glide reflection, 140
Golden ratio, 6
Great circle, 46
Group, 136

Half-plane model, 105
Hamming distance, 276
Harmonic set, 228
Hausdorff dimension, 216
Hermite curve, 82
Hilbert's axioms, 23, 293
h-inner product, 251
h-length, 251
Homogeneous coordinates, 238
h-orthogonal, 251
Hyperbola, 66
Hyperbolic,
    distance, 121, 250
    geometry, 98ff, 249
    isometries, 250
    planes, 284
Hyperboloid, 86
Hypercube, 88
Hyperplanes, 86, 164

Icosahedron, 41
Ideal point, 226
Identity, 135
IFS, 158
IFS fractal, 158
Incommensurable, 3
Independent, 31
Indirect isometry, 140
Inscribed, 17
Interior, 249, 284
Interpretation, 29
Invariant, 133
Inverse, 135

Inversion, 167ff
Inversive plane, 167
Invertible, 147
Irrational, 3
Isomer, 212
Isometry, 137ff, 161, 165
Isomorphic, 33
Iterated function system (IFS), 158

Kirkman's schoolgirl problem, 265
Klein model, 104
Klein's definition of geometry, 142ff
Koch curve, 214ff

LaGrange's theorem, 186
Law of cosines, 20
Lightlike, 214
Line conic, 242
Locus, 65
Lorentz transformations, 212, 258
Lunes, 7, 46

Median, 63
Metamathematics, 30
Metatheorem, 30
Mickelson–Morley experiment, 211
Midline, 191
Minkowski geometry, 212, 258
Mirror reflection, 138
Möbius transformation, 171ff
Mod, 279
Model, 29ff
Modulo, 279
Motif, 181

Napoleon's theorem, 94
n-gon, 17
Non-Euclidean geometry, 97ff
Nonorientable, 127, 256
Nonsingular; *see* invertible.

Octahedron, 41
Omega,
    point, 109, 250
    triangle, 109
One-to-one, 133
Onto, 133
Orbit, 186
Order,
    field, 279

Index

Order, *(cont.)*
    plane, 267, 271
Orientable, 127
Orientation, 132
Oriented point, 256
Oriented projective geometry, 256ff
Orthogonal,
    circles, 103
    matrix, 164
Oval, 282

Parabola, 66
Parallel, 12, 267, 288
Parallelogram, 18
Parametric equations, 72
Pascal's theorem, 260
Pasch's axiom, 23, 110, 293
Penrose tiling, 209
Pentagram, 6
Periodic point, 136
Perspective,
    from a line, 229
    from a point, 229
    in art, 226, 230
    in computer-aided design, 254ff
Perspectivity, 227
Pick's theorem, 93
Playfair's axiom, 12, 295
Poincaré model, 104
Polar coordinates, 73
Polygon, 17
Polyhedron, 39
Polytope, 88
Projective plane, 238, 270, 281
Projectivity, 236, 243
Proof, 26ff
Proportional, 35
Pseudosphere, 105
Pyramid, 39
Pythagorean theorem, 7, 292

Quadratic formula, 6
Quasicrystal, 209

Ratio of proportionality, 35
Ray, 25, 294
Reflection property for parabola, 69
Regular polygon, 17
Regular polyhedron, 41
Relatively consistent, 31

Relativity, theory of, 210ff
Repelling fixed point, 136
Rotation, 138
Rotatory reflection, 162

Saccheri quadrilateral, 113
SAS, 12, 295
    Euclid I-4, 289
Scaling ratio, 153
Screw motion, 164
Self-similar, 215
Semiregular polyhedron, 55
Sensed parallels, 107
Separation, 232ff
Shear, 156
Similar, 35ff
Similarity, similitude, 153
Simplex, 90
Single elliptic geometry, 124ff, 251
Special theory of relativity, 210
Spherical,
    excess, 43
    geometry, 129ff
    triangle, 46
Squaring a circle, 14
SSS, 12
    Euclid I-8, 289
Stabilizer, 186
Stable, 133, 244
Statistical design theory, 265
Statistically self-similar, 215
Steiner triple system, 275
Subgeometry, 247ff
Subgroup, 191
Summit, summit angle, 113
Symmetric,
    BIBD, 277
    group, 205
Symmetry, 179ff, 181
    group, 182
Synthetic, 2

Tangent, 76, 282
Taxicab geometry, 31, 78
Tetrahedron, 40
Theorem, 26
36-officers problem, 264
Timelike, 214
Torus, doughnut shape, 78
Tournament, 277

Transcendental, 16
Transformation, 133
　group, 136
Transformational geometry, 132ff
Translation, 137
Transversal, 12
Trilinear plots, 74
Trisecting an angle, 14

Ultraparallels, 107
Undefined terms, 25

Variety, 272
Vertex, 39

Wallpaper pattern, 190
Without loss of generality (WLOG), 4

Zeno's paradox, 4

People

Archimedes, 15
Aristotle, 5

Bernoulli, Jakob, 73
Bhaskara, 7
Bolyai, János, 100
Bravais, Auguste, 181, 204

Cayley, Arthur, 226, 248

da Vinci, Leonardo, 185, 226
Desargues, Girard, 226
Descartes, René, 61
Dürer, Albrecht, 226

Einstein, Albert, 210
Eratosthenes, 7
Euclid, 2, 9, 226
Eudoxus, 4
Euler, Leonhard, 41, 265

Fedorov, Vyatseglav, 194
Fermat, Pierre de, 60, 67
Fisher, Sir Ronald A., 265, 273

Galileo Galilei, 210
Gauss, Carl Friedrich, 17, 101

Gödel, Kurt, 25, 31

Hausdorff, Felix, 216
Hilbert, David, 24
Hippocrates, 7

Kant, Immanuel, 98
Kepler, Johannes, 226
Kirkman, Thomas, 265, 275
Klein, Felix, 104, 132, 143, 226, 247

Lie, Sophus, 132, 155

Mandelbrot, Benoit, 216
Minkowski, Hermann, 212
Möbius, August, 132, 172, 226
Monge, Gaspard, 87, 226

Newton, Sir Isaac, 210

Oresme, Nicole, 60

Pascal, Blaise, 226
Penrose, Roger, 209
Plato, 4
Plücker, Julius, 94, 143, 226
Poncelet, Jean Victor, 226, 233
Pythagoras, 3

Riemann, Georg Bernhard, 102

Saccheri, Girolamo, 98
Senechal, Marjorie, 209
Steiner, Jacob, 275

Theaetetus, 5

Viète, François, 60
von Staudt, Karl, 226

Zeno, 4

Chapter One Chapter opener: Copyright © 1996 by Allegra Fuller Snider. Courtesy of the Buckminster Fuller Institute.

Chapter Two Chapter opener: Courtesy of the Toro Company.

Chapter Three Chapter opener: Courtesy of Douglas Dunham; Figures 3.31 and 3.32: courtesy of Douglas Dunham.

Chapter Four Figure 4.3: from Thompson et al., *On Growth and Form,* Abridged Edition. New York: Cambridge University Press, 1942. Reprinted with permission.

Chapter Five Chapter opener: from Wade, *Geometric Patterns and Borders.* Copyright © 1982, Nostrand Reinhold Co., New York; Reprinted with permission. Figures 5.1, 5.3, and 5.6: from Wade, *Geometric Patterns and Borders.* Copyright © 1982 Nostrand Reinhold Co., New York; Reprinted with permission. Figure 5.8: Bentley and Humphrey, *Snow Crystals.* Copyright © 1962 by Dover Publications, New York; Reprinted with permission. Figures 5.11, 5.15, 5.16, and 5.17: from Wade, *Geometric Patterns and Borders.* Copyright © 1982, Nostrand Reinhold Co., New York; Reprinted with permission. Figure 5.18: from Crowe and Washburn, "Groups and geometry in the ceramic art of San Ildefonso," *Algebras, Groups and Geometries,* September 1985, Hadronic Press Inc., Palm Harbor, Florida; Reprinted with permission. Figure 5.19: from Wade, *Geometric Patterns and Borders.* Copyright © 1982, Nostrand Reinhold Co., New York; Reprinted with permission. Figure 5.21: from Gallian, *Contemporary Abstract Algebra.* Copyright © 1994 by D.C. Heath Publishing Co., Lexington, Mass.; Reprinted with permission from Houghton-Mifflin Publishing Co., Boston, Mass. Figures 5.22, 5.23, 5.24, and 5.25: from Wade, *Geometric Patterns and Borders.* Copyright © 1982, Nostrand Reinhold Co., New York; Reprinted with permission. Figure 5.26: from Crowe and Washburn, "Groups and geometry in the ceramic art of San Ildefonso," *Algebras, Groups and Geometries,* September 1985, Hadronic Press Inc., Palm Harbor, Florida; Reprinted with permission. Figures 5.27 and 5.28: from Wade, *Geometric Patterns and Borders.* Copyright © 1982 Nostrand Reinhold Co., New York; Reprinted with permission. Figure 5.30: courtesy of Rev. Magnus Wenninger. Figure 5.37: from Holden and Morrision, *Crystals and Crystal Growing.* Copyright © 1982 MIT Press, Cambridge, Mass.; Reprinted with permission. Figure 5.39: from Peterson, *The Mathematical Tourist.* Copyright © 1988 by W.H. Freeman and Company; Used with permission. Figure 5.47: from Mandelbrot, *The Fractal Geometry of Nature.* Copyright © 1982 by W.H. Freeman and Company. Used with permission. Figure 5.50: Courtesy of U.S. Geological Survey. Figure 5.51: from Moore and Persaud, *The Developing Human.* Copyright © 1991 by W.B. Saunders, Philadelphia; Reprinted with permission. Figure 5.52: Courtesy of U.S. Geological Survey.

Chapter Six Chapter opener: from Dodgson, *Albrecht Durer: Engravings and Etchings,* vol. 1. Copyright © 1967 by Da Capo Press, New York.

Chapter Seven Chapter opener: FPG International Corp., New York, photography by Jerry Driendl.